Building Information Modelling (BIM)
in
Design, Construction and Operations IV

WIT*PRESS*

WIT Press publishes leading books in Science and Technology.
Visit our website for the current list of titles.
www.witpress.com

WIT*eLibrary*

Home of the Transactions of the Wessex Institute.
Papers published in this volume are archived in the WIT elibrary in volume 205 of WIT Transactions on The Built Environment (ISSN 1743-3509).
The WIT electronic-library provides the international scientific community with immediate and permanent access to individual papers presented at WIT conferences.
Visit the WIT Press eLibrary at www.witpress.com/elibrary

Fourth International Conference on Building Information Modelling (BIM) in Design, Construction and Operations

BIM 21

Conference Chairmen

Juan Casares
University of Santiago de Compostela, Spain
Member of WIT Board of Directors

Lamine Mahdjoubi
University of the West of England, UK

Antonio Galiano Garrigós
University of Alicante, Spain

International Scientific Advisory Committee

Organised by

University of the West of England, UK
Wessex Institute, UK

Sponsored by

WIT Transactions on The Built Environment

WIT Transactions

Wessex Institute
Ashurst Lodge, Ashurst
Southampton SO40 7AA, UK

Senior Editors

Building Information Modelling (BIM)
in
Design, Construction and Operations IV

Editors

Juan Casares
University of Santiago de Compostela, Spain
Member of WIT Board of Directors

Lamine Mahdjoubi
University of the West of England, UK

Antonio Galiano Garrigós
University of Alicante, Spain

WIT*PRESS* Southampton, Boston

Editors:

Juan Casares
University of Santiago de Compostela, Spain
Member of WIT Board of Directors

Lamine Mahdjoubi
University of the West of England, UK

Antonio Galiano Garrigós
University of Alicante, Spain

Published by

WIT Press
Ashurst Lodge, Ashurst, Southampton, SO40 7AA, UK
Tel: 44 (0) 238 029 3223; Fax: 44 (0) 238 029 2853
E-Mail: witpress@witpress.com
http://www.witpress.com

For USA, Canada and Mexico

Computational Mechanics International Inc
25 Bridge Street, Billerica, MA 01821, USA
Tel: 978 667 5841; Fax: 978 667 7582
E-Mail: infousa@witpress.com
http://www.witpress.com

British Library Cataloguing-in-Publication Data

A Catalogue record for this book is available
from the British Library

ISBN: 978-1-78466-441-1
eISBN: 978-1-78466-442-8
ISSN: (print) 1746-4498
ISSN: (on-line) 1743-3509

The texts of the papers in this volume were set individually by the authors or under their supervision. Only minor corrections to the text may have been carried out by the publisher.

Preface

Building Information Modelling (BIM) has become established as the industry standard approach to the modelling and management of a building's lifecycle, from design and construction to maintenance, and decommissioning. BIM standards and protocols have now been adopted internationally and they are increasingly becoming mainstream in various built environment and associated disciplines. With the wider BIM adoption, there has been increased level of expectation. BIM is anticipated to become a major enabler for greater productivity, and to deliver reduced cost and risk within the construction and infrastructure industry.

The International Conference on Building Information Modelling in Design, Construction and Operations has become one of the key reference fora for research and development in BIM. Over the years it has attracted a wide range of participants, including built environment practitioners, researchers, academics, and students. This is reflected in the advanced topics presented and debated at the fourth International Conference on Building Information Modelling in Design, Construction and Operations, which took place from 27 to 29 October 2021. Participants presented cutting-edge research findings, including national and international case studies, as well as future BIM challenges for research and practice.

The first theme was focused on BIM and sustainability to determine the role of BIM novel approaches and technologies to address the challenges of climate change and reducing built-environment carbon footprint. The second theme was concerned with research and development of advanced BIM platforms, and their implications on advanced visualisation in education and practice, as well as delivering a more effective common data environment, and the challenges of implementing BIM in small and medium-sized enterprises and water treatment plans. The third theme was focused on BIM and automation in construction and was more concerned with the challenges and case studies of automating BIM implementation in education, in prefabrication of heating, ventilation and air conditioning systems, as well as generative techniques for building envelope design and structural optimisation. The fourth theme was concerned about BIM and cultural heritage and examined issues pertaining to the role of Heritage BIM (HBIM) for preventive maintenance, Roman temple management, old depot of Salma (Alabama), and European renovation projects. The fifth and final theme was focused on the emerging digital twin and artificial intelligence for buildings and infrastructures.

The Editors, 2021

Contents

Section 3: BIM and automation in construction

Section 4: BIM and cultural heritage

Section 5: Digital twin and artificial intelligence for buildings and infrastructures (Special session organised by F. Cumo)

SECTION 1
BIM AND SUSTAINABILITY

GREEN BUILDING DESIGN AND ASSESSMENT WITH COMPUTATIONAL BIM: THE WORKFLOW AND CASE STUDY

YAIK-WAH LIM[1,2], TAKI EDDINE SEGHIER[3], MOHD HAMDAN AHMAD[1], PAU CHUNG LENG[1], AZARI MAT YASIR[1], NORASLINDA ABDUL RAHMAN[1,2], WAI LAI CHAN[1] & SHARIFAH SALWA SYED MAHDZAR[1,2]
[1]Faculty of Built Environment and Surveying, Universiti Teknologi Malaysia (UTM), Malaysia
[2]Centre for the Study of Built Environment in the Malay World (KALAM), Institute for Smart Infrastructure and Innovative Construction (ISIIC), UTM, Malaysia
[3]School of Architecture and Built Environment, UCSI University, Malaysia

ABSTRACT
Due to growing concern regarding sustainability in the built environment, several green building assessments and rating tools have been established worldwide including Green Building Index (GBI) in Malaysia which was introduced in 2009. However, the current methods of measuring, analysing and documenting the green building design rely on a number of disjointed processes to meet the discrete requirements for various building systems. The development of building information modelling (BIM) together with computational programming has made it easier for complicated building modelling to be digitally constructed, generating required information to support green building design and assessment throughout various project stages. Thereby, the aim of this research is to integrate computational BIM with green building design and assessment in Malaysia, using GBI as a unique case and an office building in Kuala Lumpur, Malaysia as a case study. Match-up of BIM (Revit) functionalities and GBI (NRNC) criteria was formulated, then visual programming (Dynamo) was employed to automate the BIM data management process. The findings of this research have developed workflows and templates to assess several criteria, namely energy efficiency (EE), indoor environmental quality (EQ), sustainable site planning and management (SM) and material and resources (MR), which allow a higher level of automation in green building assessment.
Keywords: building information modelling, visual programming, office building, sustainability, Green Building Index.

1 INTRODUCTION

Green building development is an urgent move for sustainable built environment, and many green building assessments such as Leadership in Energy and Environmental Design (LEED) in the US, Building Research Establishment Environmental Assessment Method (BREEAM) in the UK, Comprehensive Assessment System for Built Environment Efficiency (CASBEE) in Japan, Green Mark in Singapore, and Green Building Index (GBI) in Malaysia have been established to assess the green building performances. Sustainability in the building sector has received concerns from various stakeholders because the building industry accounts for 36% of energy use and 39% of energy-related carbon dioxide (CO_2) emissions worldwide [1]. Furthermore, the world building energy consumption is expected to increase by nearly 50% between 2018 and 2050 if no further action is taken to improve building energy efficiency [2].

To assess the success of a green building project in achieving sustainability, assessment tools which the designers can use to predict and evaluate the building performances are needed [3]. With a predictive assessment of the building performance, the decision makers can put forward strategies and methods in their decisions to improve the building performance in a more cost-effective way [4]. Even though various green building assessments or ratings such as LEED and BREEAM have been established since the 1990s,

WIT Transactions on The Built Environment, Vol 205, © 2021 WIT Press
www.witpress.com, ISSN 1743-3509 (on-line)
doi:10.2495/BIM210011

the current methods of measuring, analysing and documenting the green building rely on a number of disjointed processes which cause regression in productivity [5].

With the development of building information modelling (BIM) process and technology, all information or data related to a building, including its physical features and performances during its life cycle, can be digitally represented in a single model [6]–[8]. Furthermore, the building information can be extracted and managed using computational algorithms such as visual programming language (VPL) to automate the performance evaluation process. Thus, integration of computational BIM and green building design can best address the unprecedented challenges in productivity and sustainability encountered by the building industry [9]–[12].

Till today there is no standard workflow for computational BIM-based modelling for green building design and assessment, more efforts are needed to further develop a model or framework for the integration of these two aspects. Therefore, this study aims to integrate GBI Malaysia assessment process and BIM functionalities with the implementation of VPL. It explores the implementation of computational BIM to achieve a higher level of automation in the green building assessment process using GBI Malaysia as a unique case and a recently constructed office tower in TRX Kuala Lumpur as a case study.

1.1 Green Building Index (GBI)

In Malaysia, GBI is one of the assessment tools being adopted by the building industry since it was jointly founded and developed by Pertubuhan Akitek Malaysia (PAM) and the Association of Consulting Engineers Malaysia (ACEM) in 2009. GBI includes rating tools for non-residential new construction (NRNC), residential new construction (RNC), existing non-residential buildings (NREB), NRNC hotel, NRNC resort, NRNC retail, township, industrial new construction (INC) and so on. Some of the criteria in GBI used the benchmarks as stated in Malaysian Standard (MS) 1525:2019 such as overall thermal transfer value (OTTV) and roof thermal transfer value (RTTV) [13].

The GBI classification is as shown in Table 1. This research focuses on a new office building, thus adopting the GBI (NRNC) tool. It has six major parts consisting of energy efficiency (EE), indoor environmental quality (EQ), sustainable site planning and management (SM), materials and resources (MR), water efficiency (WE) and innovation (IN) as outlined in Table 2.

In this research, a generic integration framework was established on the match-up of the GBI credit requirements and the functionalities available from current BIM solutions, for the purposes of this research, the Autodesk Revit software suite and supplemental software application from Autodesk. Two major tasks in establishing the framework include evaluating GBI credits requirements, and screening available functionalities of Revit and supplemental BIM solutions. The gaps in the framework where no existing functionality was available to help compliance with certain GBI credits were identified.

Table 1: GBI classification.

Points	GBI Rating
86+	Platinum
76 to 85	Gold
66 to 75	Silver
50 to 65	Certified

Table 2: GBI (NRNC) assessment criteria overall points.

Part	Item	Max. points
1	Energy efficiency (EE)	35
2	Indoor environmental quality (EQ)	21
3	Sustainable site planning and management (SM)	16
4	Material and resources (MR)	11
5	Water efficiency (WE)	10
6	Innovation (IN)	7
Total		100

2 METHOD

An office tower in TRX Kuala Lumpur completed in 2019 was selected as a case study building for BIM authoring using Autodesk Revit with Level of Development (LOD) 300. It is a 27-storey height office tower which consists of lobby and facilities at Ground floor to the 2nd floor, multi-storey car park at the 3rd to 8th floor, and office spaces at the 9th to 26th floor. It was selected due to its efficient rectangular floor layout with central core design, and the unique full glazed façade design. Though this building was not originally designed with a green building concept, the purpose of the case study building is to develop and validate the BIM-GBI workflow and model rather than to achieve GBI rating. Fig. 1 illustrates the case study building modelling in Revit.

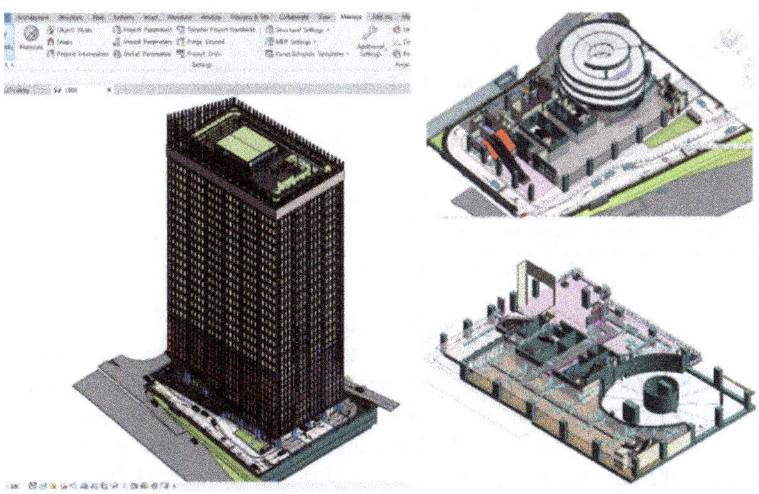

Figure 1: Revit modelling of the case study building at TRX Kuala Lumpur.

2.1 Conceptual framework development

A conceptual framework to categorise the GBI assessment criteria and to match-up with the BIM process was established to study the general workflow of the BIM-based GBI assessment process (Fig. 2). The information required for the GBI assessment were categorised according to the discipline of architectural, structural, electrical, mechanical and plumbing. Besides, the required element BIM LOD and attributes (geometrical and non-geometrical) were determined.

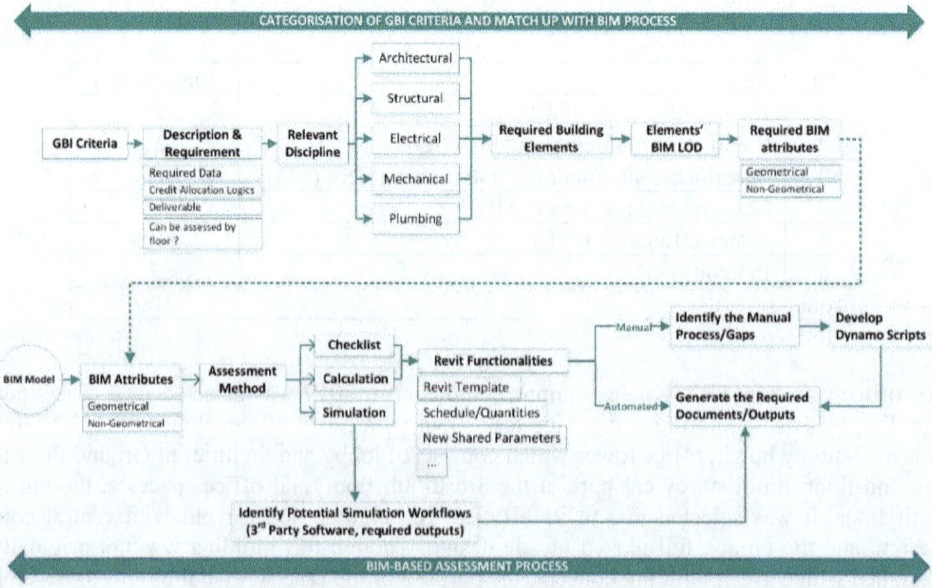

Figure 2: Conceptual framework of the BIM-based GBI assessment process.

The BIM-based GBI assessment method was divided into "checklist", "calculation" and "simulation". Various Revit functionalities such as template, schedules/quantities and shared parameters can be employed to support the "checklist" and "calculation", whereas "simulation" requires plug-ins such as Autodesk Insight or other 3rd party simulation software. Apart from that, Autodesk Dynamo, a visual programming tool which is integrated with Revit, was used to develop computational scripts to increase the level of automation for the BIM-based assessment process. This conceptual framework has been used as the basis to further develop the BIM-GBI model.

3 BIM-GBI ASSESSMENT MODEL

To determine which GBI (NRNC) criteria is potential to be used for developing the BIM-GBI use case and model, a detailed mapping between the assessment criteria and BIM functionalities was conducted. There are some GBI criteria which can be evaluated directly using Revit functionalities such as 2D drawing, schedule/quantities, material take-off, spacing and zoning, 2D tagging and the plug-in Insight Analysis. However, some criteria require external (3rd party) tools such as simulation software. Furthermore, some evaluations can be done partially in Revit, thus they are categorised as "indirectly evaluated by Revit".

The findings show that 25 points in the EE can be directly evaluated using Revit while 3 points (EE9) can be evaluated indirectly (Table 3). The EE9 requires evidence of building maintenance room and facilities in the design floor plan (1 point, which can be shown in Revit drawing); it also requires evidence of commitment to engage at least 50% of permanent building maintenance team and documented plan for at least 3-year facility maintenance (2 points); these cannot be provided using Revit. Thus, EE9 is considered as indirectly evaluated by Revit. EE consists of the highest number of points that can be evaluated by Revit.

Table 3: Summary of GBI (NRNC) points evaluation using Revit.

Part	Item	Max Points	Points can be evaluated by Revit	
			Directly	Indirectly
1	EE	35	25	3
2	EQ	21	5	10
3	SM	16	11	1
4	MR	11	7	0
5	WE	10	0	4
6	IN	7	0	0
		100	48	18

Out of the total of 21 points for EQ, 5 points can be evaluated directly while another 10 points can be evaluated indirectly. Many points are considered as indirectly evaluated by Revit because they require external tools; for instance, EQ8: Daylighting, EQ9: Daylight glare control, and EQ10: Electric lighting levels need simulation software such as IES VE and DesignBuilder.

For the SM part, a total of 11 points can be evaluated directly while 1 point can be evaluated indirectly. Many of the SM criteria can be directly shown in the Revit modelling or drawing (site plan) such as SM1: Site selection, SM3: Development density and community connectivity, SM8: Public transportation access, SM12: Greenery and roof, etc. Besides, there are some criteria that can be quantified using Revit schedules/quantities such as SM9: Green vehicle priority and SM10: Parking capacity.

The criteria of MR can be potentially assessed directly using the Material Take-off function in Revit because this function allows building materials in the BIM model to be quantified and sorted out one by one. GBI (NRNC) MR consists of a maximum 11 points; 7 points can be evaluated directly using Revit. Material Take-off can directly quantify the contents of materials for MR1 to MR4. Whereas MR5: Storage and collection of recyclables can be shown on the drawings.

The use of Revit to evaluate water efficiency (WE) is quite limited due to these criteria requires technical reports with calculations and diagrams to support the assessment. Hence, only 4 points out of the total 10 points can be evaluated indirectly using Revit. Drawings such as the roof plan can be produced by Revit to support the technical reports for WE1: Rainwater harvesting and WE2: Water recycling.

In the last part of GBI criteria, which is the innovation (IN), the requirements are depending on the types of innovative design to reduce energy or improve sustainability. IN1: Innovation in design and environmental design initiatives requires a report on each innovation such as condensate water recovery, light pipes, self-cleaning façade, etc. Though some of these innovations might be supported by Revit drawings or schedules, in general this criterion cannot be assessed using Revit. IN2: GBI facilitator also cannot be evaluated by Revit.

4 CASE STUDY RESULT AND ANALYSIS

The research has developed nine use cases for the GBI (NRNC) criteria which can be directly evaluated using BIM, namely EE1: Overall thermal transfer value (OTTV), EE5: Building energy intensity (BEI), EQ8: Daylighting, SM9: Green vehicle priority, SM10: Car park capacity, MR1: Materials reuse, MR2: Recycled material, MR3: Regional material, and MR4: Sustainable timber. However, only two of them, which use different methods and workflows, are presented in this paper as case study.

4.1 EE1: Overall thermal transfer value (OTTV)

To evaluate the OTTV, the case study building envelope was divided into 3 main parts (as shown in Fig. 3):

1. Ground to 2nd floor with partial air-conditioned (lobby and facilities) and partial natural ventilated (car park);
2. 3rd to 8th floor with natural ventilated car park (not to be included for OTTV calculation);
3. 9th to 26th floor with air-conditioned office spaces (typical floor).

As the OTTV should only include air-conditioned spaces, 3rd to 8th floors were excluded from the evaluation. With that, the building envelope was divided into Part 1 (ground to 2nd floor) and Part 2 (a typical office floor which was multiplied by 18 floors).

Building's floor Level/facade	AC level?	Included in OTTV calculation?	Comment
SAA_UM_B1017_CO_L27	No	No	Roof level
SAA_UM_B1017_CO_L26	Yes		
SAA_UM_B1017_CO_L25	Yes		
SAA_UM_B1017_CO_L24	Yes		
SAA_UM_B1017_CO_L23	Yes		
SAA_UM_B1017_CO_L122	Yes		
SAA_UM_B1017_CO_L121	Yes		
SAA_UM_B1017_CO_L120	Yes		
SAA_UM_B1017_CO_L19	Yes		
SAA_UM_B1017_CO_L18	Yes	Yes	Typical floor
SAA_UM_B1017_CO_L17	Yes		
SAA_UM_B1017_CO_L16	Yes		
SAA_UM_B1017_CO_L15	Yes		
SAA_UM_B1017_CO_L14	Yes		
SAA_UM_B1017_CO_L13	Yes		
SAA_UM_B1017_CO_L12	Yes		
SAA_UM_B1017_CO_L11	Yes		
SAA_UM_B1017_CO_L10	Yes		
SAA_UM_B1017_CO_L9	Yes		
SAA_UM_B1017_CO_L8	No		
SAA_UM_B1017_CO_L7	No		
SAA_UM_B1017_CO_L6	No		
SAA_UM_B1017_CO_L5	No	No	Parking Levels
SAA_UM_B1017_CO_L4	No		
SAA_UM_B1017_CO_L3	No		
SAA_UM_B1017_CO_L2	Partially		Facades modelled
SAA_UM_B1017_CO_L1	Partially	Yes	separately from
SAA_UM_B1017_CO_L0	Partially		the levels
SAA_UM_B1017_CO_FACADE	N/A	Yes	
SAA_UM_B1017_CO_B2	No	No	Basement
SAA_UM_B1017_CO_B1	No		

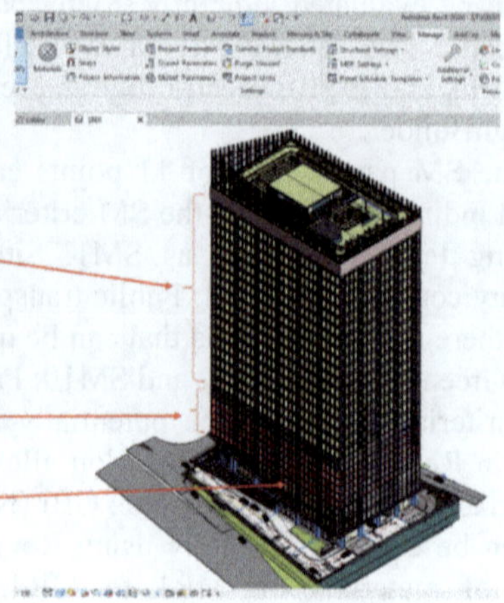

Figure 3: Division of the case study building envelope for OTTV assessment.

The BIM-GBI OTTV assessment workflow is outlined in Fig. 4. New shared parameters and schedules had been created in the Revit template for model preparation and data verification. However, some assumptions were made for the building properties data inputs due to limited information available. Then Dynamo scripts (Fig. 5) were developed to extract the BIM data automatically for OTTV calculation in "BIM-GBI Template" Excel spreadsheet. Fig. 6 shows the combination of both Part 1 and Part 2 OTTV. Due to Part 2 was only representing a single typical floor, it was multiplied by a repetition factor of 18.

4.2 SM9–10: Green vehicle and parking capacity

Under GBI (NRNC) SM9: Green vehicle priority, a building which provides preferred parking for low-emitting and fuel-efficient for 5% of full-time equivalent (FTE) occupants is entitled for 1 point. "Preferred parking" refers to the parking spots that are closest to the main entrance of the project. Apart from that, 1 point will be awarded for a building that sizes parking capacity not exceeding the minimum local zoning requirement and provides preferred parking for carpools or vanpools for 5% of the total provided parking spaces under SM10: Parking capacity.

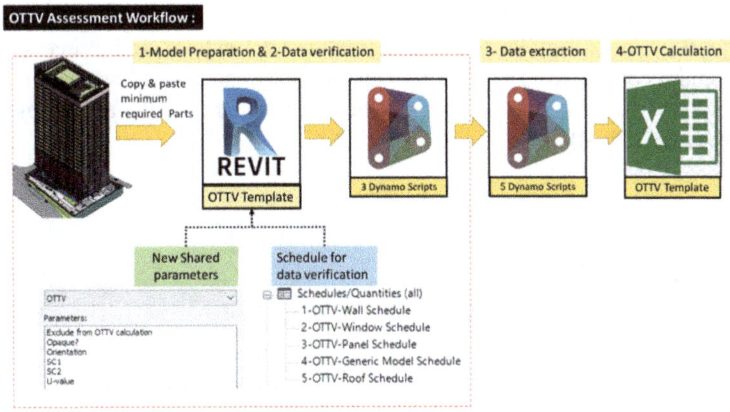

Figure 4: BIM-GBI OTTV assessment workflow.

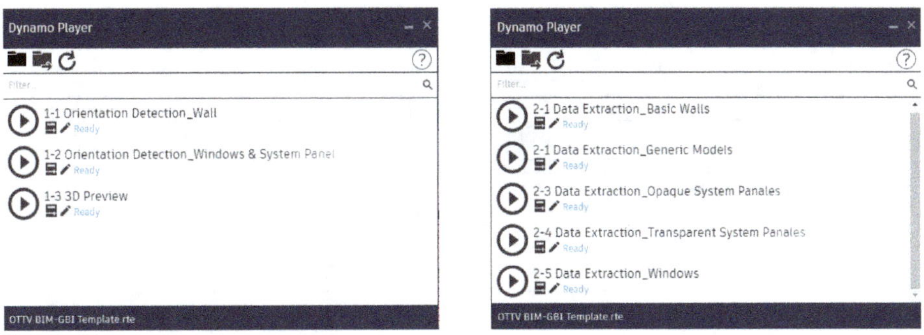

(a) (b)

Figure 5: Dynamo scripts for OTTV assessment model preparation and data extraction. (a) Model preparation; and (b) Data extraction.

OVERALL BUILDING'S OTTV (W/m²)	55.22

Building Parts	r: Repetition factor	An: Total Envelop Part Area (m²)	Overall OTTV (W/m²)
Part 1	1	2187.08	62.69
Part 2	18	791.48	54.08
Part n	0	0.00	0.00

$$Overall\ Building's\ OTTV = \frac{(OTTV\ part\ 1\ x\ A1\ x\ r)+(OTTV\ part\ 2\ x\ A2\ x\ r)...+(OTTV\ part\ n\ x\ An\ x\ r)}{(A1\ x\ r+A2\ x\ r...+An\ x\ r)}$$

Figure 6: OTTV assessment result based on input data assumption.

Fig. 7 presents the workflow of assessing SM9 and SM10 using the BIM-GBI model. It starts from data input for the green vehicle (low-emitting and fuel-efficient) parking, preferred parking, and carpools/vanpools parking in the Revit schedules. The newly created Revit schedules include SM09-All parking (Fig. 8), SM09-Only FTE occupants parking,

SM09-Parking green vehicle, SM10-Only car park and SM10-Parking carpools/vanpools. Then, Dynamo scripts (Fig. 9(a)) were formulated to extract and process the data to quantify the points achieved. The parking capacity was calculated based on the floor area and local authority requirement. Finally, the assessment results are displayed in the Dynamo Player (Fig. 9(b)).

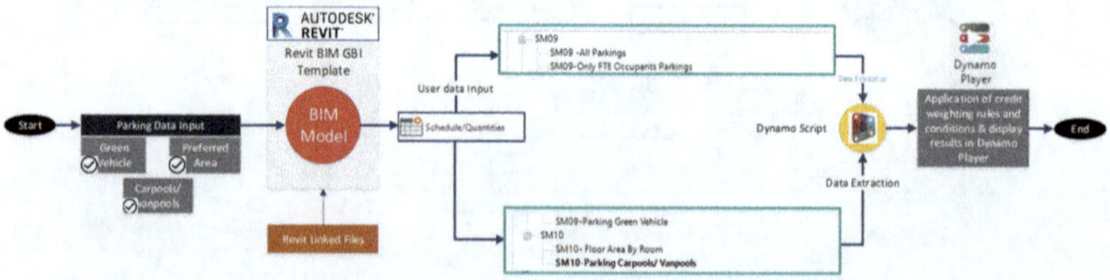

Figure 7: BIM-GBI green vehicle and parking capacity assessment workflow.

SM10-Parking Carpools/ Vanpools | (3D) | SM09 -All Parkings

<SM09 -All Parkings>

A	B	C	D	E	F	G
Family and Type	Exclude	Count	FTE Occupant	Green Vehicle	Prefered Parking Provided	Carpools/ vanpools
SAA_Carpark Line1: 2500 x 5000_CAR PARK 2		15	☑	☑	☑	
SAA_Carpark Line1: 2500 x 5000_CAR PARK 2: 15		15				
SAA_Carpark Line1: 2500 x 5000_CAR PARK SINGLE LADY DRIVE		5				
SAA_Carpark Line1: 2500 x 5000_CAR PARK SINGLE LADY DRIVER 2: 5		5				
SAA_Carpark Line1: 2500 x 5000_CAR PARK TANDEM		1	☑	☑	☑	
SAA_Carpark Line1: 2500 x 5000_CAR PARK TANDEM: 1		1				
SAA_Carpark Line1: 2500 x 6000_CAR PARK 2		7				
SAA_Carpark Line1: 2500 x 6000_CAR PARK 2: 7		7				
SAA_Carpark Line1: 2500 x 6050_CAR PARK		7				
SAA_Carpark Line1: 2500 x 6050_CAR PARK: 7		7				
SAA_Carpark Line1: 2700 x 5000_CAR PARK 2		30				
SAA_Carpark Line1: 2700 x 5000_CAR PARK 2: 30		30				
SAA_Carpark Line1: 2700 x 5000_CAR PARK TANDEM		5				
SAA_Carpark Line1: 2700 x 5000_CAR PARK TANDEM: 5		5				
SAA_Carpark Line1: 3000 x 5000_CAR PARK		4				
SAA_Carpark Line1: 3000 x 5000_CAR PARK: 4		4				
SAA_Carpark Line2: 2500 x 5000_CAR PARK 2		25				
SAA_Carpark Line2: 2500 x 5000_CAR PARK 2: 25		25				
SAA_Carpark Line2: 2500 x 5000_CAR PARK SINGLE LADY DRIVE		5				

Figure 8: Revit schedule SM9-all parking.

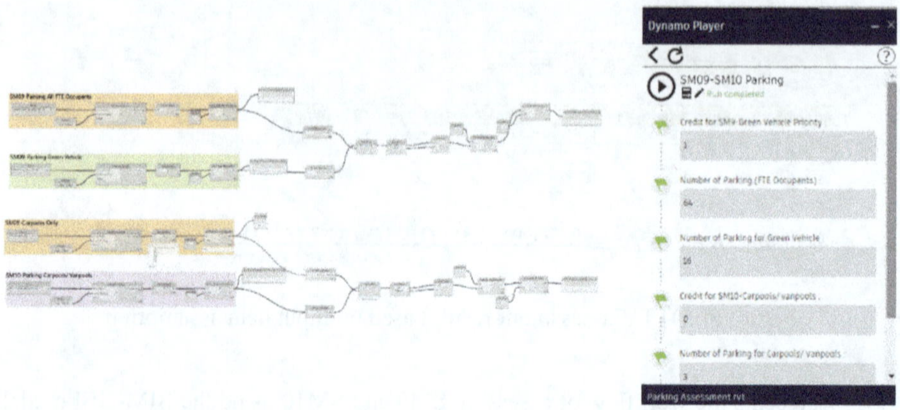

(a) (b)

Figure 9: Use of VPL for SM9 and SM10: (a) Dynamo scripts; and (b) Assessment results in Dynamo Player.

5 DISCUSSION AND CONCLUSION

BIM can support the design process and decision-making to achieve green building criteria in the early design phases. Although previous research [14], [15] has explored BIM-based method for green building assessment in different contexts, the current study has specifically applied BIM functionalities for GBI Malaysia assessment criteria with test cases developed using an actual building located in Kuala Lumpur as case study. Furthermore, the study also integrated visual programming language (VPL) as a solution for data management and higher level of automation.

Based on the detailed match-up between the BIM functionalities and GBI (NRNC) criteria, results showed that only 48 points can be directly evaluated while 18 points can be indirectly evaluated using various Revit functionalities. Thus, the development of the BIM-GBI model focused on the GBI (NRNC) criteria that can be directly evaluated using Revit and require quantifications using calculation or simulation assessment methods. As a result, the newly developed BIM-GBI model covers 26 points of the GBI (NRNC) using the calculation (schedules, material take-off, Dynamo, spreadsheet) and simulation methods as presented in Table 4.

Table 4: Summary of GBI (NRNC) points evaluation using Revit.

GBI criteria	Description	GBI point	Assessment method	Tool/functionalities
EE1	Minimum EE performance	1	Calculation	Schedule, Dynamo, Spreadsheet
EE5	Advanced EE performance – BEI	15	Calculation	Schedule, Dynamo, BEIT, Spreadsheet
EQ8	Daylighting	2	Simulation	Insight Lighting Analysis, Schedule
SM9	Green vehicle priority	1	Calculation	Schedule, Dynamo
SM10	Parking capacity	1	Calculation	Schedule, Dynamo
MR1	Materials reuse and selection	2	Calculation	Material Take-off, Dynamo, Spreadsheet
MR2	Recycled content materials	2	Calculation	Material Take-off, Dynamo, Spreadsheet
MR3	Regional materials	1	Calculation	Material Take-off, Dynamo, Spreadsheet
MR4	Sustainable timber	1	Calculation	Material Take-off, Dynamo, Spreadsheet
		26		

It is worth mentioning that the aim of this research is to implement computational BIM to achieve a higher level of automation in GBI assessment process, hence the case study office building in TRX Kuala Lumpur was employed for the purpose of developing the workflows and test cases rather than for accurate evaluation of GBI points. Due to the constraints of time and resources, some modifications on the Revit model and assumptions of building data were made for the unavailable information required for GBI assessment such as the building components, material properties as well as mechanical and electrical systems. Nevertheless, the geometrical information such as the building area, layout and volume were based on the actual building condition.

This research has made significant contributions to the implementation of BIM for green building assessment with special reference to GBI Malaysia. It has provided detailed comparison of BIM functionalities with GBI (NRNC) assessment criteria which is an advancement of knowledge in the relevant study; its findings can be further applied or adopted to different building typologies and contexts. The integration of visual programming language with BIM, also known as computational BIM method, opens the possibility to increase the level of automation for BIM data extraction and management for green building assessment. Although the current research only developed use cases for 9 GBI criteria, the similar workflow can be further extended for other relevant green building assessment criteria including other green building assessment tools apart from GBI.

ACKNOWLEDGEMENTS

The authors would like to acknowledge the research funding by the Ministry of Higher Education, Malaysia (MOHE) under Fundamental Research Grant Scheme (FRGS/1/2018/SSI11/UTM/02/4), titled "Computational Building Information Modelling-based Method for Green Retrofitting", and Technology Depository Agency (TDA) Berhad and IJM Construction Sdn. Bhd. under Contract Research Grant (4C270) titled "Building Information Modelling (BIM) for Green Building Assessment".

REFERENCES

[1] IEA, Towards a zero-emission, efficient, and resilient buildings and construction sector, 2017. Global Status Report 2017, UN Environment and International Energy Agency, 2017.

[2] USEIA, *International Energy Outlook 2019 with Projections to 2050*, US Energy, 2019.

[3] Lim, Y.W., Sediadi, E., Shahsavari, F. & Fazlenawati, N., Building information modelling for building energy efficiency evaluation: Integration with green building index (GBI) in Malaysia. *Proceedings of the 4th ACE*, pp. 42–48, 2016.

[4] Akadiri, P.O., Chinyio, E.A. & Olomolaiye, P.O., Design of a sustainable building: A conceptual framework for implementing sustainability in the building sector. *Buildings*, **2**(2), pp. 126–152, 2012.

[5] Lim, Y.W. et al., BIM and genetic algorithm optimization for sustainable building envelope design. *International Journal of Sustainable Development and Planning*, **13**(1), pp. 151–159, 2018.

[6] Cavalliere, C., Habert, G., Dell'Osso, G.R. & Hollberg, A., Continuous BIM-based assessment of embodied environmental impacts throughout the design process. *Journal of Cleaner Production*, **211**, pp. 941–952, 2019.

[7] Lim, Y.W., Chong, H.Y., Ling, P.C.H. & Tan, C.S., Greening existing buildings through building information modelling: A review of the recent development. *Building and Environment*, **200**, 107924, 2021.

[8] Santos, R., Costa, A.A., Silvestre, J.D. & LincyPyl, T.V., BIM-based life cycle assessment and life cycle costing of an office building in Western Europe. *Building and Environment*, **169**, 106568, 2020.

[9] Lim, Y.W., Seghier, T.E., Harun, M.F., Ahmad, M.H., Samah, A.A. & Majid, H.A., Computational BIM for green retrofitting of existing building envelope. *WIT Transactions on the Built Environment*, vol. 192, WIT Press: Southampton and Boston, pp. 33–34, 2019.

[10] Seghier, T.E., Ahmad, M.H. & Lim, Y.W., Automation of concrete usage index (CUI) assessment using computational BIM. *International Journal of Built Environment and Sustainability*, **6**(1), pp. 23–30, 2019.

[11] Wei, L., Liu, S., Wei, Q. & Wang, Y., Concept, method and application of computational BIM. *Advances in Intelligent Systems and Interactive Applications. Advances in Intelligent Systems and Computing,* eds F. Xhafa, S. Patnaik & M. Tavana, Springer: Cham, p. 1084, 2019.

[12] Seghier, T.E. & Lim, Y.W., A BIM-based method for building energy intensity (BEI) evaluation. *2nd International Civil Engineering and Architecture Conference*, 2021.

[13] GBI, *Green Building Index: Design Reference Guide,* 1st ed., Green Building Index Sdn. Bhd.: Kuala Lumpur, Malaysia: 2009.

[14] Azhar, S., Carlton, W.A., Olsen, D. & Ahmad I., Building information modeling for sustainable design and LEED® rating analysis. *Automation in Construction*, **20**(2), pp. 217–224, 2011.

[15] Sanhudo, L. et al., Building information modeling for energy retrofitting: A review. *Renewable and Sustainable Energy Reviews*, **89**, pp. 249–260, 2018.

EXPERIENCE IN THE FIELD OF SUSTAINABILITY ENHANCED CONSTRUCTION CLASSIFICATION SYSTEM

RODRIGO LIMA, FILIPA SALVADO, MARIA JOÃO FALCÃO SILVA & PAULA COUTO
National Laboratory of Civil Engineering (LNEC), Portugal

ABSTRACT

Currently, due to the increasing complexity of new buildings and the urgency to promote their economic sustainability, the Architecture, Engineering, Construction and Operation (AECO) sector faces the challenge of managing all the necessary information to fulfil its basic requirements. For this information management (e.g. control, organization, structuring, storage), the importance of using standardized construction information classification systems (CICS) is emphasized. In parallel, the use of Building Information Modelling (BIM) methodology is increasingly essential. In addition, there is a global challenge to rationalize the use of natural resources in line with the United Nations (UN) Sustainable Development Goals. As the AECO sector is one of the most demanding in terms of the use of these resources, there is an urgent need to develop production models that involve the sharing, reuse, repair, renovation and recycling of existing materials and products, thus extending their life cycle. The Sustainability Enhanced Construction Classification System (SECClasS) Project presents a solution that unifies these challenges. It proposes to develop and implement a CICS optimized for sustainability, based on the principles of Circular Economy, aiming to reduce construction and demolition waste (CDW), through the use of digital tools that help and promote the careful selection and management of products with less environmental impact. This paper describes the assumptions for the development of a Portuguese CICS (in accordance with SECClasS Project), that adapts classification systems already established at an international level, such as: (i) UniClass2015 (United Kingdom); (ii) OmniClass (North America); (iii) CCS (Denmark); (iv) CoClass (Sweden); and (v) CCI (Estonia, Czech Republic, Norway, Denmark, Sweden and Finland). International standards, such as ISO 12006 and ISO 81346, as well as the Action Plan for the Circular Economy defined by the European Union in 2020 are also analysed. This research work aims to obtain a CICS that streamline decision-making processes in AECO sector, reducing negative environment impacts and promoting sustainability.

Keywords: classification systems, building information modelling (BIM), standardization, sustainability, circular economy.

1 INTRODUCTION

In the context of construction design and planning, the building information modelling (BIM) methodology has imposed new challenges in the Architecture, Engineering, Construction and Operation (AECO) sector. Project information, previously transmitted through graphic elements, in two dimensions, and with instructions in text form, is now transmitted through three-dimensional models, containing data that allows several virtual simulations such as: buildings behavior, quantities, environmental impacts, costs or deadlines.

The use of smart objects in three-dimensional BIM models is capable to support different simulations that increase the predictability and reduce the risks of construction projects. The information contained in these objects causes a large data increase, requiring tools to assist in its management.

The Construction Information Classification Systems (CICS) have been developed in this context. While they were previously useful for processing, organizing and retrieving information, with the new technology they are now fundamental for managing the large volume of data. The integrated way of sharing this information is also a BIM benefit, and this sharing relies on the CICS to improve communication between those involved in the construction and operation processes, in order to ensure a uniform language throughout the

WIT Transactions on The Built Environment, Vol 205, © 2021 WIT Press
www.witpress.com, ISSN 1743-3509 (on-line)
doi:10.2495/BIM210021

building's life cycle, standardizing it through codes recognized by human actors or by machines [1].

To overcome these challenges, the Sustainability Enhanced Construction Classification System (SECClasS) Project aims to present a solution that meets the needs and peculiarities of the Portuguese AECO sector, proposing a Portuguese Construction Information Classification System optimized for sustainability, based on the principles of Circular Economy and on international experiences in developing their own systems and adapting them to the several works and materials types used in Portuguese AECO sector.

This paper aims to analyze the main existing CICS and to expose their most relevant concepts. It also intends to present the important concepts that can be a basis for the Portuguese CICS standardized structure, as well as the concepts used by European legislation and normalization for a circular economy. Finally, the paper proposes the choice of one of the international systems as a basis to develop the Portuguese CICS.

2 CONCEPTS, USE AND STANDARDIZATION OF CICS

Classification is defined as the grouping of a set of items into classes, categories or groups that have common characteristics, based on pre-established rules and according to the purpose of a process. In general, classification is done with the objective of finding the best possible order. So, after classifying an element, it is easier to find it within a given set. Although there is no absolute way to classify, the most correct would be that all trading partners use a common language to group elements [2].

There are several types of classification that can be described, and that are associated, depending on the objectives, the scope and the particularities of the system to be developed. Classifications can be divided into: (i) specialized, when it is intended to focus on a particular subject; (ii) general, when it is intended to cover a universe of information; (iii) enumerative, when it is sought to list all subclasses, including compound ones, that relate to the main class; (iv) hierarchical, when subclasses are created from a simple and particular principle of division of the main class and compound classes are defined by association of the subclasses; (v) documental, when the main objective is the classification of documents or other types of information in order to facilitate their organization, location and storage; and (vi) analytical (taxonomy or scientific), when it is intended to systematize information and provide a basis for its explanation and understanding [1].

CICS emerged in the twentieth century in order to meet the sector's needs, to organize itself in a rational manner, to facilitate storage and retrieval of information, and to exchange information that is relevant to the sector. In Europe, the United Kingdom developed the CESMM system, in 1976, by the ICE (Institution of Civil Engineers), geared to the detailed survey of civil engineering works quantities. Its great applicability to civil engineering works led it to be updated three times, with the last version (CESMM3) published in 1992 [2].

A CICS is configured in a set of interdependent elements that form an organized whole. Such systems can be developed according to the needs of each country, region or even a company, in order to meet their desires in organizing themselves, and can follow international classification standards [1]. The classification and harmonization of information through those systems have the main objective of making collaborative work more effective, even when performed at a distance or in different formats, to ensure consistency and comparability, and to enable the construction process stakeholders to be less dependent on the weaknesses of others [2].

2.1 ISO 12006

ISO has produced standards to organize the classification and exchange of information, such as ISO 12006-2 and ISO 12006-3, whose major objective is to guide the structuring of construction classification systems. The base structure proposed by ISO aims to provide the systems in development with the necessary scope to respond to the growing size and complexity of construction works. Indeed, a classification system designed in accordance with ISO 12006-2 should cover the entire life cycle of the construction project, with a wide variety of construction works [1].

However, there is no specific and closed recommendation on the structuring model to be adopted for each situation, which allows the user to choose the one that best suits the system he want to make the classification. The standard still has another role in the development of classification systems. Due to the existence of several CICS, the information exchange between them tends to be facilitated if they are developed in accordance with the proposed guidelines. Based on this function, ISO 12006-2 recommends tables, which can be used in combination or separately according to the user's needs [3].

2.2 ISO/IEC 81346

The latest series of ISO/IEC 81346 constitute the standards for the components identification belonging to any technical system. These standards deal with technical systems in the broadest sense of the word, including construction (structures and technical installations), industrial systems, as well as machinery in general and, moreover, are of considerable importance for the base documentation of the machinery and equipment European Directive. This documentation links the marking of components and associated documentation [4].

The systems based on ISO/IEC 81346 series are structured differently from the ones based on ISO 12006-2. They have a smaller group of Functional Systems, Technical Systems and Components (items corresponding approximately to the Elements, Systems and Products tables of ISO12006-2 based systems), and differentiates them by assigning different properties (for example, a port is always designated as a port, but subtypes are defined by their properties), which is reflected in the designation or classification. Additional notations are attached to the initial/root notation, as the item is progressively defined throughout the design and documentation process. This results in a stable and recognizable designation throughout the project life [5].

ISO/IEC 81346 is divided into four parts: (i) Part 1: Basic rules (IEC 81346-1); (ii) Part 2: Classification of objects and codes for classes (IEC 81346-2); (iii) Part 10: Power plants (ISO/TS 81346-10) and Power supply systems (ISO/DIS 81346-10); and (iv) Part 12: Construction works (ISO 81346-12). Parts 1 and 2 also have rules for the creation of unambiguous identifiers that define, through the Reference Designation System (RDS), a combination of well-defined relationships (part and type of relationships) and different aspects (function, product, location and type) [6].

3 CICS AND THE BIM METHODOLOGY

Based on the BIM concepts and objectives, recent CICS have been developed and adapted to this methodology. The purpose is to manage all the information produced in the process that, through smart objects, is based on a three-dimensional virtual representation of the construction project, where the information is integrated, relating to its entire life cycle, so that its results and tools bring advantages to the AECO sector, such as the entire process transparency [1].

The information is introduced in BIM models through the smart objects. Those objects are responsible for integrating valuable information about production processes, communication and analysis of digital three-dimensional models, assuming then particular importance in the creation and development of models that allow to represent virtually, in an equivalent way, the real products. In fact, the correct insertion of information in the object is fundamental for the best use of BIM methodology. For this, there is the need to adopt methods and strategies that enable the correct definition and implementation of these elements. This is the case of the CICS, which guide and organize the way of information is made available in smart objects [1]. Fig. 1 shows the types of data, in a smart object, using BIM.

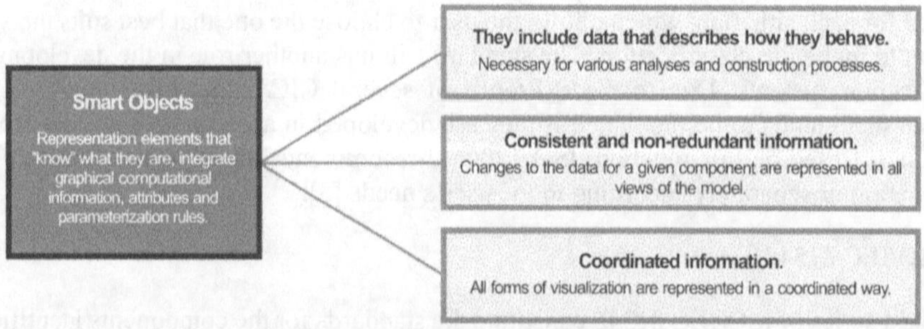

Figure 1: BIM object data [7].

BIM encompasses information about developments during their entire life cycle, at the design, planning, construction and operation phases. The classification and structuring of information thus enables the integration between the different stages and designs of the construction system, reducing errors and optimizing the process. The technology on which BIM is based allows the generated model to contain precise and detailed information about the geometry, as well as a set of data regarding the construction procedures, the technologies used and the procurement process, necessary to realize the project [8]. This presupposes a complete change in the flow of information exchanged among the stakeholders, relative to the traditional workflow, that becomes centralized in the BIM model avoiding the eventual loss of information, as shown in Fig. 2. Thus, CICS become indispensable as an information management tools.

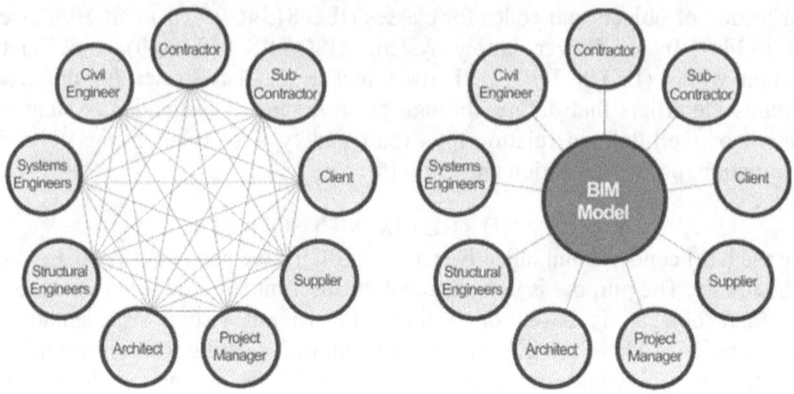

Figure 2: Traditional vs. BIM workflow [9].

4 INTERNATIONAL CLASSIFICATION SYSTEMS

Considering the analysis of CICS used in Europe and wide world, the development of the present paper is based on a study conducted by the Faculty of Civil Engineering of the Czech Technical University in Prague, Czech Republic, in association with the Czech Agency for Standardization (PS03). This research includes 18 international CICS, that through pre-established criteria were evaluated: (i) horizontal details, dealing with more complex classifications and referring to the scope of CICS; (ii) vertical details, consisting on checking the basic classifications suggested by means of ISO 12006-2 tables; (iii) chronological details, concerning the classification of the phases of a construction project and the state of the structures; (iv) the main properties, comprising ontological, functional and license; and (v) compliance with other systems, regarding compatibility with BREEAM, LEED, etc., certification systems and with BIM software (e.g. Revit, ArchiCAD, Tekla).

An "ideal construction information classification system" was created that can be used as a basis for comparison between the selected CICS. The different characteristics of the systems do not allow such a direct comparison between them, so the method is adopted to avoid a distortion of the final results.

The overall results are presented in Table 1, with the respective scores obtained by each CICS. The values presented indicate how close the best rated are to the CICS imaginary ideal, as defined based on approved criteria and weights for each aspect. The table is ranked by ranking from the best CICS to the worst.

Table 1: Comparison results. *(Source: Adapted from Faculty of Civil Engineering at the University.)*

General classification			
CICS	Range	Average	Classification
Coclass	80–90%	84.85%	1
CCS	66–82%	74.08%	2
UniClass 2015	67–77%	72.01%	3
TFM a NS6451	59–70%	64.30%	4
CI/SfB	52–75%	63.49%	5
KKS	48–74%	61.24%	6
TALO 2000	53–69%	60.90%	7
OmniClass	53–68%	60.49%	8
Natspec Worksection	45–57%	52.05%	9
Uniformat and Masterformat	42–62%	52.00%	10
SKP	39–56%	47.65%	11
SNIM	44–51%	47.43%	12
TSKP	45–49%	46.99%	13
ASA QS Elemental Class	37–54%	45.28%	14
RTS BIM	30–57%	43.69%	15
CZ-CC	32–53%	42.75%	16
BIM7AA	34–48%	41.12%	17
KSO (JKSO)	24%	23.83%	18

In the study conducted by the Czech researchers, UniClass2015 (United Kingdom), CoClass (Sweden) and CCS (Denmark) systems are the best evaluated.

In addition, it was defined that, because they are globally consolidated CICS, OmniClass (North America) and CCI (a system under development by several northern European countries) would also be included in the analyses carried out by the present research.

Besides the Czech study, the SECClasS Project takes into consideration aspects that are fundamental for the Portuguese AECO sector and that aim to facilitate the implementation and diffusion of the system in Portugal. Specific decision-making information for the choice of a new base system was analyzed: (i) basic information; (ii) structure, tables and compliance with ISO 12006-2 and ISO/IEC 81346; (iii) designation of reference codes; (iv) license and language of the system documentation; (v) Updates and expandability of the system; (vi) sustainability component; (vii) digitization of the system and interconnection with modelling software.

4.1 UniClass2015

The CICS developed by the United Kingdom, the UniClass2015, under the responsibility of Nation Building Specification (NBS) has great advantages. Its structure based on the concepts of ISO 12006-2, is wide, hierarchical and its tables make relationships with each other. The complement of its terminology and coding database is flexible, allowing the introduction of new elements. The system is adapted to the constructive process of the European market, facilitating adaptation to the Portuguese AECO sector. Its updates are systematic, and documentation is available in English. NBS also provides tools that enable interconnection with BIM modeling software. In addition, the SECClasS Project was authorized by NBS to translate and use UniClass2015 as the basic system for a Portuguese CICS.

4.2 CoClass

CoClass was created in Sweden between 2017 and 2018 in response to the digitization of the construction industry and its connection to BIM. This is a complex CICS, and its use is mainly linked to the country of origin [10]. CoClass is structured differently from the “classic CICS” therefore brings differences with respect to UniClass2015 and OmniClass. Its general structure is based on ISO 12006-2 but uses ISO 81346 series for classifications of functional systems, techniques systems, elements, and spaces. The CICS proposes to have a human-readable digital language and machines using the RDS coding system, proposed in ISO 81346, to codify references to products. CoClass has some disadvantages: the documentation is partially available in English and most in Swedish, and CICS is not fully open source, having cost of use, providing only basic tools and classifications for free. Its updates are systematic and allow development and insertion of new classifications. The system considered the state of the structures. When analysing the Property table, information is found regarding durability, mechanical durability, object state or year of reconstruction, according to real estate registration data. Due to limited access to the system, no further information can be verified.

4.3 CCS

Cuneco Classification System (CCS) originates in Denmark, under the responsibility of Molio – Construction Information Centre, BuildingSmart’s Danish partner. The development started in 2010, using ISO 81346 series, in the 2009 version. The purpose of the system is to ensure that digital data can be clearly exchanged between production chains, phases and software's, thereby helping to increase construction productivity. Provide a unique identity for each part of the building and be able to be applied to all parts of the building, from design to operation, so that all parties using the same concepts can be based on each other's information and classify the information. The documentation is in Danish, limited to a part of it and tables translated into English. CICS is available in the basic version free of charge but requires annual subscription for full use of resources. CCS provides tools for sorting objects, along with a system for defining codes and definitions that can be accessed using the CCS Navigate tool. The connection with the modelling software, Revit and ArchiCAD, is made using plug-ins [10]. No sustainability classifications were found.

4.4 OmniClass

OmniClass is a CICS developed by CSI (Construction Specification Institute), CSC (Construction Specification of Canada) and AIA (American Institute of Architects), developed to serve the AECO sector of those countries. It is considered a traditional system, as UniClass2015, which has its structure based on ISO 12006-2. The biggest difference between them is the construction process in which they use as a base, with OmniClass being adapted to the processes of the North American construction sector, but still compatible with other international systems. This CICS is an open source standard to any person or organization, promoting the exchange of information between stakeholders. The development and update have the participation of the industry, but is verified an outage, being the last occurred in 2019. The system has good connection with BIM modeling software, documentation in English and, although the study of the Czech Agency for Standardization affirm that the license terms allow editing and distribution, the information could not be verified about the commercial use. A great advantage of OmniClass, to the SECClasS Project, is the property table, capable to classify information associated with sustainability.

4.5 CCI

The origin of the development of the Construction Classification International (CCI) arises from Estonia's need to develop its own CICS that would serve that country. Soon other neighbouring nations would feel the same need and together the Czech Republic, Finland, Norway, Denmark, Sweden join Estonia for this goal. The CCI working group then decides to use CoClass and CCS as the basis for the new international CICS, believing that the implementation of a common a cross-border classification system improves cooperation, increases standardization, and improves digital working methods. As a result, the CCI core is the common element between ISO, CCS and CoClass standards [11]. The system uses ISO 12006-2 in its general structure and the ISO 81346 series for systems and components. The system aims to be developed in English with possible translations into the languages of the countries that are part of the project. According to the letter of intent signed by the developers, CICS must be free of cost, but a formal use license document as well as upgrade protocols is not available yet. The sustainability component also cannot be evaluated because of the development stage.

5 CONCLUSIONS

The BIM methodology, admittedly, brings advantages to the AECO sector. The applications generate a large amount of information needed to provide designers an increasing number of possibilities for simulation and operationalization of assets. CICS becomes an essential tool, but not the only one, in the management of this information. Therefore, and with the demand for a CICS that meets the needs of the Portuguese market, the SECClasS Project seeks to develop a system based on European and North American experiences that also adds an unexplored component in existing CICS: sustainability.

The study carried out by researchers from the Faculty of Civil Engineering of the Czech Technical University in Prague, in association with the Czech Agency for Standardization, was analysed in this paper and the conclusions were compared with the objectives of the SECClasS Project presenting the following results:

1. None of classification system is perfect, as the requests of each user group are distinct. The flexibility and adaptability of the system is essential, as each country will always have its own classification needs;
2. The CICS's purpose must be in accordance with the BIM methodology and able to increase the sustainability and circular economy, responding to current uses, but also to future uses;
3. Systems that are hyper-adapted to a specific type of construction or a life cycle phase should be avoided;
4. Correspondence with other systems is essential, so they must be in accordance with international standards, in this case ISO 12006-2;
5. More complexes CICS have greater breadth of future application and development. However, the level of complexity should always be evaluated, so as not to derail human interaction and perception;
6. The oldest CICS have advantages such as greater consolidation in the market, acceptance by professionals and the language availability;
7. The CICS's future maintenance, including, but not limited to its expansion, updating and maintenance, are factors to consider, and it is important to define a clear methodology, including and defining responsible and resources;
8. The license to use and adapt, the current dynamism of the entity(s) holder of the rights in the expansion, correction and improvement and responsiveness to external contributions should be investigated and considered in decision-making for a basic CICS.

Based on the reasons presented, the existence of an open license and the authorization for its translation and adaptation, the degree of dissemination and implementation in object libraries, the choice for the base system for the SECClasS Project falls on the UniClass2015 system, managed by NBS. The next steps are: firstly, establish a methodology for adapting and translating the system into the Portuguese language and reality; secondly, establish how to include the sustainability component. Finally, it is of the utmost importance to engage the technical environment, collecting and incorporating their feedback.

ACKNOWLEDGEMENTS

The implementation of BIM in Portugal is a reality and professionals demand for a Portuguese Classification System that facilitates the processing, communication and sharing of information and models that maintains standard communication between all agents involved and throughout the construction life cycle.

SECClasS is a project funded by The EEA Grants with the purpose of developing a Construction Information Classification System for Portugal, with a component focused on sustainability, supporting decision-making that encourages reuse, recycling, durable products, possibility of relocation of structures and minimal waste.

The project aims to facilitate the Circular Economy in Construction by introducing a CICS optimized for sustainability, oriented to the BIM methodology and that will serve not only the sustainability component, but also the other BIM uses. It will also allow unifying terminology at all scales, facilitating communication, selection of materials and components, as well as an accurate assessment of the impacts of buildings throughout their life cycle.

With the development of the SECClasS Project, it is expected to achieve the following results: (i) Portuguese classification system that will allow more advanced BIM uses based on traditional systems with sustainability component: environmental impacts, reuse, recycling; (ii) Manual and Library of BIM Objects, with traditional and sustainability components; and (iii) Online platform with artificial intelligence tools that learns from users.

REFERENCES

[1] Nunes, H.M., *Sistema de Classificação de Informação da Construção – Proposta de metodologia orientada para objetos BIM.* Dissertação de mestrado, Departamento de Engenharia Civil, FCT: Lisbon, Portugal, 2016.

[2] Pereira, R.M.S., *Sistemas de Classificação na Construção. Síntese comparada de métodos.* Dissertação de mestrado, Departamento de Engenharia Civil, Faculdade de Engenharia da Universidade do Porto: Porto, Portugal, 2013.

[3] Lima, R.T. et al., *Sistemas de Classificação da Informação da Construção – Comparação entre as realidades portuguesa e brasileira.* Relatório do Projeto P2I, LNEC: Lisbon, Portugal, 2019.

[4] Balslev, H., *DS Handbook 166:2010 – A Guide to Reference – Preparation of TAG Numbers, Letters Codes, and Interfaces Between Systems.* Danish Standards Foundation: Charlottenlund, Denmark, 2010.

[5] Friborg, G., ICIS PROJECT #3: Classification, Identification and BIM. ICIS – International Construction Information Society: Zurich, Switzerland, 2017.

[6] Balslev, H., The Reference Designation System (RDS): A common naming convention for systems and their elements. *26th Annual INCOSE International Symposium (IS 2016)*, Edinburgh, 2016.

[7] The NBS, What is Uniclass 2015? https://www.thenbs.com/knowledge/what-is-uniclass-2015 Accessed on: 17 Mar. 2021.

[8] Eastman, C. et al., *BIM Handbook: A Guide to Building Information Modelling for Owners, Managers, Designers, Engineers and Contractors,* New Jersey, 2008.

[9] ARC BIM, *BIM: What is it and how does it affect me?*, 2012

[10] Agência Checa para a Normalização, Pesquisa e Comparação dos Sistemas de Classificação dos Elementos de Construção no Contexto do Building Information Modeling (BIM) – Análise de sistemas de classificação selecionados para efeitos da Norma de Dados Checa. Universidade Técnica Checa em Praga, Faculdade de Engenharia Civil: Prague, Czech Republic, 2018.

[11] Saar, J., CCI International – Presentation of Estonia Government Agenda. Ministry of Economic Affairs and Comunications. Estónia, 2019.

[12] CSI, *OmniClass: Introduction and User's Guide.*, p. 29, 2006. http://www.omniclass.org/tables/OmniClass_Main_Intro_2006-03-28.pdf. Accessed on: 19 Mar. 2021.

[13] ISO 12006-2:2015 – Building construction – Organization of information about construction works – Part 2: Framework for classification. http://visnormas.lnec.pt/php/menu.php.html. Accessed on: 8 Mar. 2021.

[14] CSI, 2020. OmniClass. https://www.csiresources.org/standards/omniclass. Accessed on: 19 Mar. 2021.

[15] Svensk Byggtjänst (Swedish Building Centre), CoClass Website. https://coclass.byggtjanst.se/. Accessed on: 24 Mar. 2021.
[16] Molio Construction Information Centre, CCS: Cuneco Classification System Website. https://molio.dk/produkter/digitale-vaerktojer/gratis-vaerktojer/ccs-cuneco-classificationsystem.

TOWARDS SUSTAINABLE BUILDINGS USING BUILDING INFORMATION MODELLING AS A TOOL FOR INDOOR ENVIRONMENTAL QUALITY AND ENERGY EFFICIENCY

AHMED ALI SHABAN ABDELAZIM, MOHAMED ABDELAAL & WALID MOHAMED
Department of Architecture, Faculty of Engineering, Alexandria University, Egypt

ABSTRACT

Nowadays, many people suffer within the existing buildings from problems related to human comfort and energy consumption. The poor quality of the internal environment of buildings for instance uneven temperatures, inadequate ventilation, humidity, poor lighting, acoustics, and noise pollution leads people to use methods that harm the environment. The objective has always been in green thought existing from the continue sustainably. By reviewing the literature illustrating the contributions of BIM to sustainability and its applications in achieving indoor environmental quality within public buildings through conserving energy, natural resources and respecting the needs of its users. So, the field of research is How can BIM is applied to accomplish sustainable Public and housing buildings. BIM technology has created significant benefactions firstly in the design phase, then construction stage, finally in the process of operation, and other perspectives of sustainable buildings. This paper aims to understand BIM technology, its relationship to the sustainability of buildings. This is done by conducting a questionnaire on specialized organizations and companies to learn about the dimensions of BIM and the extent of its use. In addition to discussing the positive and negatives of BIM in sustainable design processes and coming up with a working method that can be applied. The consequences of the questionnaire and analysing literature show that the contribution, restrictions, and barriers of applying BIM for developing sustainable buildings have not been precisely analysed and explained yet. The overlap between BIM and building sustainability needs to be demonstrated practically. The application of BIM in a large number of examples is required to show positive results in raising the quality of the indoor environment and energy efficiency. These solutions which BIM introduces help designers increase the performance of buildings, thus maintaining the life cycle of cities and achieving a suitable quality of life for the inhabitants of the 21st century.

Keywords: BIM, energy efficiency, green building, indoor environmental quality, sustainability, sustainable buildings.

1 INTRODUCTION

Currently, sustainability is a top priority in the agenda of the international community pursued by decision-makers in all aspects of life. Architecture and cities are a reflection of society. There is a strong race among construction workers to achieve more sustainable buildings than other traditional buildings that people are accustomed to. It is necessary to clarify the meaning of sustainable building here. In the previous literature, there are many discussions and studies of the thought of sustainable buildings. These modern explanations can be summarized that sustainable building can be described as a healthy facility created and built in a cradle-to-grave resource-effective pattern, using ecological principles, social equity, life-cycle quality value, etc. promotes a sense of sustainable community [1].

To fully achieve sustainability in buildings, it is necessary here to identify the factors that affect sustainability. After extensive research, sustainability can be reduced to four main headings according to research topics such as water, indoor environmental quality, energy, materials, and waste minimization [2]. Then comes the powerful role of BIM, which is a great tool for achieving sustainability goals. It is noted that the link between

www.witpress.com, ISSN 1743-3509 (on-line)
doi:10.2495/BIM210031

BIM and sustainability has been mentioned in the literature significantly lately. The National BIM Standard-United States describes BIM as "the action of producing an electronic model of buildings for the goal of visualization, architecture analysis, interpretation analysis, code standards checking, expense engineering, as-built output, budgeting, and a lot of other purposes" [3].

BIM is understood as a potentially promising powerful driver guiding the construction and plaining sectors towards sustainability. BIM aims at a high-efficiency construction project. As one of the main applications of BIM, sustainable building construction is driven by energy-efficient [4]. The traditional architectural model faces challenges such as lack of information exchange and low work efficiency at various stages. BIM is a modern technological method that is highly recommended by the construction industry and government authorities [5]. The birth of BIM is based on the combination of construction techniques and information technologies, leading to a revolution in the modern construction industry [6].

The most progressing BIM can reach up to eight dimensions which proposes which it is a 3D visualized conception. It benefits to program and presents the projected growth along over with forecasts (4D = time and schedule). It gives in real-time all cost-data, so it draws an actual budget (5D = cost). The sixth dimension facilities management (6D = operation), then the model also provides environmental impact analysis (7D = Sustainability). The eighth Dimension is about security on site: emergency plans. It prevents security issues (8D = safety) [7].

The research focuses on the seventh purpose of BIM models and the overlay connecting BIM and sustainable design. Architects can quantify the environmental influences of systems and materials to support the determinations needed to produce sustainable buildings, using essential information about sustainable materials that are saved in the database and compatibility between the design process and analysis tools. Such data is useful for building life cycle assessments and important to enhance the internal environment in the design process, which are indoor air quality, Moisture comfort, thermal comfort, lighting comfort, and acoustic comfort.

Architects and engineers become more capable to digitally model the different components of a building (shape, structure, heating/cooling, cost, materials, etc.) in real-time and quickly understand how specific changes in design or construction models will impact other variables like structure, loads, energy efficiency and the fiscal bottom line. BIM has particularly helped empower sustainable design which providing architects with a particular path to high-tech means to carefully integrate parameters like heat gain, ventilation, solar, and energy efficiency. Hence the term BIM in Green Building, or "Green BIM," is a process that can help architecture, engineering, and construction companies increase sustainability in the construction industry. It can allow designers to incorporate and analyse environmental issues in their design over the life series of a building [8].

The benefit of a BIM process is that many types of analysis and simulation can take place as part of the design process, enabling engineers quickly to cycle through experiments and get instant feedback on project achievement. Another advantage of a BIM process is that all reports can be saved in a cloud database. Thus, all information is parametric and that way interconnected, thus making it available for all project partners. Accordingly, a BIM model could be managed holistically during the design-to-build-to-own process since any changes to the BIM model are instantly revealed, thereby promoting up the project process.

This paper is structured in six sections: Section 1 a quick review of sustainability and its relationship to BIM, Section 2 Clarification of the research method and its steps, whereas

Section 3 Presentation of the questionnaire, how to implement it, and the results reached. Section 4 Discuss the results and analyse them. Review the benefits and obstacles that BIM faces in achieving sustainable buildings and, finally, Section 5 summarizes the findings of this work (Fig. 1).

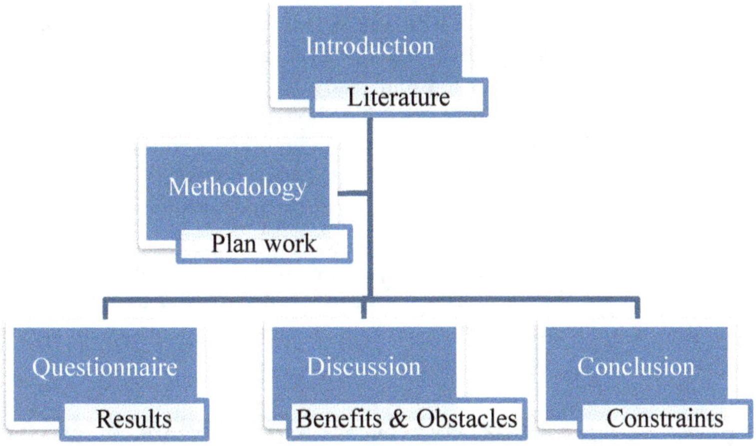

Figure 1: Research structure.

2 METHODOLOGY

The research aims to clarify the contributions of BIM to sustainable building processes, especially on the levels of indoor environmental quality and indoor energy efficiency. With the increasing use of this technology worldwide, the researcher first wants to stand on solid ground and have a clear vision of the extent of using BIM at the local level. The research begins with the conduct of a local questionnaire in Alexandria, Egypt. The extent to which BIM is used in the institutions and offices of architecture, construction, and engineering is determined by asking them a set of questions.

By analysing the results, the positive aspects are quickly highlighted. But more emphasis is placed on the negatives related to the use of BIM to achieve sustainability in the buildings. A discussion of the means and tools that are being used to achieve sustainability is held. In addition to analysing previous successful models in the literature and case studies that achieved tangible and measured successes. Finally, the conclusion comes which clarifying the results of the research. With giving a set of recommendations that help activate the role of BIM in sustainable design.

3 QUESTIONNAIRE AND DATA GATHERING

Through online interviews and site investigation, some closely related companies have selected and are ready to provide answers to the questionnaire. It's issued questionnaires to more than 30 relevant engineers. The questionnaire includes 15 questions with the subsequent viewpoints:

1. Individual Information, qualifications of the taskmaster, BIM training of users, length of serving, and which BIM software is used.
2. Application of BIM in different dimensions in the ACE sector from 2D shapes, 3D visualization, 4D schedule, 5D cost, 6D operation, 7D sustainability, and 8D safety.

3. The possible contribution of BIM to sustainable buildings. In addition, asking about the specific contributions of BIM in building water preservation, indoor environmental quality, energy efficiency, materials, and waste minimization.
4. Factors that impede BIM application to sustainable buildings are classified into technology, budgetary constraints, social environment, and resource management.

Before distributing the questionnaires, the researcher seeks to determine interviews closely linked to the building and construction shareholders and BIM. Various organizations are chosen closely related and ready to give feedback to the questionnaire through online discussions and site inventory. The honesty and validity of the questionnaire are analysed through its review of three questionnaires and opinion poll specialists. The results of the analysis show that the questionnaire results are valid and reliable after deleting some questions. Based on the research conclusions, policies have been suggested to promote effectively sustainable buildings through BIM.

However, before starting the discussion process, there is a need to know the areas of using BIM technology on the ground in the design process. And to what extent are they used by professionals in the processes of (ACE) in the eight dimensions mentioned before? This is clearly shown in Fig. 2.

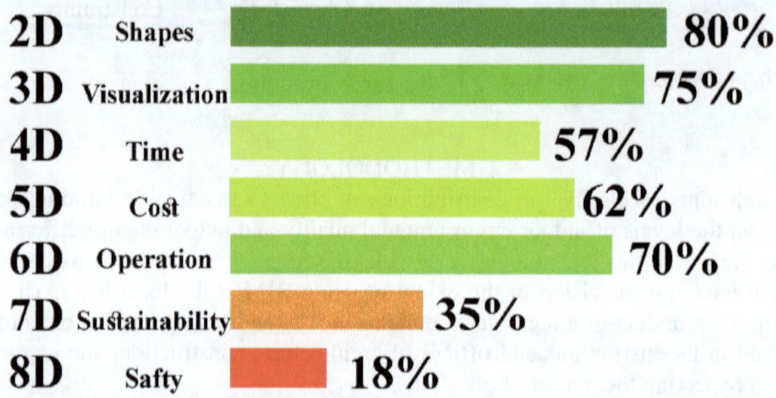

Figure 2: Dimensional frequency diagram for which BIM is used extensively in the design process.

The intensity of the use of BIM in the design process can be divided into three parts. In the foreground comes the use of BIM in two-dimensional drawing operations, as well as making 3D models and external and internal perspectives of buildings. Then it is followed by strongly above-average application of BIM in management, operation, assays, quantities, specifications, determining budgets, costs, schedules, and timelines for the start and end of projects. Finally, at a weak level, but in recent times it is considered promising, which is the use in the field of sustainability, especially in the quality of the internal environment and energy efficiency, as well as safety and security operations inside the workplace. Therefore, in the next part, the contribution and obstacles of BIM in green buildings are analysed by reviewing the questionnaire and the results deduced from it.

4 DISCUSSION, BENEFITS, AND OBSTACLES

Although researchers have discovered many benefits and hurdles of BIM, the participation of BIM to sustainable buildings is not clear. Through the questionnaire conducted by the research, a set of these benefits and obstacles that designers currently face were identified. Moreover, a search for solutions and methodologies in the literature can be applied.

4.1 BIM's contribution to increasing sustainable buildings

Participants think that within BIM software, designers can merge design numerous conveniently. BIM helps users to be more efficient in making design drawings and blueprints, promotes their interaction efficiency with the owners. BIM Technology can straight give targeted information in succession with multiple construction circumstances and put ahead reasonable recommendations for sustainable buildings. Compared with conventional buildings, sustainable building design is a complex work that needs to consider many factors, such as energy-saving analysis, daylighting analysis, sunshine analysis, etc... According to the summary of the questionnaire, it is obvious that BIM can act a significant role in energy preservation, material maintenance, and conserving, improving indoor and outside environments [9].

BIM technicality can additionally support increasing the availability area of outer windows, facilitate accurate simulation of sunlight, and promote natural lighting inside and underground. BIM technology can apprehend the demands of group analysis, centralized presentation, a quick calculation of data, and reports needed for sustainable buildings. All of these would achieve better green energy-saving and produce an extra relaxed living environment for people. In addition, the sustainable building evaluation method can be hugely optimized by BIM's characteristics as simulation and visualization. With BIM, the sustainable building valuation issues can too be recorded easily and traced.

4.2 BIM obstacles in design sustainable building

BIM has good application potential in sustainable buildings. However, many barriers remaining are still, including:

The paradox of BIM software measures and criteria are the most significant obstruction from the technology viewpoint. Not only the shortage of technical criteria but also the delivery criteria of BIM document Production without elaboration. All of these led to numerous disturbances for users in terms of management and control.

The shortage of interactivity among the multiple applications comes second. It is a really essential restriction in the technology aspect. Until now, there are more than a hundred types of BIM software, when software programmers ordinarily only think about the adaptability and flexibility among software in fields they used. Resulting in inadequate interactivity which hugely limits the communication and interchange of information amongst the software. In order to actualize sustainable mentality by using BIM, it should rely on collaboration among the software. At present, it takes a lot of time to build a new building model by importing the information contained in the BIM model into the software of each stage, which leads to the low practicability of the BIM technology [10]. For example, BIM and energy simulation takes place in three main steps: (1) BIM Generate tool, (2) BIM file, (3) Energy simulation tool. Problems often appear when going between these steps. Energy simulation software could be constituted of two major components. The first one is a Graphic User Interface and the second one is a simulation engine. GUIs such as Open Studio, BEopt, Design Builder, and EQuest promote the energy simulation method by implementing a graphical interface for designers. Nevertheless, the second component, which is the energy simulation engines as Energy Plus and Doe2 runs following these tools. BIM file schemas are the second link of the chain. IFC and gbXML extensions are two of the most common types of files, which could carry information on diverse aspects of buildings. In Table 1 shows how common energy simulation tools are aggregated under either energy simulation GUI or simulation engine. In Table 1 shows how common energy simulation tools are aggregated under either energy simulation GUI or simulation engine [11].

Table 1: Categories of energy simulation tools [11].

Energy simulation GUI only	**Energy simulation engine**	**Independent energy simulation engine with graphical interface**
OpenStudio *DesignBuilder* *Hevacomp* *Simergy* *BEopt*	EnergyPlus	Ecotect TRNSYS IDA ICE ApacheSim (used in IES VE) EDSL Tas
GBS eQuest RIUSKA	DOE2	Modelica language (can be used with graphical interfaces such as Dymola)

Based on this explanation, it appears there is still a demand for extensive research on the abilities and skills of these BIM file standards and their interoperability characteristics with various BIM and energy simulation tools to recognize all the latent shortcomings.

The high cost of software, hardware, and maintenance grant and the doubts of BIM expense returns. This high cost of software is clearly visible in the initial cost when adding the BIM to the workflow of the company. This cost includes training BIM technicians, hardware upgrades, licenses for using BIM programs, establishing standard libraries, and the possibility of duplication and returning to programs for two-dimensional graphics for use. However, when balancing the high initial cost against the cost of current tools and labour, saving time and increased working hours when making adjustments and correcting errors that appear during implementation processes. The transition to using BIM technology will save more effort and budget [12]. In the issue of public and environmental obstructions, incomplete BIM standards, the lack of proper Politicking directing and support the major obstacle. Many respondents consider that the sharp deficiency of BIM specialists in companies is also a hindrance that cannot be ignored.

The shortage of knowledge and applicability of standards such as ISO 19650 is a significant difficulty, for example, lack of necessities, minimum preparation, not examining the method of model evolution and replacement, and delaying sustainability to the end steps in the operation process, which leads to it being a complementary and non-essential step in the design process.

The inconsistency of BIM software standards, Of course, is not necessarily correct. But this happens due to the lack of knowledge, experience, and understanding of global standards for BIM, for example, but not limited to ISO 19650. It is a universal standard for managing and handling information across the whole life cycle of a constructed asset using BIM. It includes all the identical policies, principles, and essential requirements Like as United Kingdom BIM Framework, and is almost followed by the contemporary UK 1192 regulations and restrictions.

There are many other problems, but what is of interest to this research is summarized that the lack of smooth connection between BIM and sustainable building simulation tools such as Energy Plus, FLUENT, PKPM, and e-QUEST [11]. While BIM is mainly used in the building design phase, it fails to provide adequate support for subsequent phases such as operation and maintenance. However, they are considered key evaluation stages for sustainable buildings. Therefore, a clear and proven practical method must be found that achieves the desired goal of using BIM in sustainable design.

4.3 Methodology for design sustainable buildings

The application process of implementing BIM for indoor environmental performance analysis is in general terms highlighted by Fig. 3. The suggestion of process is based on the circumstances of a smart 3D augmented simulation construction form, provided by containing all features of the building information into an intelligent format. This method is used to improve the sustainable performance of a building before committing to the design process [13]. That is Meaning after evaluating an initial indoor environmental performance analysis and quality. It is reasonable to modify the building design for performance enhancements.

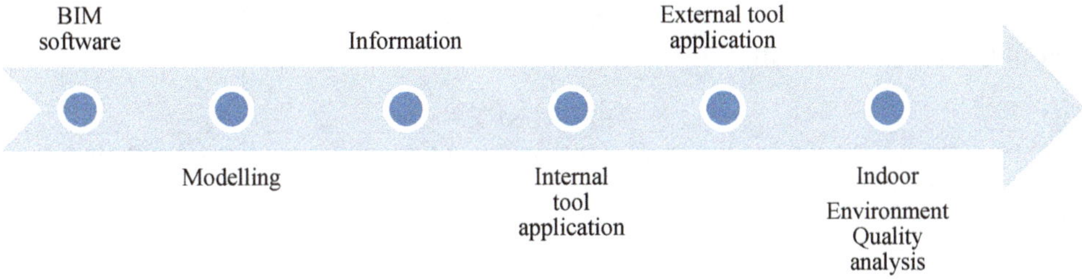

Figure 3: The BIM workflow for indoor environmental quality analysis.

Likewise, the indoor environment is related to buildings' energy consumption. Therefore, the research focuses on achieving the quality of the internal environment, which in turn will achieve energy saving and efficiency. Furthermore, the costs of a poor indoor environment are often more expensive than the expenses of regular energy used in the building due to the lower performance of the users, whether residents or workers. It is important to analyse all parameters of the internal environment in the design process, which are indoor air quality, Moisture comfort, thermal comfort, lighting comfort, and Comfortable acoustic range. With all these parameters, BIM can raise its efficiency and deal with to achieve the quality of the internal environment.

5 CONCLUSION

BIM is taking important steps to become the Ideal technology for users in the ACE industry. There is no doubt that BIM is facing many difficulties now to reach this position. However, the current and future research plans are considered promising and predict many positives in the coming periods.

This study discusses the contribution of BIM to sustainable buildings. Through the questionnaire, it is concluded that because the usage of BIM increases in its first dimensions, the seventh dimension of sustainability is nevertheless in its cradle. BIM can raise indoor environmental quality and achieve energy efficiency by improving indoor air quality, Moisture comfort, thermal comfort, lighting comfort, and acoustic comfort. The problems facing the use of BIM in designing sustainable buildings are (1) The inconsistency of BIM software standards is not necessarily true, and if it appears, it may be due to a lack of knowledge of international standards, (2) the lack of interaction between different applications, and (3) the expensive price of software, hardware, and maintenance as an initial cost. Access to a BIM workflow for internal environmental quality analysis that can be used in sustainable design.

This research is a step in the way of identifying BIM as a powerful tool and a framework to achieve sustainability, which opens the way for more research on this topic. The next step by the researcher will be a case study on an already existing building. And how to convert it from a traditional building to a sustainable building using BIM and building energy simulation modelling (BEM).

ACKNOWLEDGEMENTS

Thanks and appreciation to everyone who helped collect information for this research and the researchers who reviewed their literature searches. As well as all those who participated in the questionnaire from co-workers and companies related to the design, construction and implementation processes. Special thanks to the body responsible for publishing the research, Wessex Institute at this international conference on BIM technology, which constitutes a sustainable future.

REFERENCES

[1] Berardi, U., Clarifying the new interpretations of the concept of sustainable building. *Sustainable Cities and Society*, **8**, pp. 72–78, 2013.

[2] Khaddaj, M. & Srour, I., Using BIM to retrofit existing buildings. *Procedia Engineering*, **145**, pp. 1526–1533, 2016.

[3] NBIMS-US, National BIM Standard-United States®, National Institute of Building Sciences, Washington, DC. http://mddb.apec.org/Documents/2013/SCSC/WKSP5/13_scsc_wsp5_007.pdf. Accessed on: 15 Jan. 2021.

[4] Reizgevičius, M., Ustinovičius, L., Cibulskienė, D., Kutut, V. & Nazarko, L., Promoting sustainability through investment in building information modelling (BIM) technologies: A design company perspective. *Sustainability*, **10**(3), pp. 600, 2018.

[5] Li, L. & Wang, L.X., Application of BIM technology in green building engineering construction, pp. 1301–1305, 2013.

[6] Haruna, A., Shafiq, N. & Montasir, O., Building information modelling application for developing sustainable building (multi-criteria decision-making approach). *Ain Shams Engineering Journal*, **12**(1), pp. 293–302, 2020.

[7] Smith, P., BIM implementation–global strategies. *Procedia Engineering*, **85**, pp. 482–492, 2014.

[8] Krygiel, E., & Nies, B., *Green BIM: Successful Sustainable Design with Building Information Modelling*, Wiley: Indianapolis, pp. 26–35, 2008.

[9] Huang, B. et al., Contribution and obstacle analysis of applying BIM in promoting green buildings. *Journal of Cleaner Production*, **278**, 123946, 2021.

[10] Shehzad, H.M., Ibrahim, R.B., Yusof, A.F., Khaidzir, K.A., Iqbal, M. & Razzaq, S., The role of interoperability dimensions in building information modelling. *Computers in Industry*, **129**, 2021.

[11] Kamel, E. & Memari, A.M., Review of BIM's application in energy simulation: Tools, issues, and solutions. *Automation in Construction*, **97**, pp 164–180, 2019.
[12] Shin, M., Lee, H. & Kim, H., Benefit-cost analysis of building information modelling (BIM) in a railway site. *Sustainability*, **10**(11), p. 4303, 2018.
[13] Lim, Y., Building information modelling for indoor environmental performance analysis. *American Journal of Environmental Sciences*, **11**(2), pp. 55–61, 2015.

SECTION 2
ADVANCED BIM PLATFORMS

VIRTUAL REALITY STORIES FOR CONSTRUCTION TRAINING SCENARIOS: THE CASE OF SOCIAL DISTANCING AT THE CONSTRUCTION SITE

FELIPE MUÑOZ-LA RIVERA[1,2,3], JAVIER MORA-SERRANO[1,2] & EUGENIO OÑATE[1,2]
[1]International Centre for Numerical Methods in Engineering, Universitat Politècnica de Catalunya, Spain
[2]School of Civil Engineering, Universitat Politècnica de Catalunya, Spain
[3]School of Civil Engineering, Pontificia Universidad Católica de Valparaíso, Chile

ABSTRACT

The architecture, engineering, construction and operation industry has high accident rates, and its workers are susceptible to certain risks due to the variability of their worksites, the tasks they perform and the equipment they use. Of particular interest is the coexistence of workers and heavy machinery in the same space, which is one of the main causes of high accident rates. Increasing digitisation also incorporates the use of new systems (e.g. UAVs, autonomous vehicles, robotic devices), which must function while preserving worker safety. As such, worker training is also essential in these contexts. Research indicates that occupational risk prevention training in construction is not satisfactory. Traditional methods based on lectures (where the student does not take an active role) and traditional test-type assessments do not result in significant learning and fail to create a culture of prevention among workers. In the context of the COVID-19 pandemic, the construction industry has had to incorporate new prevention measures to continue projects while protecting workers' health and safety. Virtual reality has demonstrated advantages in adapting to different contexts, allowing for the agile development of training experiences with immersive characteristics and more significant learning than traditional methods. This research shows the application and agile development of a training experience for social distancing at construction sites to prevent COVID-19 transmission, based on virtual reality and building information modelling and using serious games as a teaching strategy.

Keywords: virtual reality, BIM, construction site, safety in construction, storytelling, job safety analysis technicians (JSAt), COVID-19.

1 INTRODUCTION

The construction industry is highly relevant to countries' development, accounting for about 6% of the world's gross domestic product [1]. It is also one of the most dangerous industries worldwide, with more than 30% of total recorded accidents in the sector being fatal [2]. Construction sites are complex and demonstrate diverse work dynamics, including workers with different training levels, disciplines and activities [3]. In addition, working outside subjects construction workers to unpredictable weather conditions This, coupled with the use of heavy machinery, dangerous tools and large materials, makes it difficult to manage safety on construction sites [4].

Given this complexity, events like the COVID-19 pandemic present even greater challenges for a sector where health and safety management was already difficult to manage in normal times. The importance of the construction industry as an engine of global development – and its relevance in terms of the jobs it provides (especially to low-skilled workers), accounting for 18 million jobs in Europe alone – challenges the sector to reinforce its preventive health and safety measures to enable workers to continue working even during the pandemic.

In addition to these aspects, several authors have pointed to the lack of real awareness or significant training regarding health and safety issues in construction. The

WIT Transactions on The Built Environment, Vol 205, © 2021 WIT Press
www.witpress.com, ISSN 1743-3509 (on-line)
doi:10.2495/BIM210041

implementation of a safety culture is one of the most crucial needs in this sector. However, traditional safety training for construction workers has not had the expected effects. Lectures with little active student participation, unattractive training material and simple evaluations assessing memory are not enough. Thus, immersion in Building Information Modelling (BIM) models and training in virtual reality (VR) environments have emerged as strategies to bridge these gaps. VR allows for first-person immersion in realistic scenarios, allowing workers to interact with dynamic and risky environments [5]. The incorporation of serious game (SG) elements provides a layer of gamification, which captures workers' attention and keeps them challenged in the virtual simulations.

Although VR, virtual scenarios and SGs have proven effective in safety training in association with learning tasks, they have not succeeded in leading workers to fully internalize the message of a safety culture [6]. This goes beyond learning how to safely perform a task to extend into the realm of emotionality – that is, the worker should become truly aware of the effects of accidents and the need to behave responsibly. Under this approach, storytelling allows researchers to study the role of emotions and cultural factors as determinants of people's behaviour and a means of deepening users' awareness; users can experience the content as a protagonist who is responsible for the outcome of the story, which offers emotional meaning beyond theoretical content on safety protocols [7].

Thanks to these technologies and methods, job safety analysis technicians can quickly and easily build customised VR experiences to improve construction workers' safety awareness. Topics might include awareness of new, much more robotic-based building sites; safe ways to interact with new heavy machinery; improved sensor-based warnings for working at heights; and the use of ladders, scaffolding and ropes alongside dangerous drops, among others [8]. A last the focus of this work is social distancing adapted to the new post-COVID-19 era.

As a specific example of this versatile and fast technique, this work shows the step-by-step procedure of creating social distancing learning material for construction workers. This research shows the results of the development of a virtual experience for social distancing awareness at construction sites in the context of the COVID-19 pandemic. The tool draws on the benefits of development environments for VR experiences and their link to BIM models, which are associated with realism and immersion. In addition, scenarios for worker decision-making aligned with the generation of storytelling were developed. The aim of this research is to raise awareness of and evaluate workers' behaviour with regard to social distancing and how this new aspect can trigger the occurrence of other risky actions.

2 RESEARCH METHODOLOGY

The Design Science Research Methodology (DSRM) was used to guide the entire process of justification, development and testing [9]. In general, the DSRM involves four stages: (1) identifying observed problems and motivations; (2) defining the objectives of a potential solution; (3) design and development; and (4) demonstration. Fig. 1 shows a diagram of the research methodology. In the first stage, a literature review was carried out to identify the characteristics of construction safety; the impact of COVID-19; the forms of construction safety training; and how VR, SGs and storytelling improve training. The literature search was conducted on the Scopus and Web of Science databases. In the second stage, the objective of a potential solution was defined: use of VR, SG and storytelling can help safety training to raise awareness about social distancing. In the third stage, the conceptual proposal was created, and the scenario and algorithms for virtual interaction were developed. Finally, in the fourth stage, the virtual experience was implemented and initially evaluated.

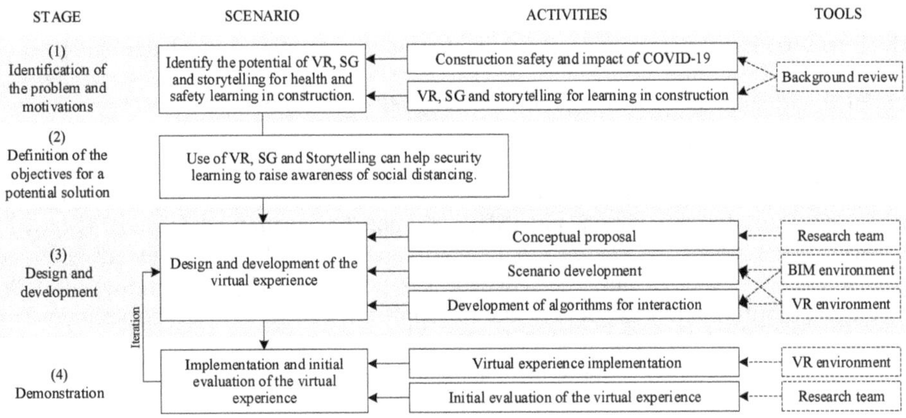

Figure 1: Research methodology.

3 BACKGROUND

3.1 Health and safety in construction

The construction sector worldwide has a high rate of occupational accidents due to high levels of risk and major safety challenges. Construction workers tend to have a higher probability of suffering an accident in the workplace because of physical overexertion, falls and trips, collisions or blows against moving objects, contact with sharp materials or being run over by construction machinery [10]. Other causes of fatal occupational accidents include heart attack and stroke, traffic accidents, entrapment and amputations, collisions with moving objects and falls [11]. In addition, the most important risk factors influencing safety performance in the construction industry are associated with poor work site setup and organisation, lack of coordination, poor communication, constant workplace changes, the different languages, cultures and educational backgrounds of the workforce, inadequate training, and the absence of a culture of prevention and safety [12].

The COVID-19 pandemic has undoubtedly impacted the construction sector. In particular, in the area of safety, the pandemic has forced the industry to incorporate new prevention measures, which are not easy to implement and control in a sector that already experienced difficulties in normal times [8]. Masking and social distancing, combined with the stressful and challenging situations that workers on construction sites already experience, increase the variables that workers must consider when carrying out their routine tasks [13].

3.2 Health and safety training in construction

Training in occupational health and safety in construction is considered crucial in reducing accident rates and moving towards a zero-accident culture [14]. Traditional training includes classroom lessons on theoretical concepts, manuals and digital syllabi. In this type of training, the worker plays a passive role and learning is verified through (inefficient) final questionnaires [15]. Studies show that the level of information retention decreases over time, so it is important to use tools that generate lasting learning. Traditional methods are limited to complying with the norms associated with prevention, with no intention of taking action or implementing training measures that lead to meaningful learning (for which real-world experience is required in addition to theoretical knowledge) [16].

3.3 BIM, virtual reality, serious games and storytelling for training in construction safety

In line with the trend of digitalisation in the construction sector and the increasing use of BIM, new tools for prevention and safety training have emerged [17]. BIM-based VR scenarios make it possible to generate a fully immersive, lifelike environment wherein the user experience relates to how the real world is perceived [18]. In addition, through the integration of information technology and the application of games, SGs make it possible to recreate real danger zones in a virtual environment to promote learning by configuring video games to establish a series of challenges and cognitive-behavioural changes via reward and penalty mechanisms to achieve the desired objectives [16].

Incorporating storytelling into these virtual training experiences has indicated a little-Incorporating storytelling into these virtual training experiences shows a preliminary promising series of series of characteristics that open new ways to improve the significance of the learning [19]. Storytelling refers to the art of presenting a narrative with elements that generate emotionality and transport recipients, enabling them to in a meaningful way and learn from the events being described [20]. Storytelling promotes the development of effective and selective attention, reasoning skills and cognitive skills. Its application in accident prevention has been observed in the context of road safety education. Storytelling can be used to study the role of cultural factors and emotions as determinants of people's behaviour [21]. Thus, in virtual immersive environments, the users experience content as protagonists responsible for the outcome of the story and thus receive emotional meaning beyond the theoretical content on safety protocols and standards [22].

4 DESIGN AND DEVELOPMENT OF THE VIRTUAL EXPERIENCE

For the development of the virtual experience, the conceptual design route considers 3 elements: (1) Learning goals, (2) VR experience goals, and (3) Decision tree linked to learning options depending on the Safety head. The learning objectives are related to: awareness of social distancing, the importance of wearing a mask, and the identification of unsafe work areas.

On the other hand, VR experience goal is to get the worker to a safe zone. In order to do so, many kind of conventional missions can be assigned, in accordance with the project requirements. The decision tree is designed to allow the user to experience the following options: the worker enters the construction site. In order to reach the defined safe zone, the worker must traverse a large part of the construction site. On the route (something to be decided by the user), the worker encounters groups of workers who must be avoided (so as not to violate social distancing). In addition, unsafe zones are placed near these workers to expose the worker to other potential risks. In other words, the worker must maintain awareness not only of people but also of unsafe typical areas of the construction site. Fig. 2 shows a simplified schematic of the virtual experience route.

Fig. 3 shows the steps to develop the virtual experience: (1) Technical project and occupational risk prevention information is obtained. With this information, (2) a BIM model of the building is designed, and its textures are improved (to obtain higher levels of realism in the virtual environment), (3) in the VR environment, physics, gravity, context elements, building site objects are added, along with programming for the user and

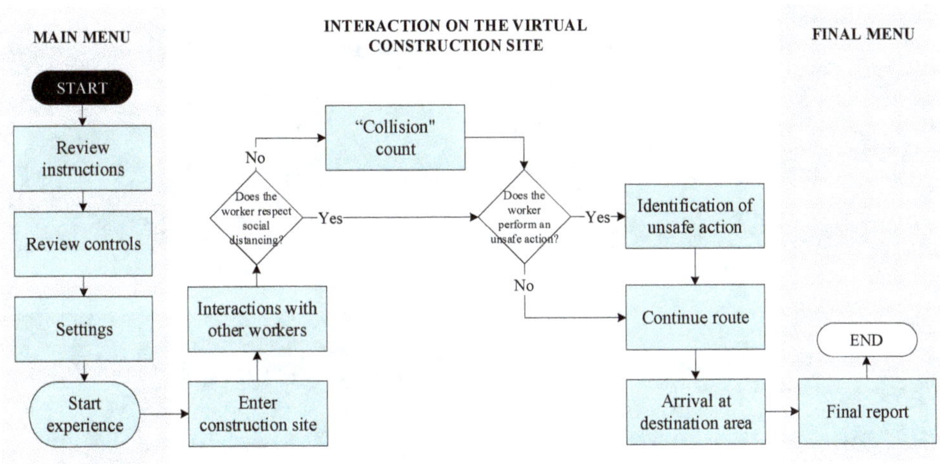

Figure 2: Diagram of virtual experience route.

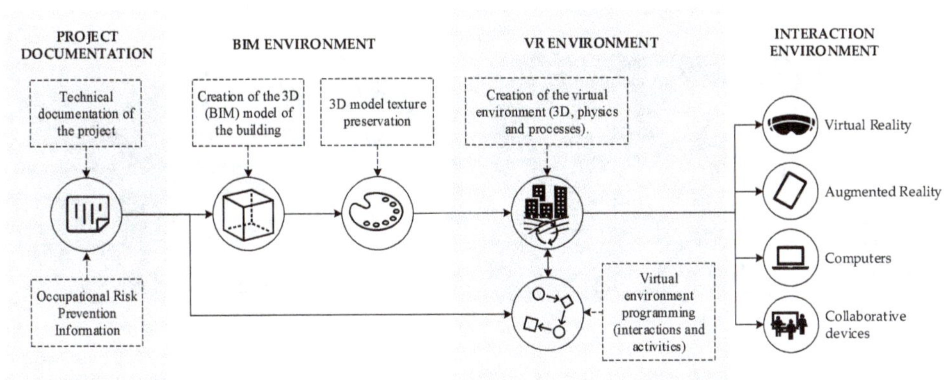

Figure 3: Tool flow for creation of virtual experience.

environment interactions. Finally, (4) the virtual experience is exported for viewing on different devices. For this particular development, the tools used included: Autodesk Revit, 3D Studio Max, Unity 3D, Visual Studio (C#) and computer visualization of the experience.

In accordance with the training objectives, different elements were created to generate realism in the virtual environment based on the project information and security plan. Fig. 4 shows the objects in the virtual construction site: Fig. 4(a) shows the plan view of the construction site; Fig. 4(b) shows the context of the city where the construction site is located (trees, surrounding buildings, streets and cars, among others); Fig. 4(c) shows the main building (construction phase); Fig. 4(d) shows excavation areas and details of safety fences; Fig. 4(e) shows workers and material stockpile area; and Fig. 4(f) shows machinery, safety elements and secondary buildings.

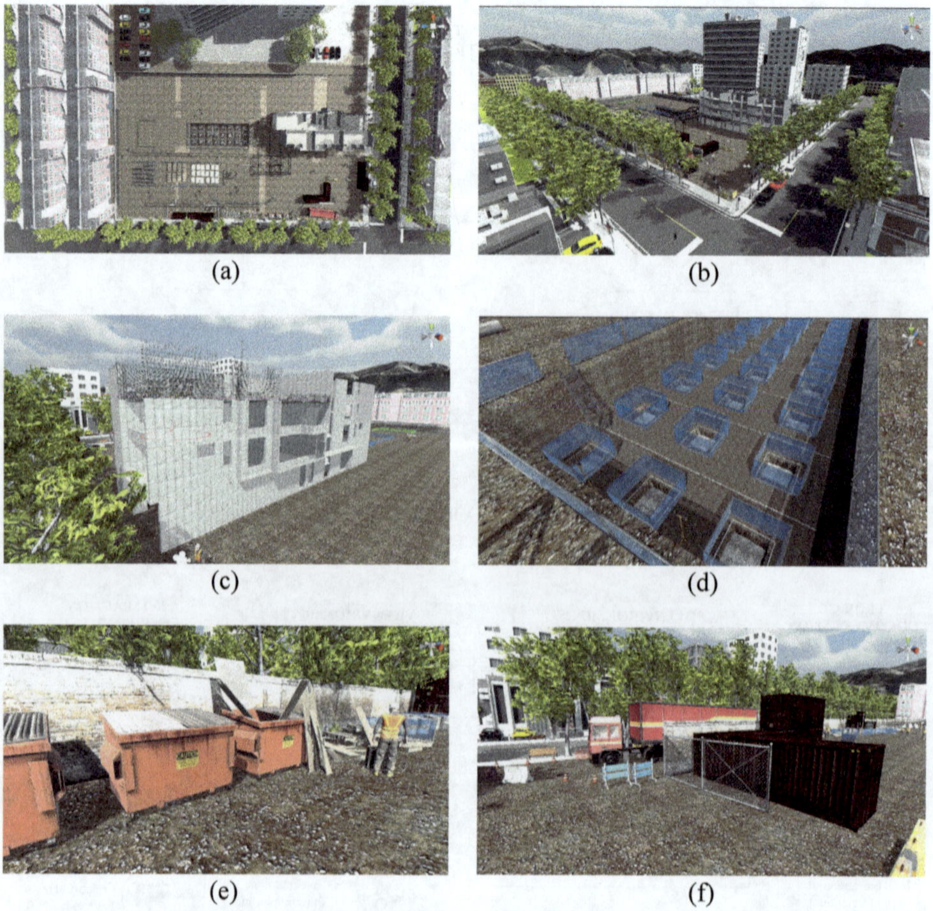

Figure 4: Virtual scenario elements. (a) General view of the construction site; (b) General site view and general elements; (c) Building view in the construction phase, imported from building information modelling environment; (d) Excavation areas; (e) Stockpile areas and other workers performing tasks; and (f) Auxiliary buildings, safety objects and machinery.

Characters were created to enable the worker (player) to interact with other (virtual) workers and to generate the social distancing aspects of the virtual experience. Based on the design of the main bot representing the player [23], elements were added to detect collisions with other workers (i.e. minimum social distance). Fig. 5 shows the collision zone and interactions between workers. Different options for prevention measures for COVID-19 may arise. One of these is the wearing of face masks: how important is the way the people wear the mask? Are we able to identify more risky situations depending on how the people wear the mask? In response to these and other questions, and to study the behaviour of workers in different cases, Fig. 6 shows the different mask-wearing options that were implemented.

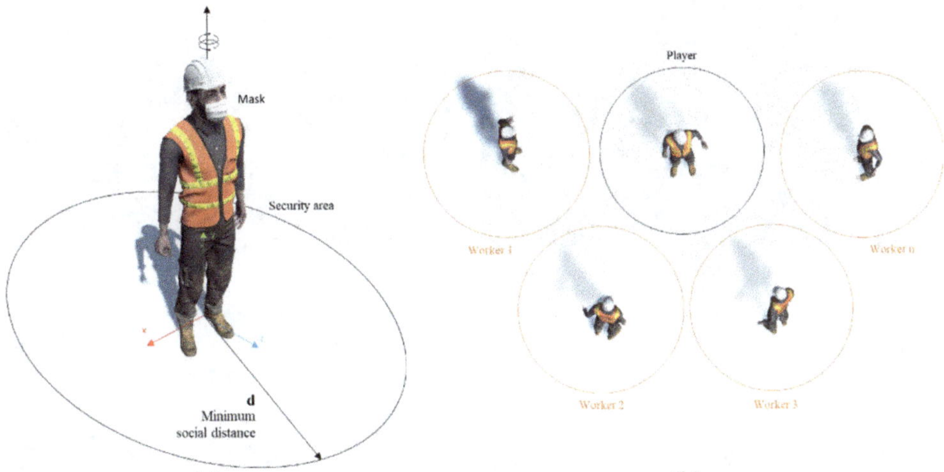

(a) (b)

Figure 5: Safety areas for minimum social distancing. (a) Player's personal safety area; and (b) Player interaction with other workers in the virtual environment.

(a) (b) (c) (d)

Figure 6: Options for mask wear by workers. (a) Correct wearing of mask; (b) Incorrect wearing of mask; (c) Incorrect wearing of mask; and (d) No mask.

Table 1: Algorithms developed for interaction in the virtual environment.

N	Algorithm name	Description
1	Start menu	Allows the player to access or exit the virtual experience: instructions, settings, credits and start of the experience.
2	Player animations	Enables the movement and interaction of the main bot associated with the player.
3	Worker animations	Enables the movement and interactions of the bots located in the different parts of the construction site, both 'human-like' movement and construction-related activities.
4	Collision	Establishes interaction zones for each virtual worker; detects collisions between the main bot and the rest of the workers (when social distance is violated).
5	Collision counter	Enables traceability of the main bot's behaviour provides information on the cumulative number of collisions.

Figure 7: Scenes in virtual experience route. (a) Start menu; (b) Game start at entrance to construction site; (c) Example of virtual interaction with another worker; (d) Example of actions performed by other workers; (e) Interaction with group of workers; (f) Interaction with workers and unsafe environment; (g) Arrival at 'end zone'; and (h) Final menu.

Different algorithms were created to control the interactions of the main player, the activities of the construction site workers and the gamification aspects. Table 1 shows the five main algorithms used in the development of the virtual experience.

5 DEMONSTRATION OF THE VIRTUAL EXPERIENCE

Fig. 7 shows the different 'scenes' of the route (not fixed, since the worker is free to move around the construction site). Until the defined goal area is reached, the player is able to see the start menu, the worker's trajectory and the worker's interactions with the surroundings and other workers. In Fig. 7(a), the main menu is shown, with buttons for Instructions, Configuration, Credits, and Start Experience. Fig. 7(b) shows the start of the experience and the user interface, composed of an interaction counter and a mini-map, where it is possible to see the interaction zones of the user and the workers. Fig. 7(c)–7(f) show possible interactions of the user with other workers (using the mask incorrectly, performing routine work, group worker chats, and interaction with the environment in general). Fig. 7(g) shows the worker arriving at the safe zone, and Fig. 7(h) shows the final menu, where information on the number of interactions where the minimum social distancing is exceeded is shown.

Within the experience, which is focused on health and safety, the worker's goal is not only to maintain social distance from other workers but also to avoid risky behaviour more generally. The experience thus allows for the evaluation of how the new condition of social distancing could lead to incorrect decision-making with regard to safety matters on construction sites. For example, Fig. 8 shows part of the route taken by the player where he encounters a group of workers who must be avoided. In this context, the worker has two options: a) avoid the workers by walking to the left, which is the quickest route but would force him to walk too close to an unprotected excavation; or b) avoid them by walking to the right – a longer, but safer, route.

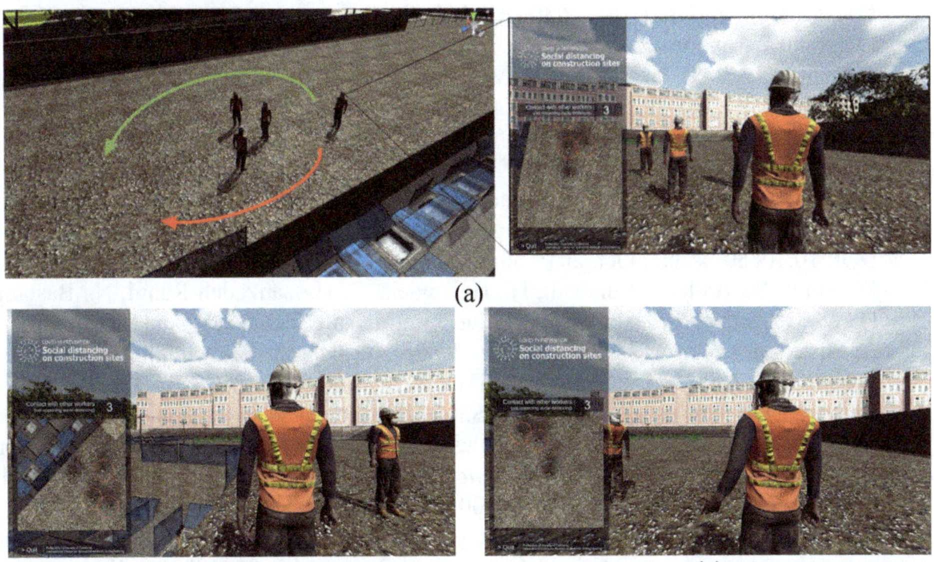

(a) (b) (c)

Figure 8: Example of decision-making: maintaining social distance and/or choosing a safe route. (a) Player faces best route choice decision; (b) Wrong decision (risk of fall); and (c) Correct decision (risk-free, safe area).

6 CONCLUSIONS

This research successfully created a virtual experience to raise construction workers' awareness of COVID-19 prevention issues with a focus on maintaining social distancing. The developed game incorporated the integration of BIM models, the generation of realistic construction site environments and the programming of interactions and gamification aspects. It was thus demonstrated that VR can be adapted to different context, enabling the agile and simple development of customised VR experiences to raise awareness among construction workers in order to improve their safety. Thus, this research shows guides, tools, procedures, and examples for the development of virtual learning experiences.

ACKNOWLEDGEMENTS

This work has been supported by the CONICYT for its economic support to Felipe Muñoz, beneficiary of a pre-doctoral grant (CONICYT – PCHA/International Doctorate/2019-72200306) and by the Ministry of Science, Innovation and Universities of Spain (MICIU) through the BIMIoTICa project (RTC-2017-6454-7). The authors also acknowledge the financial support from the Spanish Ministry of Economy and Competitiveness, through the "Severo Ochoa Programme for Centres of Excellence in R&D (CEX2018-000797-S)". The research team would like to thank Ignacio Valero (professor at the School of Civil Engineering of the Polytechnic University of Catalonia and researcher at CIMNE) for his constant support regarding health and safety issues in construction.

REFERENCES

[1] Global Construction Perspectives and Oxford Economics, *A Global Forecast for the Construction Industry to 2030*. Global Construction Perspectives and Oxford Economics: London, 2015.

[2] Jannadi, O.A. & Bu-Khamsin, M.S., Safety factors considered by industrial contractors in Saudi Arabia. *Building and Environment*, **37**(5), pp. 539–547, 2002. DOI: 10.1016/S0360-1323(01)00056-7.

[3] Ahmed, S., Causes and effects of accident at construction site: A study for the construction industry in Bangladesh. *International Journal of Sustainable Construction Engineering Technology*, **10**(2), pp. 18–40, 2019. DOI: 10.30880/ijscet.2 Oct. 2019.003.

[4] Khosravi, Y., Asilian-Mahabadi, H., Hajizadeh, E., Hassanzadeh-Rangi, N., Bastani, H. & Behzadan, A.H., Factors influencing unsafe behaviors and accidents on construction sites: A review. *International Journal of Occupational Safety and Ergonomics*, **20**(1), pp. 111–125, 2014. DOI: 10.1080/10803548.2014.11077023.

[5] Bhagwat, K., Kumar, P. & Delhi, V.S.K., Usability of visualization platform-based safety training and assessment modules for engineering students and construction professionals. *Journal of Civil Engineering Education*, **147**(2), 04020016, 2021. DOI: 10.1061/(asce)ei.2643-9115.0000034.

[6] Joshi, S. et al., Implementing virtual reality technology for safety training in the precast/prestressed concrete industry. *Applied Ergonomics*, **90**, 103286, 2021. DOI: 10.1016/j.apergo.2020.103286.

[7] Muñoz La Rivera, F., Mora Serrano, J., Delgado, C., Briceño, C., Vera, R. & Núñez, T., Storytelling in mixed reality experiences to develop a culture of occupational risk prevention in construction, *VIII Convención Internacional de la Edificación*, 2020.

[8] Ogunnusi, M., Hamma-Adama, M., Salman, M. & Kouider, T., COVID-19 pandemic: The effects and prospects in the construction industry. *International Journal of Real Estate Studies*, **14**(2), p. 120, 2020.

[9] Peffers, K., Tuunanen, T. & Gengler, C., The design science research process: A model for producing and presenting information systems research. *Journal of Management Information Systems*, pp. 83–106, 2006.
[10] Winge, S., Albrechtsen, E. & Mostue, B.A., Causal factors and connections in construction accidents. *Safety Science*, **112**, pp. 130–141, 2019. DOI: 10.1016/j.ssci.2018.10.015.
[11] Yap, J.B.H. & Lee, W.K., Analysing the underlying factors affecting safety performance in building construction. *Production Planning & Control*, 2019. DOI: 10.1080/09537287.2019.1695292.
[12] Abas, N.H., Yusuf, N., Suhaini, N.A., Kariya, N., Mohammad, H. & Hasmori, M.F., Factors affecting safety performance of construction projects: A literature review. *IOP Conference Series: Materials Science and Engineering*, **713**(1), 2020. DOI: 10.1088/1757-899X/713/1/012036.
[13] Wang, Z., Liu, Z. & Liu, J., Risk identification and responses of tunnel construction management during the COVID-19 pandemic. *Advances in Civil Engineering*, 2020. DOI: 10.1155/2020/6620539.
[14] Teizer, J., Right-time vs real-time pro-active construction safety and health system architecture. *Construction Innovation*, **16**(3), pp. 253–280, 2016. DOI: 10.1108/CI-10-2015-0049.
[15] Yilmaz, M. & Kanit, R., A practical tool for estimating compulsory OHS costs of residential building construction projects in Turkey. *Safety Science*, **101**, pp. 326–331, 2018. DOI: 10.1016/j.ssci.2017.09.020.
[16] Gao, Y., Gonzalez, V.A. & Yiu, T.W., The effectiveness of traditional tools and computer-aided technologies for health and safety training in the construction sector : A systematic review. *Computers & Education*, **138**, pp. 101–115, 2019. DOI: 10.1016/j.compedu.2019.05.003.
[17] Muñoz-La Rivera, F., Mora-Serrano, J., Valero, I. & Oñate, E., Methodological - Technological framework for construction 4.0. *Archives of Computational Methods in Engineering*, **28**(2), pp. 689–711, 2021. DOI: 10.1007/s11831-020-09455-9.
[18] Kassem, M., Benomran, L. & Teizer, J., Virtual environments for safety learning in construction and engineering: Seeking evidence and identifying gaps for future research. *Visualization in Engineering*, pp. 1–15, 2017. DOI: 10.1186/s40327-017-0054-1.
[19] Pedro, A., Le, Q.T. & Park, C.S., Framework for integrating safety into construction methods education through interactive virtual reality. *Journal of Professional Issues in Engineering Education and Practice*, **142**(2), pp. 1–10, 2016. DOI: 10.1061/(ASCE)EI.1943-5541.0000261.
[20] McHugh, K. & Klockner, K., Learning lessons from rail safety storytelling: Telling safety like it is. *Safety Science*, **122**, p. 104524, 2020. DOI: 10.1016/j.ssci.2019.104524.
[21] Oschatz, C. & Klimmt, C., The effectiveness of narrative communication in road safety education: A moderated mediation model. *Communications*, **41**(2), pp. 145–165, 2016. DOI: 10.1515/commun-2016-0003.
[22] Cullen, E. & Fein, A., Tell me a story: Why stories are essential to effective safety training. *Report of Investigations*, 2005. DOI: 10.1242/jcs.02992.
[23] Mixamo, Worker character. https://www.mixamo.com/#/?page=1&query=worker&type=Character. Accessed on: 10 Jan. 2021.

CHANGING LEARNING LANDSCAPE: FROM CAD TO BIM AND BEYOND

DAVID EDWARD MORTON[1,2] & IGNASI PÉREZ ARNAL[1,2]
[1]Department of Architecture, Northumbria University at Newcastle, UK
[2]BIMAcademy.es, IBIMA, Spain

ABSTRACT

This study considers how and when students actively move between analogue (traditional), and digital methods (CAD/BIM) of design exploration. It maps the design process to explore how traditional methods with the convention of abstraction are becoming intermixed with the opportunities of synthetic simulation, via digital methods, during the students' design process. The studies original contribution to knowledge focuses on a deeper understanding of how and when analogue and digital methods are used in the design process, collated to design out comes. It is, foremost, the first study in this subject area to analyse the use of CAD and BIM in parallel with traditional methods of abstraction and simulation in the design process.

Keywords: computer aided design, building information modelling, design process, academia, learning.

1 INTRODUCTION

The use of CAD/BIM in the design process allows the opportunity to query the roles and rules of the traditional architectural conventions of visual communication. The universal use of two-dimensional, orthographic projection has become the universal tradition conventions of visual communication which are now understood by contemporary architects originated in the fifteenth century [1]. The increasing use of digital methods, such as CAD, being used in the design process are now changing both drawing conventions and the very design process itself. Extant literature suggests recognises that there are cognitive processes that run concurrent with activities that can consciously and subconsciously aid or inhibit the design process [2].

In the modern era, with the invention and subsequent use of digital technologies, the design process has become even more difficult to conceptualise and define. Therefore, students of design subjects are even more likely to find it difficult to navigate through their design process to a successful outcome. Educators are not always equipped to help, having often been educated to use analogue methods [3]. There is thought that designing on the computer may involve different cognitive processes to that of analogue methods [4]. It has been argued that the computer has aided an architect's ability to operate creatively [5] others are far less positive, especially in the educational environment. McVie and Paton-Walsh [6] referring to a study undertaken in the US, concluded that students found it 'far more difficult and less interesting to learn from the screen than through analogue methods, such as drawings and physical model making'. It is therefore difficult to fully comprehend the impact of the computer on the design process generally. The computer undoubtedly speeds up certain activities, especially those that are repetitive, while there is a suggestion that the computer may inhibit the creative process required to advance a design project [7]. Designing is very specific activity that resides within concepts of praxis. In theoretical terms, praxis is our capacity to analyse ideas, apply, make and develop them, while simultaneously reflecting as we are doing. In an educational setting, this natural way of thinking and doing is considered vitally important, especially within the creative disciplines. The effects that the computer has

WIT Transactions on The Built Environment, Vol 205, © 2021 WIT Press
www.witpress.com, ISSN 1743-3509 (on-line)
doi:10.2495/BIM210051

on design-specific praxis have not yet been considered in any depth in extant literature. However, what discussion there is suggests that it may well disrupt this natural process.

It has been argued that existing Computer Aided Design (CAD) packages were not conceived for designing in a learning environment (or operating creativity) but are aimed at professional practice as a production tool [8]. The use of such 'production tools' results in good design skills being replaced with good computer skills [9]. This is despite the word design appearing in the title (CAD). This type of software is extremely powerful and persuasive and there is concerned raised by educationalists that if it is introduced too early in an architect's education, or at the wrong stage of a project, then it could inhibit the design process thus resulting in a more linear approach to designing, or negating opportunity to develop complex design ideas [10]. It is also recognised that a students' creativity is often limited to their knowledge of the software with those who can exploit the software's full potential tend to operate more creatively than those with limited knowledge. In this case the possibilities and the potential of praxis is very much constrained by an inability to properly use the computer.

In summary, the extant research on the use of CAD/BIM in the design process is focused on skills and activities within that are or have been augmented by the use of CAD/BIM. When considering the use of CAD/BIM in the architectural design process the extant research is considerably smaller in scale and written by a limited number of researchers, some of which have been noted in this chapter. Therefore, this study will observe and map the entire design process of a randomly selected group of students, in order to understand how and where CAD/BIM activities occurred and what are the effects of CAD/BIM being exploited in these design processes. Potential patterns and occurrences will then be categorised and assessed in order to suggest a digital framework to tutors teaching in this subject.

2 IMPLEMENTATION OF CAD AND BIM INTO ARCHITECTURAL EDUCATION

The perception of CAD as merely a tool to be used in architectural design remained a persistence conviction, previously Friedman [11] posited that to consider CAD as a tool was to undermine the additive skills and conceptualisation advantages of working with an intrinsically accurate digital model. More recently, in Shelbourn et al. [12] studied students' perceptions of BIM education, finding that frameworks in degree programmes focused only on using computer technologies to teach modelling for visualisation within the architectural discipline, they did not consider possible wider advantages, such as using BIM for design collaboration with the other disciplines.

The traditional method of visualisation for a proposed design is through abstraction, in plan, section and elevation. The use of CAD realigns these conventions, displacing Schöns theory of the 'critical practitioner', creating visualisation of a design that addresses the complexities and simultaneities of the design solution in parallel. Knowledge of the design solution becomes displaced, with CAD simultaneously offering detailed visualisation at multiple levels of scale, thus increasing the opportunity to gain a richer understanding of the design problem and therefore potential solution.

When considering CAD and its impact on the design process early research of Lawson [13] with recent research of Taleyarkhan et al. [14] suggested that using CAD in the design process requires an understanding of how it impacts on the learning of a student. Lawson deemed the use of CAD, now over two decades ago, as a new and interesting discourse in the academic architectural studio, with the 'CAD tool' being the new exploration methodology. Taleyarkhan et al. [14] paralleled the proposition determined by Lawson [13] that to understand the impact of CAD on the design process was critical in its successful application within academia. Findings of Taleyarkhan et al. [14] suggest that the use of CAD achieves a significantly increased understanding of the design problem and subsequently it's potential solution. Therefore, CAD offered greater levels of detail to be perceived in a dimensionally accurate virtual model. The traditional conventions of communication as we know and understand it in the academic architectural studio of two-dimensional sketching requires abstraction and representation of the design as opposed to the CAD model allowing direct presentation.

When CAD is used the architectural intent can be visualised in three dimensions in both form and space. The communicative potential of CAD to visualise and understand a design solution realigns the traditional methodology of abstraction and representation displacing this to literal re-presentation of the complete (virtual) design model. The use of CAD within the design process establishes displacement of the traditional abstraction methodology. It creates discrete and fragmented representation, into a singular modelling convention allowing convergence between abstraction, literal representation, and simulation of the design. Extant literature suggests that traditional analogue method of sketching and manual model making in the design process is increasingly becoming replaced by the tacit convention of using CAD and as such requires a revision of the traditional methodologies used in architectural academia.

Research completed by Abdirad and Dossick [15] suggested caution noting that BIM allows far more power to assess the design, but in parallel also created a dependency on the software itself and additional time investment required in learning software. The research outcomes suggested that the designer would evolve into merely an operator, with the dumbing down of the creative process becoming the sceptical outcome. Ambrose [16] advocated that the studio structure of our architectural schools and curriculum needs to be redeveloped around the innovative approaches of BIM. Furthermore, Ambrose and Fry [17] conclude that BIM must become integrated within all design taught modules. Ambrose championed the instantaneous testing of design solutions using BIM allows a richer understanding of design solutions in the design via exploration of the virtual building model. His studies indicated that analysis in the design process using BIM would develop skills, not hinder them, and remove the concern that other academics and researchers have of students becoming mere operators of BIM.

Despite there being a large increase in the number of publications on the subject of Building Information Modelling (BIM), there remains a lack of critical review in the implications of BIM on the design process undertaken by architectural students. Therefore, this study intends to understand the 'how' and 'when' of using CAD in the design process. From this analysis it is proposed to create an informed digital framework suggesting an updated methodology for use of CAD in the academic studio context.

3 METHODOLOGY

This study presents rich data based on the immersion of the researcher in the everyday lives of master degree architecture students and proposes a series of recommendations addressing these via the development of a digital framework to be used in teaching within the architectural studio. The data was collected through weekly interviews with each student. The observation of their design processes included a review of their submitted work at each monthly design review over the entire final academic year (a period of nine months). This collection of data amounted to over 600 interviews and 48 design review submissions being analysed in order to understand the impact of CAD/BIM on the design process. The methodology of this study encompasses the 'rationale and the philosophical assumptions that underlie a particular study' [18]. Therefore, the methodology of a research project very much informs the research methods that are used to determine the research design (collection, analysing and interpreting of data) and investigate the initial question of the research, in this case, to understand the impact of CAD/BIM on the design process. As such the researcher must decide upon the structure of the research methodology and formulate the adopted research strategy in a way that is in accordance with the aims and objectives of the study whilst, at the same time, ensuring retention of an original contribution to an existing body of knowledge [19].

This research employed a combination of inductive and deductive analysis in its approach. The early stages of the research used deductive analysis within the literature review, "as the literature review helps provide a source for the focus of the research" [20]. The literature review exploited research publications from journals and conference proceedings and provided an understanding of historical and current research in the field of architectural design process including methods used within these processes (analogue sketching and modelling making to more current use of CAD and BIM).

4 THE STUDY

The approach taken in this research was a case study involving 16 Masters Architecture Degree Level students in their final year of study. Semi-structured interviews were undertaken over the longitudinal study period of 9 months, leading to over 600 interview records. Data generation methods included interviews, observation, and document review, while themes were inductively generated through thematic analysis. Bourdieu's concepts of field and habitus were used to guide the conduct of the study and interpretation of the findings. The study indicates that it is both desirable and often necessary for a student to use a blend of analogue and digital techniques in the presentation and representation of their design project. However, it must also be cognisant that an over-reliance on digital methods in the design process for visualisation can distance the student from the physical realities of the actual design. This thesis does not argue for one position over another, instead it endeavours to understand the use of both analogue and digital methods and determine how both can be used in conjunction to aid the student in their design process.

The researcher chose final year masters architecture students, with several years' design experience of using both analogue methods of design exploration and the computer. This approach afforded the study the ability to observe students who were less CAD/digitally inhibited due to their knowledge of the software and allowing for more consistency across design process and methods used and reducing the impact of these variables within the research. These students were chosen randomly from the year group who all had similar design training, this included a mix of analogue and digital approaches within their design process. The design processes of all 16 students studied were thoroughly mapped across the academic year. From this inquiry, the research observed patterns and similarities in the practice, considering amongst other questions:

- When the students start to use the computer?
- How is the computer used?
- When the computer offers advantage in the design process?
- When are the students are using the computer incorrectly?
- What impact the computer has on the traditional analogue methods?
- Can the computer lead to creative insights?
- Does the computer inhibit the creative design process?

5 THE FINDINGS

At the culmination of the case study the students sketch books and design projects were analysed in detail. The design work was assessed for use of analogue or digital activities used in design problem resolution and mapped to the interviews that took place during the design process. For the 9-month duration of the case study, a weekly interview with each student was undertaken to capture data that would facilitate the final assessment of the overall design process. This data allowed the alignment of commentary between student and tutor, observations made during tutorials and general observations made on modes of design inquiry (analogue or CAD/digital).

6 OVERALL IDENTIFICATION OF THEMES IN DESIGN PROCESS

At the culmination of the case study the students sketch books and design projects were analysed in detail. The design work was assessed for use of analogue or digital activities used in design problem resolution and mapped to the interviews that took place during the design process. For the 9-month duration of the case study, a weekly interview with each student was undertaken to capture data that would facilitate the final assessment of the overall design process. This data allowed the alignment of commentary between student and tutor, observations made during tutorials and general observations made on modes of design inquiry (analogue or CAD/digital).

6.1 Identification of themes: All students – Month 1

During the first month of the students' design processes all of the students within the study undertook site and contextual analysis. These studies included exploration of their proposed site via sketching or making models in order to process their design approach and formulate initial ideas. The sketchbooks from this period mainly contained initial thoughts of proposed schemes without yet commenting or understanding where this analysis made lead them.
All 16 students studied undertook site context and historical research to establish their thesis area. Activities used during this analysis mainly consisted of sketching, modelling, and assembling data from various sources, such as digital maps. In addition to general information gathering, physical modelling was used to explore and record initial design thoughts that were yet to be refined. The use of physical models gave the students the opportunity to gain a better understanding of their site without knowing the consequences and final outcome of these initial design decisions. These models were a mixture of semi-accurate site and context models or in some cases more conceptual in form.

Students refined and amended these models in a reflective act during this stage of the design process. The models afforded the students the ability to consider contextual issues and generate initial design ideas to develop further. These initial studies became a constant source of reference for the majority of the students in the study being continuously revisited during months 2–8 prior to the final month of production activities in readiness for the final submission of their design projects (month 9).

When assessing the use of CAD against matching cases within the data during month 1 of the study the findings indicate that 14 of the 16 students undertook no CAD during this period of their design process. When this data was further analysed, those students completing more analogue tasks were focused on understanding the design problems that their brief would need to meet as they began to formulate their initial design brief. When reviewing the two outlying students in this data query, the tasks undertaken one student remained focused on understanding the overall design problem and their development of the design brief. The remaining student, only 1 from the 16 students studied used CAD from the outset of their design process, however shared the overall aim of focusing on the development of the design brief by understanding the design problems that their brief would need to meet. These findings are shown in Fig. 1, in which the use of analogue methods used for understanding or defining the design problem can be seen to be significant and CAD use is comparatively very low.

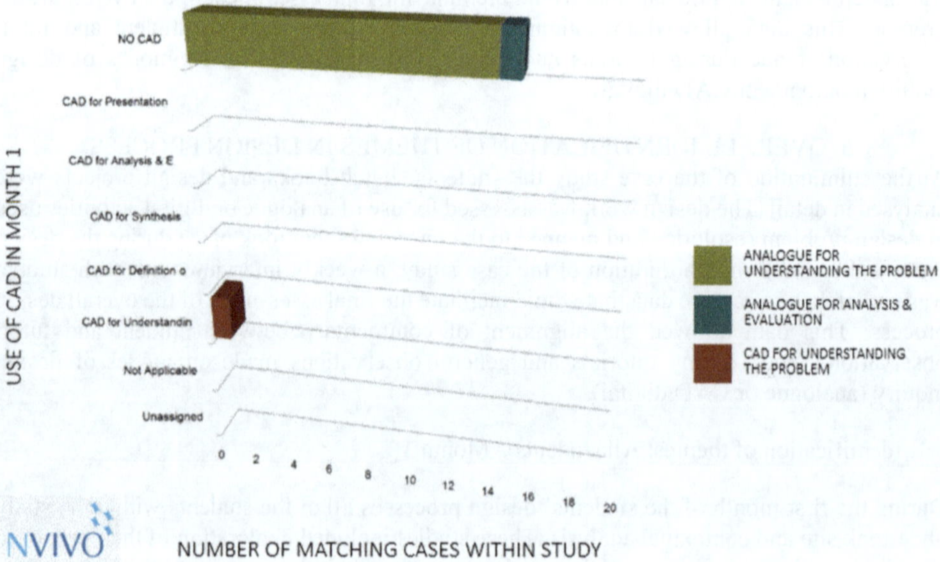

Figure 1: Comparison of CAD and analogue methods used during month 1.

6.2 Identification of themes: All students – Month 2

The design processes of the students studied during month 2 generally followed a period of focused enquiry. The activities used by each student began to differ during this month, however, the overall focus remained the development of initial site responses and how these could be used to develop the design brief. Students who began to use CAD to model their proposed site, such as S-2C and S-8H to create 3D models of their sites to assess, reflect upon and illustrate routes through and around their chosen sites. The level of data within these digital site models varied, with student S-8H also adding his analysis of the building heights, as this aspect was deemed important to his building's contextual response. Fig. 2 shows 10 of the 16 students retaining a purely analogue approach to their design process. The remaining 6 students used digital methods to understand and reflect upon their design choices in order to define the design.

All students began to increase analysis of their chosen site and surrounding context to begin to 'problem set'. Those students who continued to use analogue approaches (physical models) photographed these physical models in order to create generative sketches and complete diagrammatic outputs that summarised their analysis of the brief and site.

When creating a query of CAD use during month 2 of the study the findings indicated a continued non-CAD approach, with 10 of the 16 students using analogue methods in their design process, these included sketching, painting and photography in the recording of their

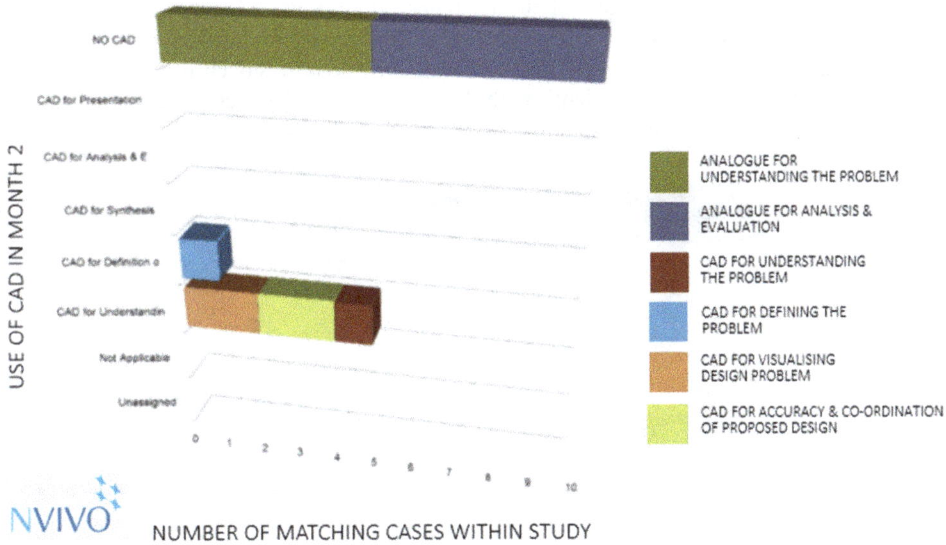

Figure 2: Comparison of CAD and analogue methods used during month 2.

proposed sites. The remaining 6 students began to use CAD during month 2. Just one student used CAD to begin to define his design brief and initially analyse the design of his proposed scheme. The remaining five students continued to refine their understanding of the site and proposed scheme by resolving their initial design brief proposals. These findings were further queried and established that within this sub-group of five students, two students used analogue methods in parallel to CAD to refine their design brief and two used analogue methods in parallel to CAD to increase definition of design brief proposals. The remaining student used CAD to collate numerous maps of their site in order to create historical overlays of the site and its development. The historical mapping of the site allowed them to refine their design brief.

In month 3, the students were exclusively focused on activities that informed concept development that reinforced initial thesis proposals and developing the design brief. Digital models of sites completed during the initial 2 months were now used to produce basic renders of sites onto which analysis of routes, levels or other significant contextual features were added in editing software to use in design brief documentation. This month was a period of focused research that consolidated analysis into initial drafts of design briefs. The students who had used solely analogue activities up to this stage in their design process spent considerable time during this month drawing and diagramming their findings in order to present site analysis at the monthly design review.

During this month of the study, the students began to take divergent routes in their design processes. Five of the students began to use digital models to gain a greater comprehension of the surrounding site topography. The example shown in Fig. 3 shows how student S-2C modelled her site and surrounding buildings to a high level of detail. The student geo-located her digital model in order to accurately plot the sun path around the site. The shading analysis from this exploration was used to inform her design brief. Further abstract studies were carried out this month by student S-2C to develop the design brief in response to findings from the digital modelling. These combined activities were used to conclude her response to context and inform the next phase of the design process. Fig. 3 depicts the reduction in analogue methods in the design process, with 8 of the 16 students retaining only analogue methods to define design problems. The increase in CAD activities occurred in the synthesis stages of the design process (1 student), with the remaining students continuing to use CAD for definition and increased understanding of their design choices.

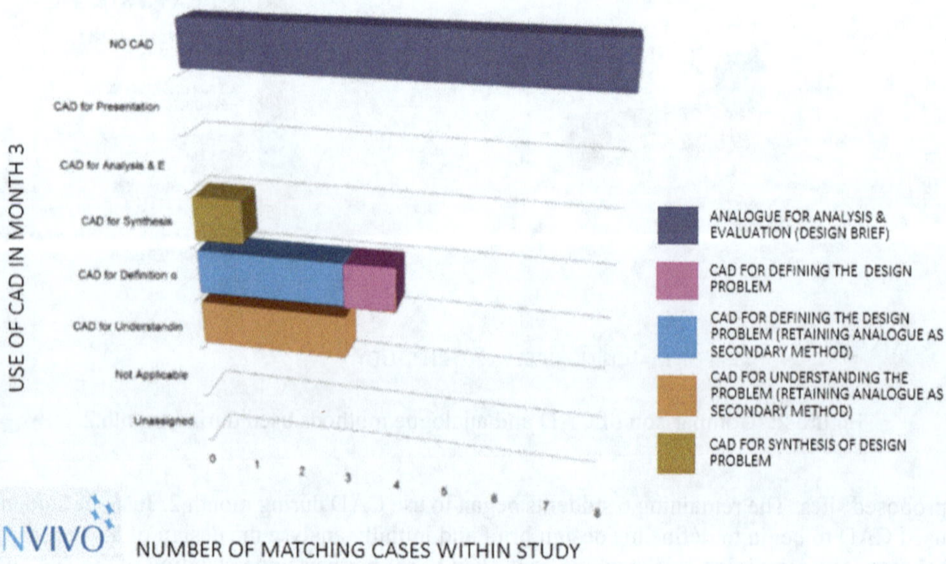

Figure 3: Comparison of CAD and analogue methods used during month 3.

6.3 Identification of themes: All students – Month 4

During month 4 the majority of students studied reached a point in the development of their design proposals where the spatial adjacencies and overall form of the building was beginning to become established. Using the digital models, the students began to assess their design brief in parallel with the site context to confirm how these related to the contextual constraints. The nine students who had been using CAD in parallel to their analogue activities during their process, now solely used the CAD models to check for constraints that may impact on their design iterations. This group of students seemed to possess a confidence in their decision making that emanated from the accuracy by which they could 'test' these site conditions.

During month 4 the data reveals that the majority of the students studied moved into synthesis and definition stages of their design process, with 12 of the 16 students focusing on these tasks. From this sub-group, 5 of the students used both CAD and analogue methods in the synthesis stage of their design process. Those students who were employing analogue approaches to synthesis in their design process, were also using CAD in parallel to accomplish definition of their design solutions. Just one student in the group studied focused on solely using CAD in during their definition stage of their design process during this month. In Fig. 4 the continued increase of CAD was observed in month 4, with only 2 of the 16 students retaining solely analogue approaches to their design process.

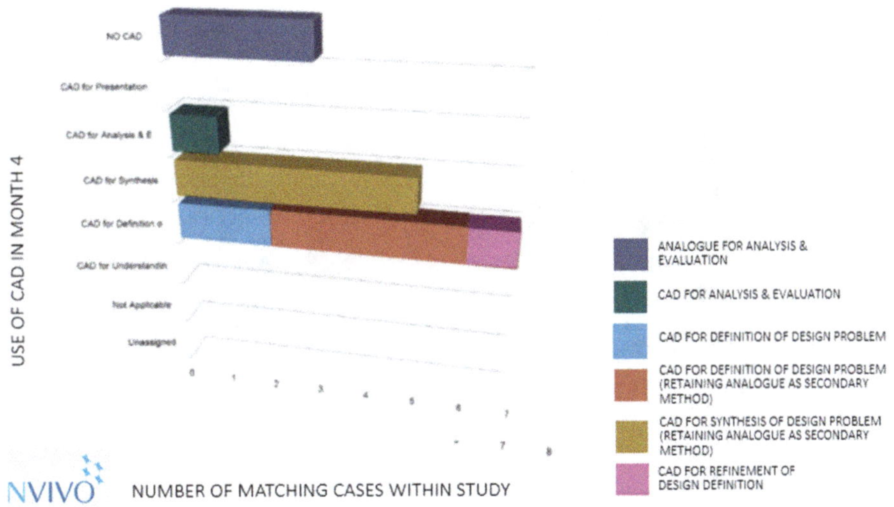

Figure 4: Comparison of CAD and analogue methods used during month 4.

6.4 Identification of themes: All students – Month 5

During month 5 the students design processes became increasingly diverse in terms of the activities used in design exploration. The examples shown in Fig. 5 shows how developed the building form is for student S-8H (a tower block) when compared to the students (S-4B and S-5J) design development where form finding was still ongoing. However, this finding from the case study requires further contextualisation, as the design of student S-8H did not develop from this architectural form for the next 2 months (Months 6 and 7) and seemed almost 'fixed' in comparison to student S-5J. The design process of student S-5J used analogue and digital activities cyclically throughout the majority of the 9-month period, he was not fixed to a solely digital approach and developed his scheme to a higher level of detail than student S-8H by month 8.

The findings indicated that CAD use during month 5 of the study converged on the synthesis stages of the design process for the majority of the students studied. During month 5, four students used only analogue methods in the synthesis stage of their design process. When considering the 8 students using CAD methods in the synthesis of their design, the data also indicates that these students also used analogue methods in parallel to CAD. The comments made by this group students suggests that CAD was used as a approach to gain increased accuracy to this period of synthesis in terms of scale, dimensional co-ordination and spatial planning of their proposed schemes. The data also indicates a small but continuing trend to retain analogue methods in analysing and evaluating design solutions but retain CAD as a parallel method to synthesise these solutions. In Fig. 5 the students in the study focused on synthesis in their design process using analogue (3 students) and digital methods (8 students) in their design process during month 5.

6.5 Identification of themes: All students – Month 6

In month 6 many students focused on the architectural form and materiality of proposed schemes. Those students using analogue activities in their design processes increased

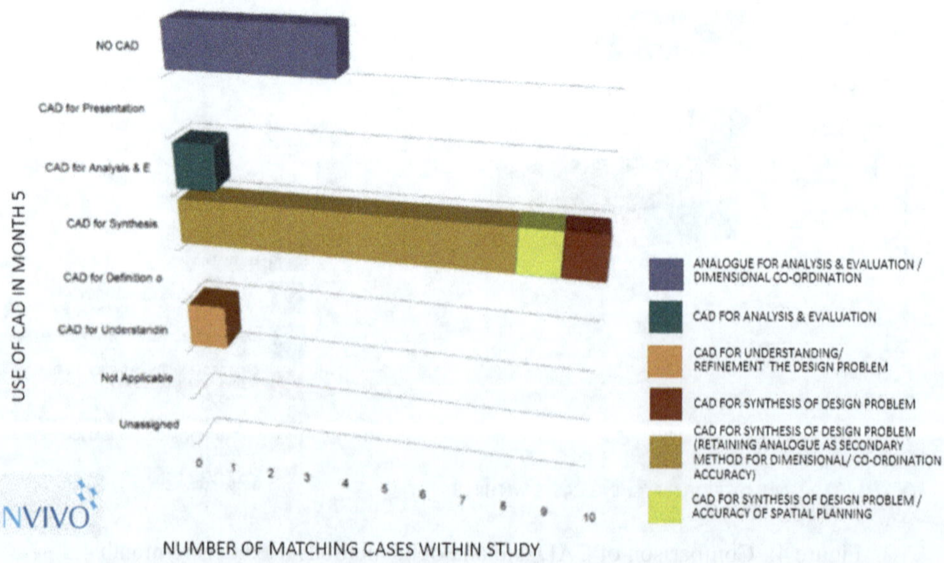

Figure 5: Comparison of CAD and analogue methods used during month 5.

physical model making and began to vary the scale and abstraction of these models in order to explore form for the external envelope or cladding for their proposed schemes. The physical model making brought together previous design ideas merging these into a more refinement of schemes. Those students within the study who used digital activities during this period also undertook to explore external forms and the design of the envelope of their schemes. These digitally driven explorations were undertaken in the virtual environment, using geolocation of models in order to accurately assess sun path, shading, and environmental inquiries.

When assessing the data from month 6 of the study it was apparent that the students were focused on analysing and sythesising their design solutions. Although there were two outlying students, the majority of the group continued to pursue analysis and synthesis in their design process. The findings also indicate the continued trend to use both analogue and CAD methods in parallel during the same design stages. During month 6, six of the students studied used analogue and CAD methods for the same design stage of analysis and evaluation in their design process. The data for this month also indicated a continued trend to use analogue methods in a specific design stage together with CAD methods in a complementary approach for another design stage. An example of this were the 4 students who used analogue methods in their analysis and evaluation stages in parallel to using CAD to synthesise the design solutions. During month 6 the significant focus in the design process was analysis and synthesis of design ideas as students refined and reflected upon each design decision. The findings indicated that CAD was the primary method of analysis and synthesis for 1 student. The use of analogue methods of analysis and synthesis was used by 14 students. This suggested that the majority of the students in the study retain analogue methods as the primary method of design exploration at this point in the design project, see Fig. 6.

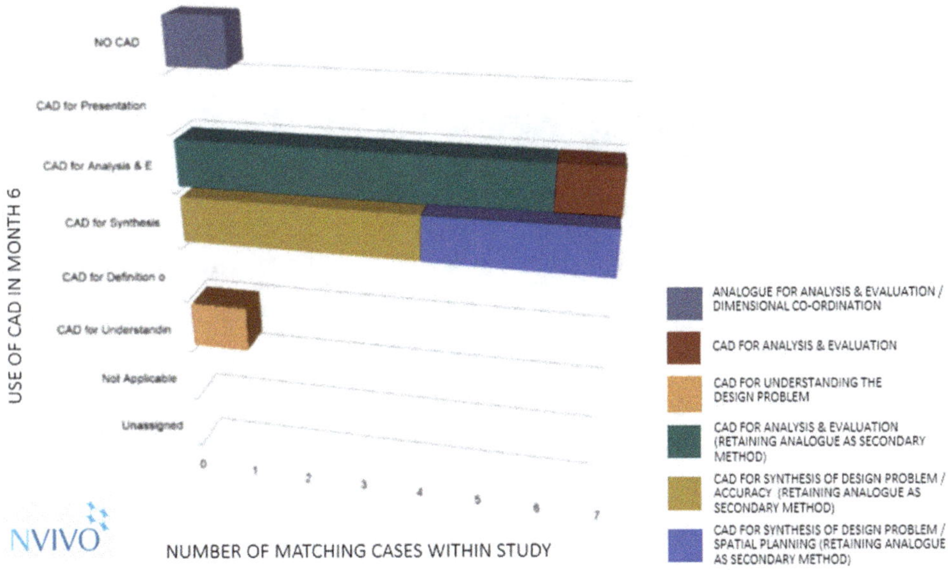

Figure 6: Comparison of CAD and analogue methods used during month 6.

6.6 Identification of themes: All students – Month 7

During month 7 the students studied continued to focus on design progression by undertaking activities that moved the space, form, and initial plans of schemes into more formalised and c coordinated plans, sections and elevational proposals. These aspects were refined over this month. The higher achieving students included displacement activities that afforded further design exploration and testing, such as physical and digital model making. Students S-1J and S-5J began to increasingly oscillate between analogue and digital activities to refine design solutions, with each iteration now occurring at a focused and refined level. By the end of the seventh month of the study the proposed design projects of the student group were becoming increasingly resolved.

The data indicates a very focused period for the group who's design processes were now focused on analysis and evaluating their proposed designs. The findings indicated that 11 of the 16 students remained in a reiterative loop of amending and re-assessing their design ideas and refining design solutions. The trend of retaining both CAD and analogue approaches in parallel can also be seen in Fig. 7. This confirms the use of analogue and CAD methods for the same design stage, analysis and evaluation. When relating this finding to the discussions with the students the use of CAD at this stage seemingly gave the students the advantage of both rapid and accurate testing of design ideas (see Fig. 7).

6.7 Identification of themes: All students – Month 8

The findings in the study for students during month 8 indicated a significant increase in the use of digital activities in the design process of all students' sub-groups. The analysis of the design processes indicated that these students oscillated between analogue and digital activities during this month continuing to use both activity modes during final design

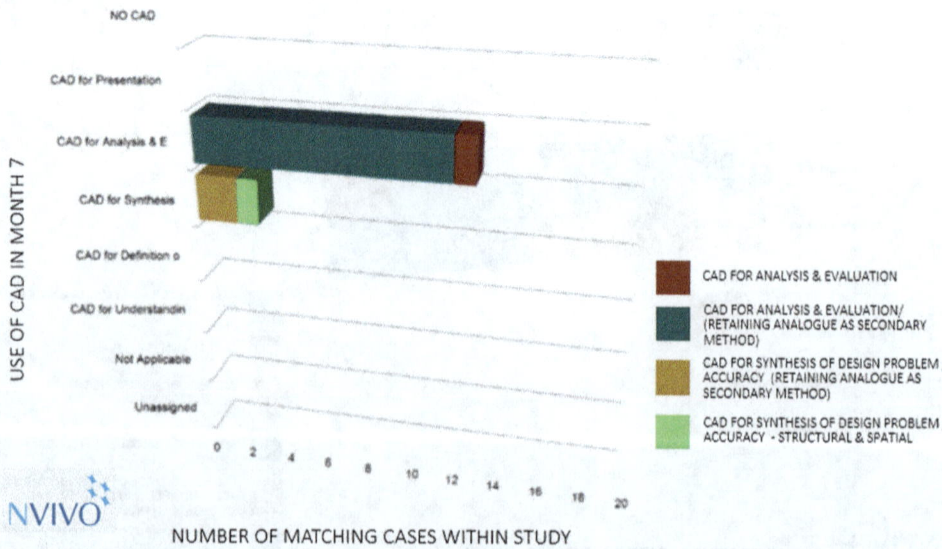

Figure 7: Comparison of CAD and analogue methods used during month 7.

resolution of schemes. The findings also indicated that those students who focused mainly on using analogue activities in months 1 to 7 began to significantly increase use of digital activities during this month. When this increase of digital activities was examined, the findings suggested that the majority of students used digital activities to rapidly test design solutions and increase levels of accuracy and detail in proposed schemes.

As the study moved into the final two months of the overall design project analysis and evaluation was still a major stage being undertaken in the design process of the student group. However, the presentation stage was now beginning to be integral in the design process of a small number of students (2 out of 16). When evaluating this initial change, the data indicated that these two students were those who had continuously used CAD throughout the majority of their design process, with the 'presentation' stage being a period of creating initial renders of their proposed schemes in order to evaluate them. The trend to use both CAD and analogue methods in a singular design stage continued this month with 10 of the 16 students employing this approach for analysis and evaluation of their design (see Fig. 8).

6.8 Identification of themes: All students – Month 9

The final month became a period in the design processes studied where the remaining students who were not already engaged in 'production' activities began to rapidly increase production of digital models in order to render their schemes and produce images for the final submission of their design projects. The digital models were used to create the plan, section and elevational outputs of their schemes from a singular dimensional accurate source.

Sections through proposed schemes were produced directly from the CAD allowing alignment of details and sections. The details were created as 'call outs' from the CAD model. During month 9 the students were focused solely on 'production' activities to produce renders and images in readiness for submission. Significantly some students within the group studied

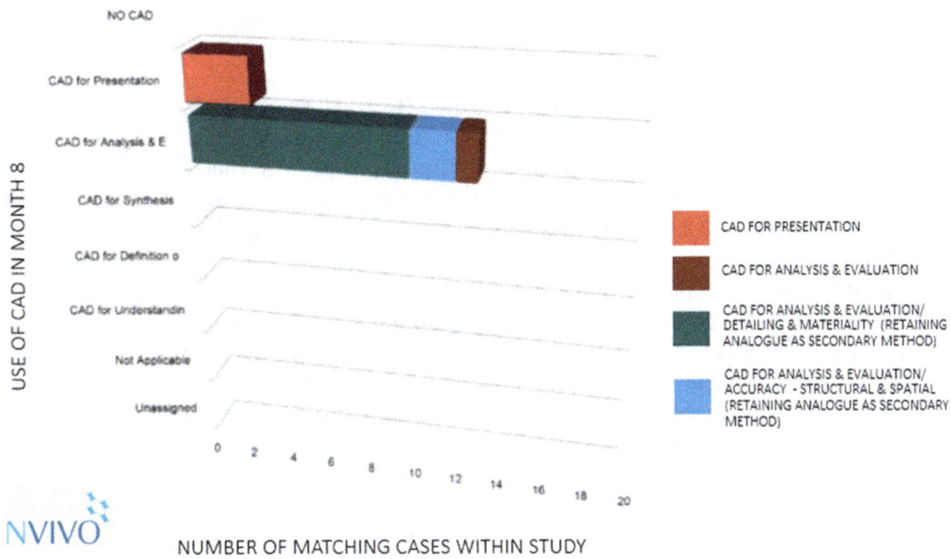

Figure 8: Comparison of CAD and analogue methods used during month 8.

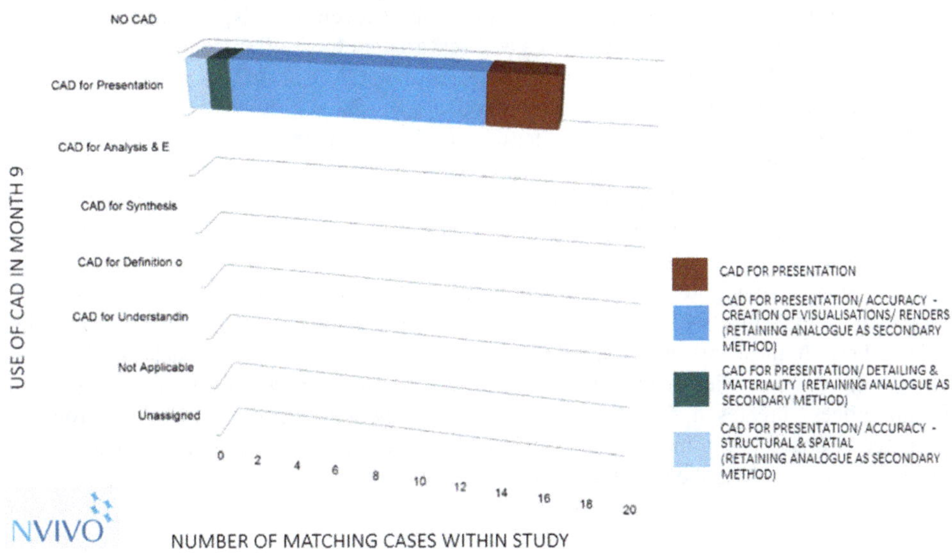

Figure 9: Comparison of CAD and analogue methods used during month 9.

continued to amend their designs using CAD to enhance visualisation of their schemes during this final month. These final amendments at this late stage in the design process created a period of intense activity where students repeatedly amended digital models to change aspects such as internal lighting in order to 'enhance' the visual appearance of the final images. Technological changes in the proposed design projects were minimal during this period for all students in the study, final changes were more for aesthetics, can be seen in Fig. 9.

7 FINDINGS FROM THE STUDY

Whilst this study was conducted current studies on the use of CAD have begun to emerge focusing on the outcomes of using CAD in terms of design resolution and transfer of representation of design ideas from manual sketches to digital models. Before the findings are discussed in detail it is important to situate this study's findings in relation to existing literature.

One of the most unexpected findings of this study was the relegated way in which CAD was used in the design process. With the majority of the 16 students studied using CAD as a re-presentation method of that focused more on the visualisation and presentation of a design ideas rather than refinement of the idea through several iterative cycles.

Extant literature suggests that CAD can be used productively to explore a design idea at multiple scales in a highly accurate environment. The findings of Haliburton et al. [21] suggested that using CAD in the design process allows the student to explore their design ideas to levels of detail not afforded by more conventional methods such as sketching. Later research by Ambrose and Fry [22] concluded that the use of BIM in the design process would allow more integrated thinking of a design problem in order to consider material, structure and environmental factors in parallel with the overall design goal. Ambrose championed the instantaneous testing of design solutions using BIM allows a richer understanding of design solutions in the design via exploration of the virtual building model. His studies indicated that analysis in the design process using BIM would develop skills, not hinder them and remove the concern that other academics and researchers have of students becoming mere operators of BIM. However, this study indicates that CAD and BIM remain more of a visualisation device that design solution method.

Previous studies, considered in the literature review, have been empirical in their method of analysis with the objective of finding a universal schema to the design process that all students adhere. Historical research completed by experts in the field of CAD use on architectural design and associated theories of design such as Schon, Bruner, Dubberly, Lawson, Robertson and Allen amongst others referred to in this study have suggested that students studied navigate individual approaches that are underpinned and guided by knowledge and skillsets obtained previously and refer to these as the students 'learning styles'. Few of these historical studies have focused on mode of approach (analogue or digital) and the activities used in the design process. The completion of an activity in the design process creates a form of learning as the new knowledge is attained through a sequence of observation, inquiry, and reflective phases. The outcome is often considered by the student but rarely is the process that led to this learning.

8 CONCLUSIONS

The design process as a theoretical construct is contested in the literature. There is extensive debate regarding what activities comprise it, and in what order they are undertaken. Designers often move between and return to previous stages; what Bruner calls spiralling [23]. Generally, the design process follows this format: concept/ideation, scheme design, design development, detail design, communication [24]. Within these stages a designer would be likely to sketch, draw, paint, draft, sculpt, model make, model on the computer, etc. as singular activities and in combination. However, there is also recognition that there are cognitive processes that run concurrent to these activities that can consciously and subconsciously aid or inhibit the design process [25]. Each designer undertakes a different process and developing a preference for particular activities over others [26]. The creative design process in architecture is not a linear or didactic, but cyclical and iterative [27].

There is an inextricable link between creativity and the design process; indeed, the design process has been referred to as the engine of creativity [27]. Those students who understand how best to navigate and construct their design process are more likely to operate creatively [28]. The importance of education in developing creativity has been stressed by many commentators [29]–[31], new thought suggests that there may be value in exploring the processes and methodologies in architectural design education to increase the likelihood of creative design outcomes [32]. There is very little investigation on the design process in architecture. In the past architectural design was seen as some sort of language that could be taught based on the use of proportion and archetypal forms. This approach was perhaps satisfactory in a past, however, in the modern era, this could be considered anachronistic, with society requiring architecture with a complexity of form. This also includes use of increased scales during the design process and designing with technical requirements that were not required in the past.

In the modern era, with the invention and subsequent use of digital technologies, the design process has become even more difficult to conceptualise and define. Therefore, students of design subjects are even more likely to find it difficult to navigate through their design process to a successful outcome. Educators are not always equipped to help, having often been educated to use analogue methods [33]. There is thought that designing on the computer may involve different cognitive processes to that of analogue methods [34]. It has been argued that the computer has aided an architect's ability to operate creatively [35] others are far less positive, especially in the educational environment [36] referring to a study undertaken in the US, concluded that students found it 'far more difficult and less interesting to learn from the screen than through analogue methods, such as drawings and physical model making'. It is therefore difficult to fully comprehend the impact of the computer on the design process generally. The computer undoubtedly speeds up certain activities, especially those that are repetitive, while there is a suggestion that the computer may inhibit the creative process required to advance a design project [37]. Designing is a very specific activity that resides within concepts of praxis. In theoretical terms, praxis is our capacity to analyse ideas, apply, make and develop them, while simultaneously reflecting as we are doing. In an educational setting, this natural way of thinking and doing is considered vitally important, especially within the creative disciplines. The effects that the computer has on design-specific praxis have not yet been considered in any depth in extant literature. However, what discussion there is suggests that it may well disrupt this natural process [38].

It has been argued that existing Computer Aided Design (CAD) packages are not conceived for designing in a learning environment (or operating creativity), but are aimed at professional practice as a production tool [39]. The use of such 'production tools' results in good design skills being replaced with good computer skills [40]. This is despite the word design appearing in the title (CAD). This type of software is extremely powerful and persuasive and there is concerned raised by educationalists that if it is introduced too early in an architect's education, or at the wrong stage of a project, then it could inhibit the design process thus resulting in a more linear approach to designing [41], [42]. It is also recognised that a students' creativity is often limited to their knowledge of the software with those who can exploit the software's full potential tend to operate more creatively than those with limited knowledge. In this case the possibilities and the potential of praxis is very much constrained by an inability to properly use the computer.

In conclusion, the extant research on the use of CAD/BIM in the design process is focused on skills and activities within that are or have been augmented by the use of CAD. When considering the use of CAD in the architectural design process the extant research is considerably smaller in scale and written by a limited number of researchers, some of which have been noted in this chapter. Therefore, this study will observe and map the entire design process of a selected group of students, in order to understand how and where CAD activities occurred and what are the effects of CAD being exploited in these design processes. Potential patterns and occurrences will then be categorised and assessed in order to suggest a digital framework to tutors teaching in this subject.

9 NOTE

The term computer-aided design (CAD) has now been used within architectural design for over three decades. The use of this software allows the designer to accurately represent floorplans and sections in a dimensionally accurate digital environment. This digital approach of the design process has evolved, and the use of Building Information Modelling (BIM) has become the successor to CAD. The term 'Building Information Model' first appeared in 1992, in a published paper by van Nederveen and Tolman. In this paper the terms 'Building Information Model' and 'Building Information Modelling' (became the acronym BIM). However, BIM did not become more widely used until a decade later in the early 2000s.

The use of CAD is no longer the only digital option for design work, with BIM becoming increasingly used as a collaborative environment tool that has proven to be increasingly important in the design of buildings. The two terms CAD and BIM can be readily confused as they both involve the use of a digital model in the design process. However, these similarities do not mean that they are essentially the same instrument. The use of CAD transfers the lines on paper to lines on a screen, these do not have any values or attributes. When using BIM the digital model is made of components, not lines as in CAD. These components have attributes that allow the model to be used to test aspects of the design in real world conditions, essentially allowing the model to be used to test design solutions in a virtual environment that can accurately assess heating, cooling, and other such environmental factors of a proposed building design.

To ensure clarity, as CAD and BIM are both digital approaches to design exploration in the design process, these acronyms are considered the same for the purpose of this study. When specific software is being discussed the terms CAD or BIM will be used to specifically reference those software approaches.

REFERENCES

[1] Lonna, F.C., Pérezgómez, A. & Pelletier, L., Prelude. Mapping the question: The perspective hinge. *Architectural Representation and the Perspective Hinge*, MIT Press: Cambridge, pp. 2–87, 1997.

[2] Lawson, B.R., Design in architecture. *Architecture and the Human Sciences*, ed. G. Broadbent, John Wiley and Sons Inc.: London, 1980.

[3] Till, J., *Architecture Depends*, MIT Press: Cambridge MA, 2009.

[4] Vasudevan, C. & Tirumala, R., Digital representation of physical artefacts: The effect of low cost, high accuracy 3d scanning technologies on engineering education, student learning and design evaluation. *ASME 2013 International Design Engineering Technical Conferences and Computers and Information in Engineering Conference. American Society of Mechanical Engineers Digital Collection*, 2014.

[5] Snodgrass, A., Coyne, R. & Martin, D., Metaphors in the design studio. *Journal of Architectural Education*, **48**(2), pp. 113–125, 1994.

[6] McVeigh, T. & Paton-Walsh, N. Computers kill pupils' creativity. *The Observer*, Sep. 24. http://education.guardian.co.uk/news/story/0,,372428,00.html. Accessed on: Nov. 2020.

[7] Garber, R., *BIM Design: Realising the Creative Potential of Building Information Modelling*, John Wiley & Sons, 2014.

[8] Paszkowski, Z.W. & Gołębiewski, J.I., International design workshops as an intensive form of architectural education. *World Transactions on Engineering and Technology Education*, **18**(1), pp. 51–56, 2020.

[9] Lawson, B., *How Designers Think*, Routledge, 2006.

[10] Thurlow, L. & Ford, P.B., Ideal ideation: A framework for the management of sketch inhibition among undergraduate designers, 2018.

[11] Friedman, D.S., Architectural education and practice on the verge, *American Institute of Architects: Report on Integrated Practice,* eds M. Broshar, N. Strong & D.S. Friedman, The American Institute of Architects: Washington, DC, pp. 3–7, 2006.

[12] Shelbourn, M., Macdonald, J., McCuen, T. & Lee, S., Students' perceptions of BIM education in the higher education sector: A UK and US perspective. *Industry and Higher Education*, **31**(5), pp. 293–304, 2017.

[13] Lawson, B., Towards a computer-aided architectural design process: A journey of several mirages. *Computers in Industry*, **35**(1), pp. 47–57, 1998.

[14] Taleyarkhan, M., Dasgupta, C., Garcia, J.M. & Magana, A.J., Investigating the impact of using a CAD simulation tool on students' learning of design thinking. *Journal of Science Education and Technology*, **27**(4), pp. 334–347, 2018.

[15] Abdirad, H. & Dossick, C.S., BIM curriculum design in architecture, engineering, and construction education: A systematic review. *Journal of Information Technology in Construction (ITcon)*, **21**(17), pp. 250–271, 2016.

[16] Ambrose, M., Agent provocateur: BIM in the academic design. *International Journal of Architectural Computing*, **10**(01), pp. 53–66, 2012.

[17] Ambrose, M.A. & Fry, K.M., Re-thinking BIM in the design studio. *6th ASCAAD Conference*, p. 71, 2012.

[18] Dainty, A., Methodological pluralism in construction management research. *Advanced Research Methods in the Built Environment*, **1**, pp. 1–13, 2008.

[19] Grix, J., Social capital as a concept in the social sciences: The current state of the debate. *Democratization*, **8**(3), pp. 189–210, 2001.

[20] Gray, D.E., *Doing Research in the Real World*, SAGE, 2013.

[21] Haliburton, J., Clayton, M., Ozener, O. & Francisco Farias, W.J., Parametric modelling and BIM: Innovative design education for integrated building practices, 2011.

[22] Ambrose, M.A. & Fry K.M., Re: Thinking BIM in the design studio-beyond tools… approaching ways of thinking, 2012.

[23] Bruner, J., *The Process of Education*, Harvard University Press: Cambridge, MA, 1960.

[24] Goldschmidt, G. & Sever, A.L., Inspiring design ideas with texts. *Design Studies*, **32**(2), pp. 139–155, 2011.

[25] Lawson, B., *How Designers Think: The Design Process Demystified*, Routledge: London, 2006.

[26] Goldschmidt, G., On visual design thinking: The vis kids of architecture. *Design Studies*, **15**(2), pp. 158–174, 1994.

[27] Kowaltowski, D., Methods that may stimulate creativity and their use in architectural design education. *International Journal of Technological Design Education*, **20**(3), pp. 453–476, 2010.

[28] Casakin, H., Associative thinking as a design strategy and its relation to creativity. *DS 68-7: Proceedings of the 18th International Conference on Engineering Design (ICED 11), Impacting Society Through Engineering Design, Vol. 7: Human Behaviour in Design*, Lyngby, Copenhagen, Denmark, 15–19.08.2011, 2011.

[29] Amabile, T., *The Social Psychology of Creativity*, Springer-Verlag: New York, 1983.

[30] Robinson, K., *All Our Futures: Creativity, Culture and Education*, National Advisory Committee: Suffolk, 1999.

[31] Cox, A., What are communities of practice? A comparative review of four seminal works. *Journal of Information Science*, **31**(6), pp. 527–540, 2005.

[32] Cropley, A.J., *Creativity in Education and Learning, A Guide for Teachers and Educators*, Kogan Page: London and Sterling, USA, 2001.

[33] Till, J., *Architecture Depends*. MIT Press: Cambridge, MA, 2009.

[34] Vasudevan, C. & Tirumala, R., Digital representation of physical artefacts: The effect of low cost, high accuracy 3D scanning technologies on engineering education, student learning and design evaluation. *ASME 2013 International Design Engineering Technical Conferences and Computers and Information in Engineering Conference. American Society of Mechanical Engineers Digital Collection*, 2014.

[35] Snodgrass, A., *Interpretation in Architecture*, Routledge: New York, 2006.

[36] McVeigh, T. & Paton Walsh, N., Computers kill pupils' creativity, 2000. http://education.guardian.co.uk/news/story/0,,372428,00.html. Accessed on: 21 Apr. 2021.

[37] Vasudevan, C. & Tirumala, R., Digital representation of physical artefacts: The effect of low cost, high accuracy 3D scanning technologies on engineering education, student learning and design evaluation. *ASME 2013 International Design Engineering Technical Conferences and Computers and Information in Engineering Conference. American Society of Mechanical Engineers Digital Collection*, 2014.

[38] Bonnardel, N. & Zenasni, F., The impact of technology on creativity in design: an enhancement? *Creativity and Innovation Management*, **19**(2), pp. 180–191, 2010.

[39] Radclyffe-Thomas, N., White heat or blue screen? Digital technology in art & design education. *International Journal of Art & Design Education*, **27**(2), pp. 158–167, 2008.

[40] Lawson, B., *How Designers Think: The Design Process Demystified*, Routledge: London, 2006.

[41] Brown, M., The teacher-tool relationship: Theorizing the design and use of curriculum materials. *Mathematics Teachers at Work: Connecting Curriculum Materials and Classroom*, ed. Remillard, Routledge: New York, pp. 17–36, 2009.

[42] Robertson, J. & Nicholson, K., Adventure author: A learning environment to support creative design. *Proceedings of the 6th International Conference on Interaction Design and Children*, ACM, pp. 37–44, 2007.

IMPLEMENTING COMMON DATA ENVIRONMENTS IN ARCHITECTURAL TECHNOLOGY STUDIES

JOAO LUIS PEREIRA DE SA & PABLO HECTOR GONZALEZ ALFARO
Architectural Technology and Construction Management, Copenhagen School of Design & Technology, Denmark

ABSTRACT

The Construction Industry's digitalisation is an iterative and ongoing process where digital tools have been primarily focused on the design and modelling and lacking traction in the adjacent stages. Communication methodologies between parties are not evolving at the same rate compared with digitalisation production. Similarly, educational institutions lack the resources to facilitate the students' growth associated with the BIM process. Different commercial entities have been tackling the issue for some time, producing tools that significantly reduce this effect. Implementing a CDE (Common Data Environment) in a pedagogical environment can improve multidisciplinary feedback across lecturers teaching different building specialities. Using digital solutions with an intuitive human interface reduces the distance between students and teachers. A CDE tool such as Dalux® provides a platform aimed for the AEC industry with an amplitude of instruments that serve the building's whole information cycle. The Copenhagen School of Architectural Technology and Construction Management has consistently been implementing in their programs the most recent BIM solutions to provide new graduates with competences to cope with industry challenges. In doing so, innovative applications of didactics strategies have been successfully applied in the light of favourable results. Students need a comprehensive understanding and application of communication flows when solving their school assignments from an industry practice perspective. In 2020, the school decided to implement Dalux in the program as a pilot project to analyse the feasibility of using the software on a major scale with promising results. This paper presents the Dalux implementation at the school, exploring the didactic and professional approach based on factual data from case studies and learning outcomes during the period of investigation.

Keywords: AEC, CDE, Dalux, documentation, digitalisation, communication.

1 INTRODUCTION

The application of digital solutions to enhance the Danish building industry productivity has been performed for more than 15 years, fostered mainly by the public sector. However, in the last years, the private sector understood the benefits of using BIM and, by synergy, become more integrated into the development of the industry. The ICT (Information and Communication Technology) regulations that are in force in public projects now became the standards to coordinate BIM performances. Within its processes, fluid communication amongst the different actors involved in a building project has become the most discussed topic to be the next step to develop building projects in a more efficient fashion. All the tools are at the disposal of the building industry. Still, it misses the capability of coordination and collaboration, especially when it comes to involving small companies due to the lack of BIM maturity.

Clients in Denmark have an increased interest in using sophisticated digital platforms for improving coordination and a better understanding of levels of collaborations. Therefore, it is vital to find flexible systems that can offer solutions and support the challenges that the interactions amongst project participant have in a project. Amassing a design and construction team can be done differently where the success of a project depends on the selected system of BIM and communication platform [1]. The main domains that drive clients to decide the right solution are based on a product that can offer optimal coordination, a quick overview of the project's status, and digital applications that

WIT Transactions on The Built Environment, Vol 205, © 2021 WIT Press
www.witpress.com, ISSN 1743-3509 (on-line)
doi:10.2495/BIM210061

can ensure quality. The industry is awaiting the next step in CDE applications – yet to come. On the other hand, educational institutions in Denmark contribute with different resources to accompany the rapid pace that the development of CDE technologies is showing recently.

2 APPLICATION OF CDE IN THE INDUSTRY AND PERSPECTIVE FOR EDUCATION

Better productivity rates in the industry require a strong linkage between cooperation and the technology that makes it possible. One tool used for planning digital communication is a BIM execution plan that sets the information level and the type of data traded amongst the project participants and designates a BIM manager who leads the team. To succeed in a project, it is key to plan the team's coordination at upfront stages using BIM to connect design, budget, schedule, planning, and facility management. Software vendors consider these basic requirements as opportunities to expand and increase their share in a tough and competitive market. The primary strategy consists of providing clients with flexible platforms and streamlining all project participants' interaction through a standard digital structure. This leads to a market that offered multiple software solutions with different capability ranges. Today some communication platforms cover most of the stages in the life cycle of a building, and its usage in Denmark has become widespread.

2.1 Levels of collaboration in Denmark

When having a look at the level of complexity of BIM applied for different building projects, the range varies from 2D BIM (drawings), 3D BIM (virtual elements with nested data), 4D BIM (time operation), 5D BIM (cost performance), 6D BIM (sustainability approach) and 7D BIM (operation and facility management) [2]. The definition of the boundary conditions where the complexity of BIM and the levels of maturity interact in the context of a specific project is not an easy task. In this regard, through several initiatives, the public and private sectors have understood that coordination and collaboration via clear definitions of roles are crucial for the sustainable growth of the construction activity. Clients are the key stakeholders in a procurement organisation that thrusts the level of collaboration in a team. They can exert power over cooperation from the early stages in the design – this is especially critical for Integrated Procurement Delivery organisations and set the degree of integration and expectations from all the team members.

In Europe, countries that understand BIM as the present and the future in the building industry have developed actions and means to realise a feasible approach of digital construction: public procurement, standardisation; research and development; and education [2]. In Denmark, the Executive Order number 118 of 2013 and Executive Order 119 of 2013 constitute the legal frame that regulates projects subject to mandatory ICT agreements in the public sector and projects in social housing. The Danish Association of Consulting Engineers and the Danish Association of Architectural Firms has developed the "Description of services for building and landscaping 2018", which outlines roles, divisions of services and responsibilities for all the professionals involved in building projects through works stages from brief to handover. The document includes a dedicated section that addresses ICT matters and gives general guidelines about digital communication and the establishment of a communication platform. The non-profit organisation MOLIO provides the building industry with standards and instructions that connect regulations and practice to coordinate ICT. Besides this, one of the most critical strategies that the Danish Government has put in place the last 15 years has been fostering research and development related to BIM via educational institutions and public offices to increase the expansion and application of BIM at all levels to obtain a shared digital language regardless the discipline.

In the light of the increased demand for solutions to face the communication and coordination challenges in the industry, different software vendors have developed products that cover, to a lesser or a greater extent, the practical demands that digital collaboration requires. The reality is that there is no one software and communication platform that can provide solutions for all types of projects. Nevertheless, there is a need to find "one single ecosystem" that can combine applications supplied by different software [3]. Digitalisation and efficient applications of CDE have become critical competitive factors in the industry due to the increased chances of success offered to all project's participants. Clients in Denmark are investing in digitalisation and CDE yet lagging compared to other sectors.

2.2 How a CDE facilitates collaboration in building projects

As per ISO 19650, a Common Data Environment CDE is an agreed Source of information for any given project asset [4]. This data aggregation enables different parties to participate and contribute to the project in all its steps through its life cycle. CDE can be simple digital storage like Dropbox or OneDrive – what defines its scope and form is the ICT agreement. Several other forms of CDE are present in the construction market in Denmark, some non-dedicated to AEC and others adapted and bound to the construction industry as Byggweb and Dalux as digital storage means for building projects documentation. CDE integration with BIM tools such as Revit, Navisworks and ArchiCAD is of the utmost importance since it is in these software where all the geometric and non-geometric data gets created and updated.

Platforms based on cloud storage enables all members to access the latest information and cooperate in a manner where all reach the same documents, improving data handling and reducing confusion in newer versions or duplicates. When a file is updated, the previous versions are stored in a straightforward interface with a shared directory. One of the most capable solutions is the possibility of seeing the newest version even with a printed document; for this Dalux uses QR codes printed in the documents, giving certainty on site and the potential to revise the alterations that occurred concerning the version printed. This tool being dedicated to the construction sector can have a file structure following national directives, and in the case of Dalux, can be adapted to any company standards. In the Danish context, Dalux defaults to the file structure Molio A104, a standard widely adopted file storage and naming in the country.

2.3 Communication platforms in building projects enabling non-BIM to BIM

The pursuit of efficient digital communication and establishing a collaboration platform try to solve cross-disciplinary coordination amongst the project participants. Effective interaction ensures the proper distribution of data in a structured manner throughout project design, construction and operation. Proper digital communication allows systematic registration processes of exchanged data and determines the responsibility and quantity of information delivered by the members of a project team. To achieve all these goals, the client has to decide who will set up the communication platform and (through respective consultancy services) delimit the extent of the system's performance. The management of metadata, versions and revisions; protocols for the definition of a folder structure, application of systematic processes via templates, and access to data through the usage of mobile devices, are critical factors to be considered when defining the type of project web to be used in a building enterprise [5].

The most prominent suppliers of complete digital project platforms solutions in the Danish market are Dalux, RIB Software and Ajour System – all of them based on cloud technology. All these products serve the same purpose in attracting public and private clients by offering solutions that are becoming more flexible due to high competition and continuous development resulted in an output of adaptation as the market evolves. The file management platform, and its means for ensuring a safe and sturdy hosting environment, are standard key parameters displayed by these products. All the data generated along with the project phases (from tender until handover and operation) can be managed across the different disciplines allowing quick access to information and providing features for 3D BIM visualisations. Previous definitions of hierarchy levels within the BIM project team, folder structures, communication workflows, and open industry standards as IFC permit all the participants included in the communication platform to use digital production. Since proper collaboration is promoted at early stages as the key to success, the available software focuses on coordinating design outcomes where the input that stems from various disciplines can be organised internally and externally through the generation of federated BIM models.

2.4 Usage of communication platforms in education: A fragmentary approach

As in other academic institutions, The Copenhagen School of Architectural Technology and Construction Management, hereafter ATCO school, uses Learning Management Systems (LMS) such as Blackboard, Docebo and others. The ATCO school uses the tool called Fronter. The platform is adequate. It can create and upload several class resources, in different types of file extensions, with varying types of access according to the courses the students should have to gain access to. The management of this tool is done before the semester starts. All students are given suitable access to their semester's courses, and the teachers are also incorporated in the respective semester disciplines. Fronter is a resource aimed for broad usage in education topics. It has been effective and proven to be a good communication tool. Still, with the intensification of digital documentation resulting from the BIM process, the assessment of the teaching staff to all the new digital submissions has been challenging. The tool relies on a fragmentary approach where different lecturers have access to further deliveries from the students.

Since multiple disciplines are aggregated in the BIM model, the documentation produced needs to be uploaded several times for the different assessments by the staff. In a final stage, the students have to assemble all comments and revise their designs in the 3D model. This workflow is problematic. Not every time the different teachers have access to the same version, and some comments can be redundant as intercommunication between the staff is not possible via this LMS. This problem is not new, the AEC industry has been struggling with it, and CDE tools have evolved to the specific industry needs. Software as Dalux has been adapting its features to the construction needs and its rapid digitalisation. It is not rare that the ATCO school students, after their internship period, refer that this tool has been in use in architectural practices and engineering offices with success.

2.5 Implementation of Dalux at ATCO school

In Denmark, BIM implementation and regulative parties have influenced the way the industry is developing. Educational institutions must work hand in hand to go along with such developments providing the industry with professionals that understand processes and work as a team. Teaching BIM is an immense challenge that requires the proper framework to align different approaches and understandings of BIM. Curriculums have to be adaptive to allow an appropriate integration at all levels of BIM complexity when teaching. Therefore, pedagogics through well-designed learning methods are essential to bringing an institution the tools for solving problems in an industry that is technical and based on practice. The ATCO school has been implementing BIM content in its curriculum for more than ten years. Great success has been achieved due to continuous improvement of the didactical and pedagogical approach. Graduates obtain multiple competences to cope with a wide range of tasks and situations presented in the professional practice and are given the capacity to produce new knowledge and innovation and transfer these fundamental factors of development to the industry. The teaching method and its BIM incorporation provide an adequate environment for students to be accountable for their learning process by setting activities to promote self-engagement and motivation.

At ATCO school, the teaching is didactically founded on Project Based Learning (PBL). Groups of students withstand a real-life project and correlate a feasible solution with their available resources in order to learn [6]. There is a set of instructions provided by the teachers (facilitators) where relevant problems related to a building project are presented to start the cycle that ignites the learning process. Through cross-disciplinary consultation and feedback, the students will connect the different learning outcomes for the topics included in the semester, delivering at the end of the course a completely designed project including primary BIM assignments. The proposed technology to make all these possible is based on assuming that communication and collaboration through digital technologies are crucial if the school seeks to depict real-life scenarios. Although the current LMS platform functions appropriately in terms of being an efficient resource in the learning process, there is a question of whether it has the flexibility to be used as a tool for simulating a real-life situation as a communication platform amongst stakeholders. In 2019 the ATCO school initiated a collaboration with Dalux, which lead to some isolated pilot projects by some lecturers. In the light of preliminary results, the staff involved in these projects decided to unify outcomes, reason and conclude. Afterwards, the team proceeds with an extensive field test and research to analyse the implementation of Dalux as a communication platform on a large scale in the program.

2.5.1 Brief description of Dalux

Dalux provides a unique ecosystem where the students can prepare their deliverables, simulate work stages and communicate with peers and lecturers. The information in the platform arrives and flows in a variety of different formats, such as drawings, documents, cost calculations, schedules and BIM models. The system includes a series of modules: Dalux Tender, Box, Field and FM, where the progression and evolution of the project can be easily tracked. Additionally, the opportunity of setting communication channels through the creation of workflows and levels of hierarchy allows the students to interact together and have access to valuable 3D views and reports with attached pictures that can be combined with 2D drawings. Dalux Tender allows to simulate any kind of competition, sets levels of communications (with or without restrictions), allows replying to questions from the competitors, and submitting the proposals. Dalux Box provides file management and organisation that can eventually be fed by information already existing in Dalux Tender. Dalux Field supports processes happening on the construction site. Together with the Dalux BIM Viewer for AR (Augmented Reality), mobile devices combine and compare physical construction with BIM models. Finally, Dalux FM is a BIM and cloud-based Computerised

Maintenance Management System (CMMS) to assist complete overview of all the building assets and manage work orders. These modules have been successfully tested and implemented at the ATCO school through different semesters, adapting learning outcomes to levels of complexity as progression goes on.

2.5.2 Dalux at KEA workshop

To show the software's capabilities, the implementation team organised an online workshop in early 2021. The lecturers and Ole Jetmundsen – (Dalux consultant) introduced several software capabilities to teachers and students at the KEA buildings department. In total, 64 attendees have attended, 57 students and seven lecturers, representing a good ratio of the KEA's academic community.

The workshop covered the basics of the software, explaining the primary User Interface (UI), creating a project in Dalux following the Molio A104 file structure standards, and adding the different contributors with the correct access to the information folders. During the workshop, a demonstration showed how to export an IFC model to the CDE via the Dalux Revit plugin – uploading of Revit sheets to the platform also possible (Fig. 1).

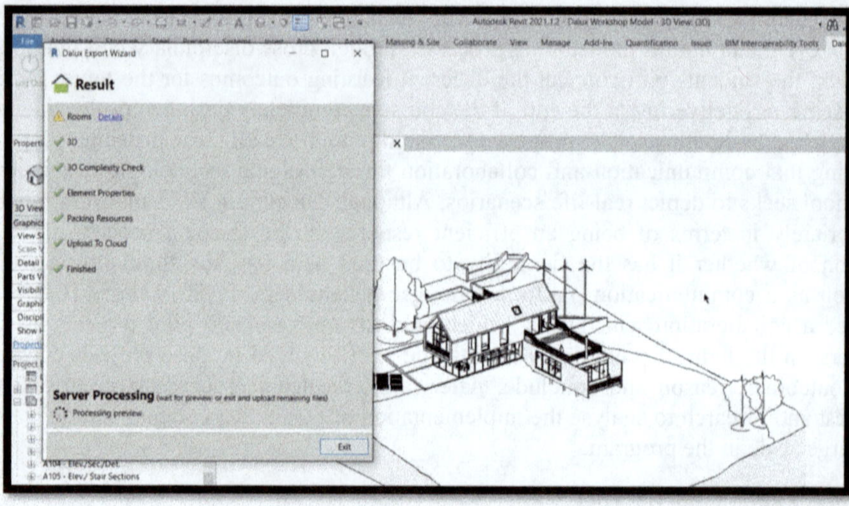

Figure 1: Exporting data from Revit using Dalux plugin.

When uploading from Revit, the sheets are automatically aligned with the correct levels and merged with the 3D. It was also shown a Dalux view mode, where both 2D and 3D are aligned – this is a solid way to inspect any discrepancies between the visual project data (Fig. 2). The software 3D visualisation is robust, allowing the user to navigate/walk virtually in the project. This feature is a keystone for success in the implementation of Dalux – it will enable non-proficient BIM staff to evaluate and create comments that are available for the students to correct afterwards.

Once the issues/comments have been created in the Dalux web interface, the students can revise them directly in Revit. The plugin includes a comment browser that attaches to the right of the drawing area in Revit (Fig. 4). This ability to review comments in Revit UI is essential, as the comments are directly shown and revised in the students' core tool.

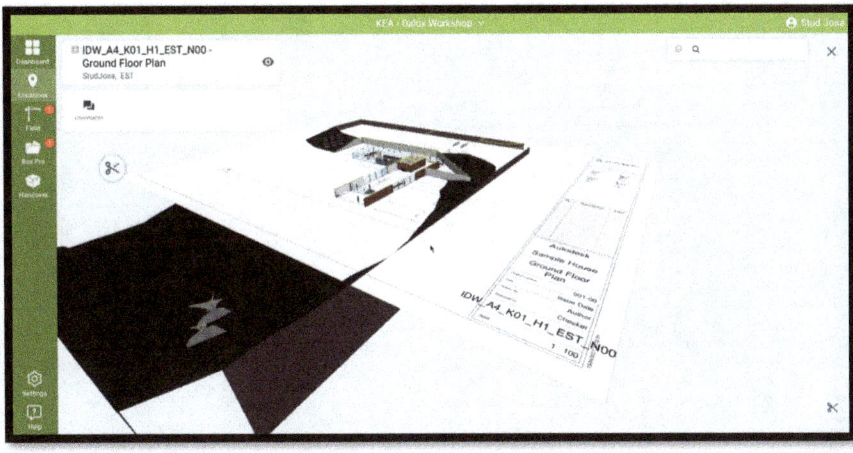

Figure 2: Merge 2D with 3D.

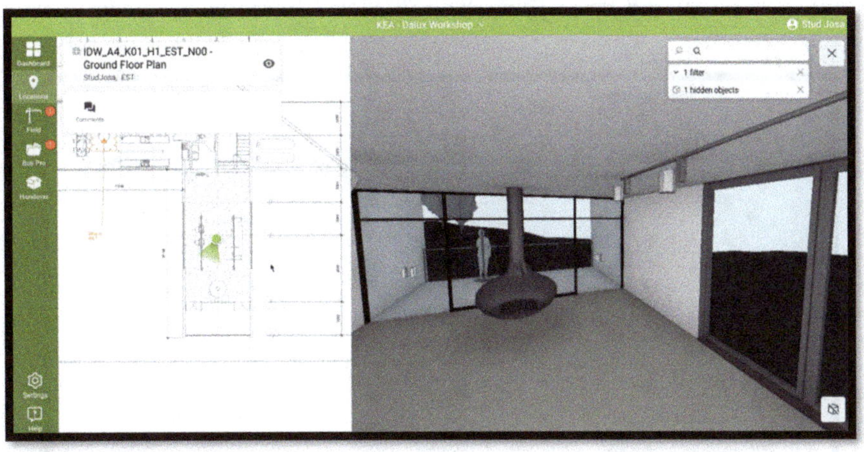

Figure 3: 3D navigation in Dalux.

2.5.3 Advanced BIM for digital tender: Elective course (third semester)

The elective course aims to work with digital tender in construction projects using ICT technologies. The program's focus is on the potential improvements that BIM offers to the construction sector by using advanced digital technologies, optimising the tender phase and the core deliverables of contractors in the construction phase. It is expected that students will acquire knowledge, skills, and competences about using ICT in digital construction projects delivered in different procurement forms. There is a strong focus on applying ICT standards, specifications and advanced digital tools for performing consistency and collision control in the 3D models, handling digital tender and digital platforms. The course includes applying 5D BIM processes for preparing model-based quantity take-off, bid

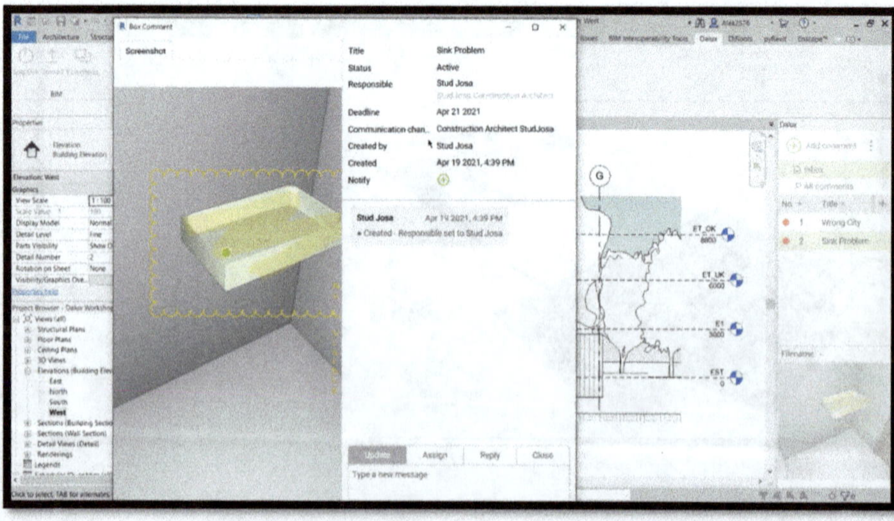

Figure 4: Issue/comment review in Revit.

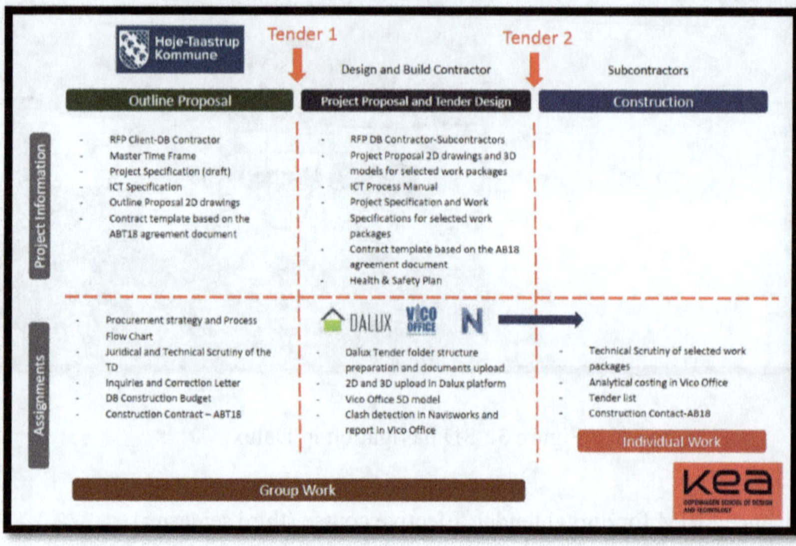

Figure 5: Project information: Procurement forms – Main assignments.

estimates and bids submission, and 4D BIM process simulation for selected phases of the construction process. The students are confronted with a series of exercises to obtain in-depth skills in planning and managing digital tendering processes, including preparation of a detailed tender phase schedule, management of documentation, specification of participating actors, services and interfaces, as well as the organisation and contracting related aspects. The students will work with a tender documentation package that corresponds to a real-life project – a new public administration building category. The released tender documentation package (handed out in a digital form) will be the core material used for the primary assignment in the course. Students scrutinise the tender documentation and prepare selected deliverables of the Tender Managers and Contractors performing in the Tender and Contract stage. Students upload the tender document in Dalux, which is the platform used as a communication tool for tenderers. Additionally, the students will use Navisworks and Vico Office BIM tools as a part of the scrutiny and costing assignments – all the outputs are uploaded in Dalux.

The course presents several challenges to the students since the project, during design stages, involves two tender phases where dynamics between assignments can be complex. The groups are given two tender packages that have to be scrutinised. Once the review is finished, the groups transfer questions to enquiry letters and perform cost calculations, schedules and contracts. Tender packages are including architectural, structural and ventilation systems BIM models. One relevant activity included in the scrutiny process is to use the BIM models as the primary resources to prepare inquiry lists and correction letters. Vico Office is the software that the students use for costing (5D) and scheduling (4D) purposes. Vico Office itself included several modules where the project information can be stored, processed and updated. One of these modules consists of a report engine that allows communication amongst the different disciplines. The students perform a clash detection control by using Navisworks, obtaining a report and exporting the results to the Vico Office module called Managed Issues. When the scrutiny of documents and BIM models has been finished, the groups will upload the original tender packages in Dalux Box for simulating a tender stage. To do that, the groups have to create a project, set rights and hierarchies and establish communication workflows where the client will communicate with bidders to reply to questions, receive proposals, negotiate and sign a contract. The course ends with an exam, where the students have to show all their outcomes in an explanatory presentation that includes opening Dalux Box performing demonstrations and display results.

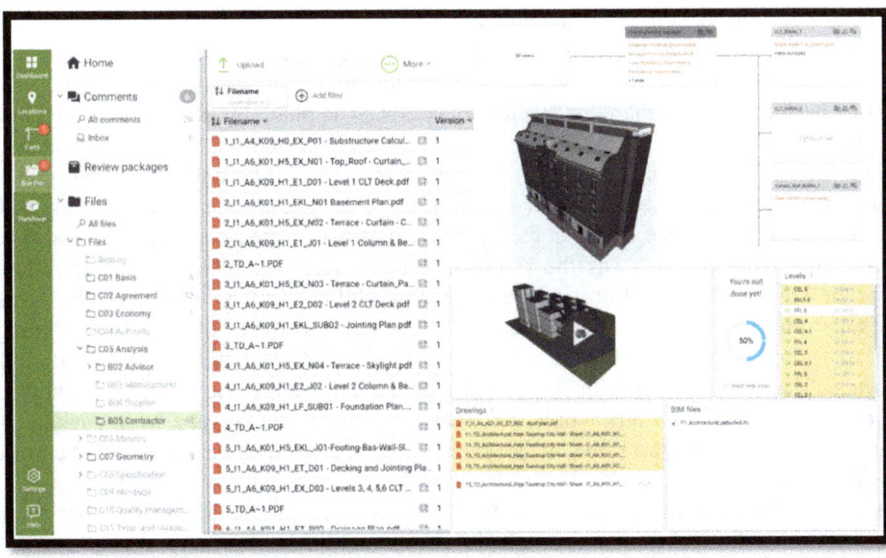

Figure 6: Dalux box: Digital tender.

2.5.4 Construction stage: Elective course (fifth semester)

The course includes selected construction management deliverables of the Client, Contractors, Design Professionals, Fabricators, Suppliers and Agencies performing in the Construction Stage. Students should strengthen their knowledge within Planning and Construction Management and develop specific Construction Management knowledge and skills about the process and core deliverables of the Construction Stage. One of the topics discussed during the course is the Handover Protocol with a general approach and afterwards more focused on the Danish scope with a specific view taking as a reference the rules set by the "General conditions for building and construction works and supplies (AB18 – ABT18)". The handover stage is an essential process in a building project. Defects associated with the performance of different trades relate to this stage. The protocols used to register and remediate imperfections constitute crucial processes that affect the productivity and the quality of the building. Additionally, the handover (with its milestones) is a legitimate critical period perfectly regulated by the agreed documents AB18 and ABT18. The parties involved in a project may conduct a pre-inspection meeting before the formal handover event, where there is a registration and discussion about the quality of the product that, in most cases, will lead (amongst others) to a request for rectification of any defect found during the actual inspection. Since the registration of defects can be a complicated process, it is essential to set the strategies, workflows and a platform where client, contractors and subcontractors will communicate [7]. So far, in the scope, the preparation of a defect list has been based on forms to be filled out manually and to a good extend by using cloud services that allow photo recording by using mobile devices. However, these methods lack proper communication channels amongst project participants and the inclusion of an engine that permits comparisons between BIM models and reality. Dalux Field is one solution to this requirement. It poses all the tools to combine project information, set communication workflows, and digital documents with excellent usage flexibility. A 3D model can be correlated with the physical completed building providing capabilities of recording defects or sending requests – all these indispensable applications to improve efficiency and quality.

In this elective course, the students work on a refurbishment project. All the design and planning information related to the project used for the exercises are the outcome of the students' performance during their mandatory course. The planning subjects included in the program are the definition of the project procurement connected with the site organisation, detailed demolition plan, construction technology analysis, health and safety plan, 5D and 4D performances, payment plan, and a defect list disaggregating trades by using Dalux Field. For the performance of the defect list, the groups will create a project in Dalux and add as many work divisions as the refurbishment project will include. The first step is setting the 3D model in the Locations and combine sections by levels with 2D drawings – eventually, the students will include views and elevations as a part of the technical design. Back in Dalux Field, the students will create work packages, and they will align the performance of a defect list with some of the trades included in the project – most of the time carpentry works, masonry works, steelworks and painting. To set a work package, the students have to create workflows where the communication channels amongst the project participants have to be defined.

Once the work packages are ready, the students can start creating tasks that for the performance of the defect list are called "issues". Other tasks that can be added are Deviation (DV), Quality Assurance (QA), Registration (REG) and Test (TEST). Regardless of the type of preset task that Dalux Field includes, there is always the possibility of

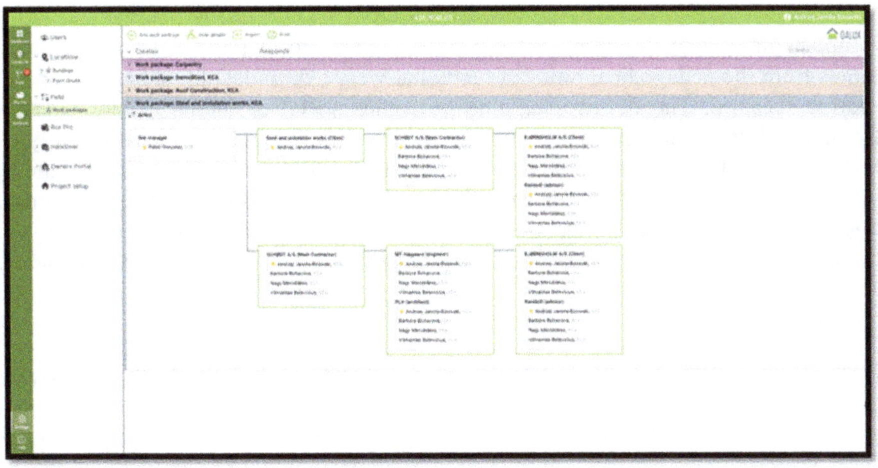

Figure 7: Dalux field: Work packages and communication channels.

creating other types of tasks, such as health and safety inspections monitoring tools. The students have to research and find the most common defects in refurbishment projects and find pictures that show those issues. Once the research is finished, the students can create issues in Dalux Field, assign a responsible (connected to the work packages) and connect the 3D model with a 2D drawing pointing exactly where the defect is located in the building and attach a picture. At the end of the exercise, there will be a list of defects in Dalux Field, with the possibility of tracking remediations and navigating the 3D model at the same time.

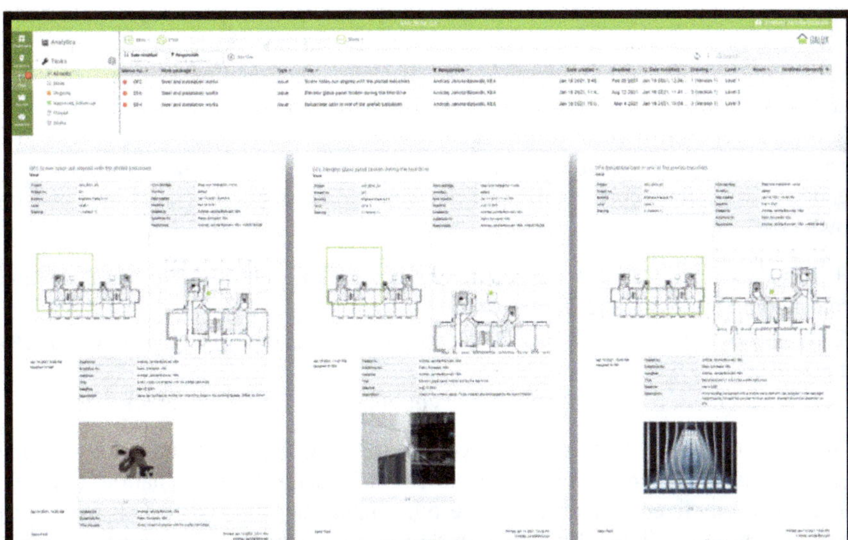

Figure 8: Dalux field: Issues, defect list and location.

3 OBSERVATIONS

The implementation of Dalux in the different courses resulted in a series of analysed outcomes from the pedagogical point of view that were extrapolated to a broader and more integrative scenario. The defined learning outcomes specified in the lecture plans were confronted with the students' presentations of results and complemented with surveys. The students managed to use all the modules included in Dalux and showed an understanding of the software as a versatile tool for data management. While going through the different exercises, there was a clear connection between how information has to be arranged and communicated and technical requirements and guidelines given by the Danish standards.

3.1 Outcomes

Collaboration via Dalux helped the groups' members work together and innovate in workflows and coordination. The usage of Dalux facilitates motivation, profound learning processes and understanding of the different stages involved in the resolution of tasks in building projects. Several disciplines can communicate via BIM models in the same ecosystem by using CDE, where the integration of inputs accomplishes the teachers' expectations.

Students were asked their opinions about having used Dalux, and answers included, "It is a very efficient tool, maximising the use of time and binding together core actors involved in a project also during the construction site. Very easy to understand and work with it and a great platform for communication. Real-life software for creating, assigning and solving problems"; "My experience working with Dalux is extremely positive, from all the new software we worked with at KEA it is the most efficient and I enjoyed working with it creating a real-life situation for preparing, assigning and solving issues. I can see immense potential for using Dalux. Also, I would like to mention that it is already being used at Metroselskabet & Hovedstadens Letbane where I am doing my internship"; "I would say that advantages are that it is user-friendly, fast with communication and also that you can use it directly on site and send immediate notes to the relevant people and have immediate action after that." In the light of more comments like those shown above, an assessment indicates that the implementation of Dalux was seen as a proficient mean for communicating, learning and working at ATCO school.

The results below demonstrate the significance of introducing a CDE tool at ATCO. The survey on implementing Dalux at ATCO corresponds to the Dalux Workshop, totalling 64 attendees with a fair representation of docents and students. The questions and results are shown below:

- Q1 – In your opinion, KEA BYG should have a CDE platform like Dalux to enhance project communication/evaluation?
- Q2 – In your opinion, a tool like Dalux can improve communication between teachers and students?
- Q3 – Do you think that a tool like Dalux could be used for Exams at KEA BYG?

To notice in Q3 results that the maturity level at ATCO is not yet at a point where either lecturers or students feel comfortable to use the tool for semester examination; only 50% of the attendees are inclined to use it. On the other hand, to the Q1 and Q2, over 70% are favourable to the queries, which shows optimism in the work and allowance to further develop the implementation at ATCO.

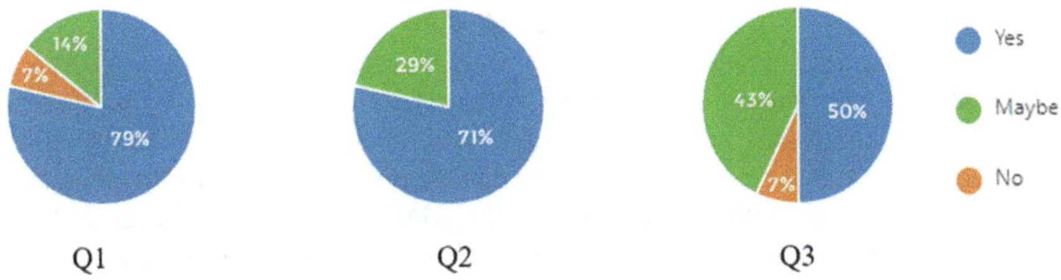

Figure 9: Survey results – Dalux Workshop in KEA.

3.2 Industry view

One of the most important activities that lectures have at ATCO school is networking with the AEC actors in the industry. To keep the education program updated, it is necessary to follow the local and international trends in a business subject to continuous development and change. On the other hand, the teachers have a relevant role as the link between companies and the students when it comes to the internship period. The school has to make sure that interns are ready to cope with a series of tasks based on learning outcomes set by the students themselves – intended to be used in the final dissertation report and bachelor project. Companies in Denmark can see the potentials of interns and graduates when it comes to implementing new technologies and processes and are open to allow them gradually incorporate their ideas and innovations in a cycle that has been accelerating in the last years.

In the autumn of 2020, the Dalux implementation team at ATCO school conducted interviews with several stakeholders in the industry (Dalux users) to obtain input about what could be the most important features and processes to teach in the courses. The AEC actors involved in the interviews stressed the importance of including at upfront stages a clear description of the communication platform in the ICT Specification and Process Manual. The usage of the MOLIO A104 is key to allowing success in setting a proper framework for data management and, in combination with a defined classification system, should allow the performance of disciplines tasks by integration of Dalux modules with special attention to BIM viewer. Setting rights and workflows within the communication platform is a particular competence that the professional should have incorporated for the proper usage of Dalux.

3.3 Limitations

The implementation path of a CDE at ATCO is far from concluded since Dalux has several shortcomings. A professional CDE tool is used in an educational context, and the lack of information to fully deploy a CDE might be difficult, as the teaching methodology is built in project-based learning and the semester assignments do not have the necessary scale to see the full potential of a CDE deployment. Competitors such as BIM 360 from Autodesk have the advantage of holding the latest data by default, without the need of uploading data or installing plugins, this Autodesk tool has the advantage of communicating directly with the model hosted in the cloud. The future holds the answers to this study, but having Dalux onboard cooperating with ATCO may have fruition. There is an expectation that the software can be adapted to an educational context allowing for better integration.

4 CONCLUSION

The increase of data associated with the BIM process has led to more precise projects and better simulations; even though the computational power is aligned with that progression, the individual workers have been struggling to handle this volume of information. The ATCO school faces the same issues; lecturers cannot revise the amount of documentation that the students generate with the inclusion of the BIM process, either because it is stored in software that they are not familiar or because the sheer number of documents is just too large. By using a CDE such as Dalux, can help to close the gap created by digitalisation. The software Dalux has a friendly user interface and a strong component in communication between BIM and non-BIM personal. Dalux is a proper tool for AEC educational institutions such as ATCO school, facilitating the linkage between the new generation, highly capable in digital tools and the senior experts that otherwise struggle in helping the students.

The introduction of a CDE in an educational context also increases its adoption by the industry, accelerating digitalisation in companies, moving forward with the national and international directives, minimising waste of resources allowing for better projects. The Danish building industry has seen the potential of CDE, and it is now under preparation for a more integrative usage of communication platforms, resulting in an increased interest for graduates with the right competences to do the job and work as a team in a digital environment. Educational institutions are willing to contribute by setting policies in terms of research and development and transferring the knowledge to the students and the industry simultaneously by bilateral collaboration. ICT agreements and the implementation of standards is vital to set the context and the framework that allows students to implements CDE in their schools' project. At the same time, it shows a tremendous potential to assist lecturers in their teaching by replicating real-life multidisciplinary work scenarios. In the light of the results after the implementation of Dalux in pilot projects at ATCO school, it seems to be evident the advantages that the application of Dalux could bring to educational institutions connected with the building industry. A wholistic cross-disciplinary pedagogical and didactical discussion is underway to decide the suitable methods and approaches at study situations. That will allow all the lecturers to communicate appropriately via CDE amongst peers and with the students and set a new benchmark to contribute to the industry development to increase productivity.

ACKNOWLEDGEMENTS

The authors would like to thank Mattias Straub and Jan Lambrecht for the vision, enthusiasm and remarkable collaboration in implementing CDE in KEA, James Harty for all the wisdom and guidance, and finally, Claudio Testa for the opportunity and trust.

REFERENCES

[1] Thomassen, M., BIM and collaboration in the AEC industry. Thesis. Department of Mechanical and Manufacturing Engineering: Aalborg, p. 120, 2011.

[2] European Commission, *Building Information Modelling in the EU Construction Sector.* European Construction Sector Observatory, p. 22, 2019.

[3] Eastman, C., Teicholz, P., Sacks, R. & Liston, K., *BIM Handbook. A Guide to Building Information Modeling for Owners, Managers, Designers, Engineers, and Contractors*. John Wiley & Sons: New Jersey, p. 506, 2008.

[4] ISO 19650 Series of International Standards, 19650-1:2018, Organization and digitisation of information about buildings and civil engineering works, including building information modelling (BIM) – Information management using building information modelling – Part1: Concepts and principles, 2018.
[5] Bygningsstyrelsen, Digital kommunikation og etablering af kommunikationsplatform. Bygningsstyrelsen: Copenhagen, 2021. https://bygst.dk/byggeri/ikt/vejledning-til-ikt-ydelser-og-implementering/digital-kommunikation-og-etablering-af-kommunikations platform/. Accessed on: Apr. 2021.
[6] Prince, M.J. & Felder, R.M., Inductive teaching and learning methods: Definitions, comparisons, and research bases. *Journal of Engineering Education*, **95**(2), pp. 123–138, 2006. https://www.researchgate.net/publication/220017716_Inductive_=Teaching_and_Learning_Methods_Definitions_Comparisons_and_Research_Bases.
[7] Værdibyg, The Hanover Process. Værdibyg: Copenhagen, 2018. https://vaerdibyg.dk/vejledning/the-handover-process/. Accessed on: 8 May 2021.

CONCEPT DEVELOPMENT FOR ADOPTING 5D BIM IN SMALL AND MEDIUM-SIZED ENTERPRISES OF THE AEC INDUSTRY

ANNA NAST & CHRISTIAN KOCH
Bauhaus University Weimar, Germany

ABSTRACT

In order to achieve multiple benefits of building information modelling (BIM) in the construction project, the BIM methodology should not only be implemented in a project-related manner but also in a fully enterprise-related manner. Thereby model-based costing is one of the most important BIM use cases, which can be found in any project stage and is relevant to all project stakeholders. But the small and medium-sized enterprises (SMEs) of the architecture, engineering, and construction (AEC) industry in the particular face numerous problems and difficulties in adopting the BIM methodology in their business processes. This research aims to investigate the process of the 5D BIM implementation in small and medium-sized planning, engineering offices, construction companies, and project developers. The main body of the work is based on a review of 24 implementation projects of the model-based costing in SMEs that the authors have carried out as part of their practice as BIM consultants and trainers in Germany. These projects are analysed to develop a concept for a smooth, individually adaptable implementation of 5D BIM in SMEs. The results are summarised in the concept framework for a sustainable 5D BIM implementation approach that illustrate the priority and hierarchy of the main activities in the implementation process. The designed concept can be used by different stakeholders of the AEC industry and provides a solid basis for organisations to make informed decisions in adopting 5D BIM in the body of the organisation structure. Thus, this research is an essential contribution to the knowledge about the enterprise-related implementation of 5D BIM.

Keywords: 5D BIM methodology, BIM implementation, adoption, small and medium-sized enterprises (SMEs), organisation, quantity take-off, cost estimation, cost calculation, concept framework.

1 INTRODUCTION

In the wake of the emergence of the coronavirus pandemic at the beginning of 2020, several construction companies started to invest more in digital resources. In this context, the BIM methodology represents an important part of the digitalisation of the construction industry. In the German AEC industry, e.g., one-third of all building companies are currently planning to align their business model more strongly than before with digital services and to expand the digitalisation of their business processes [1]. In 2017, in contrast, only 9.6% of German construction companies used BIM methods to manage construction projects [2].

BIM methodology is not spreading rapidly in the AEC sector because many construction companies still think of BIM implementation as just a simple exchange of tools and therefore find it difficult [3]. Although, a fundamental part of the BIM methodology is to address the processes by using and sharing digital building information. BIM aims to transfer project-relevant information as comprehensively and completely as possible [4]. In this case, digital tools can only bring added value and be accepted by the AEC actors if the processes are clearly defined. Technical tools must help to create a collaborative, usable, digital representation of a physical object and its functional characteristics [5].

The linking of the digital representation of the building elements with time and cost data forms one of the significant BIM use cases. Model-based scheduling and cost estimation are classified as the 4D and 5D BIM dimensions [6]. The article focuses primarily on model-based cost planning and related activities, regardless of whether model-based scheduling is

WIT Transactions on The Built Environment, Vol 205, © 2021 WIT Press
www.witpress.com, ISSN 1743-3509 (on-line)
doi:10.2495/BIM210071

performed. According to Eastman 5D BIM intends to link cost items with specific, measurable features of the model object, so that project costs can be forecasted and actual costs can be tracked about the BIM model [7]. Therefore, model-based cost estimation depends significantly on the quality of the model data, information deliveries and is linked to the other BIM use cases, such as design and execution planning. 5D BIM itself can be subdivided into several specific use cases, such as cost estimation in the early project phases, calculation of construction work for contractors, budgeting for project developers, etc.

As cost planning is a fundamental activity in the AEC industry, 5D BIM is relevant for all project phases as well as for all project participants. One of the most crucial benefits of the BIM methodology, cost assurance in construction projects, can be realized by the 5D BIM application. If to use the advantages of the BIM methodology and especially 5D BIM for the safe cost in multiple projects, BIM is to be implemented both in the project and in the company. The project-based implementation in the context of a pilot project or case study is only a part of the enterprise-related digital transformation.

The implementation of BIM depends on various influencing factors, e.g., organisation culture, education, information management [8]. For adopting BIM, construction project participants usually have to define BIM goals and use cases, develop a process road map, design a BIM execution plan (BEP), and chose a technical infrastructure, as well as clarify a lot of strategic, operational, tactical, and technical questions [9]. In most cases, SMEs are overwhelmed with all these additional tasks, the required implementation steps, and volumes of incoming information. However, SMEs often do not dispose of the financial resources to get any external support into the company. For this reason, SMEs are still lacking in the implementation of BIM. In the European Union, SMEs are a significant part of the economy, and account for approximately 22.6 million in 2021 [10]. Thereby 99.8% of all European companies are SMEs, which act as the main drivers of innovation and digitalisation [11]. In the German AEC industry, e.g., more than 93% of planning and engineering offices employed fewer than 50 professionals [12]. Hence, the comprehensive implementation of the BIM methodology in the AEC sector depends on BIM adoption in SMEs.

The focus of this research is, therefore, on developing the concept for better 5D BIM adoption and implementation in the SMEs of the AEC sector. The impulse for this scientific work is the professional experience of the authors as BIM trainer and consultant primarily for the issues of 5D BIM implementation in different companies, such as planning, engineering offices, construction companies, and project developers.

2 METHODOLOGY

The research provides a synthesis of the reviewing of scientific papers and the analysis of the authors' practical expertise and attempts to bridge a gap between theoretical fundamentals and practical implementation of the 5D BIM methodology. The literature review is a qualitative review of scientific papers in the research areas of BIM implementation, cost planning in the AEC projects, and management theories. The most relevant terms for the literature survey can be divided into three groups, which have been combined among each other: (1) BIM, implementation, adoption; (2) 5D BIM, costing, cost estimation, cost estimate, cost calculation, cost control; (3) SMEs, organisational, enterprise, construction company, contractor. It has been noticed that the research papers addressed often change management in the context of the company-related implementation of the BIM methodology. Hence, innovation management and business process change management are briefly explained.

To assess the practical applications of the theoretical findings, 24 implementation projects of the model-based costing of 5D Institut GmbH in Germany have been investigated. The article provides a qualitative review of the 5D BIM implementation projects, summary, and analysis of the preliminary project meetings, includes die results of not structured interviews with participants. The meaningful features and issues encountered are highlighted.

Due to professional confidentiality, the executed implementation projects cannot be mentioned by name. A distinction is made between the implementation packages for planners and clients as well as for construction companies and contractors. The authors are also limited in their choice of the model-based costing software and work exclusively with RIB iTWO 5D. However, the preliminary BIM model can be created in any BIM-capable software that is certified for IFC export. A BIM implementation project starts when the decision about the IT infrastructure has been made, and a technical realisation of the BIM goals must be carried out. Nevertheless, the authors often have to consult the clients on the entire BIM implementation plans and refine the BIM use cases. The general procedure of a standard consulting project for clients and contractors is shown in Fig. 1.

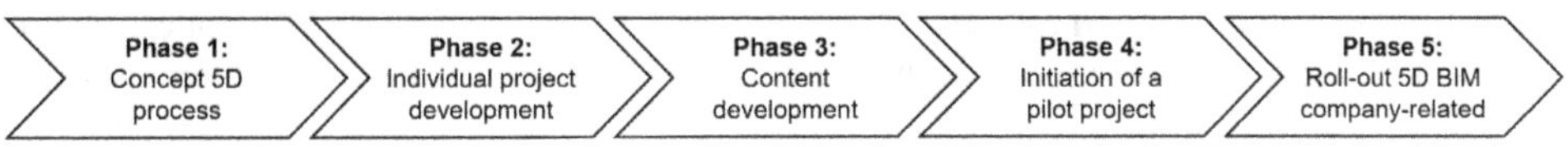

Figure 1: The general procedure of a consulting project by 5D Institut GmbH.

3 LITERATURE REVIEW

To provide a solid basis for organisations to make informed decisions in adopting 5D BIM in the body of the organisation structure, the general steps for the BIM implementation should be considered in the context of the enterprise-specific cost calculation processes. Model-based cost planning and control could be used by different AEC stakeholders, e.g., government, owner, designers, contractors, operation organisations, cost engineers, and during various project stages, e.g., in the conceptual, design, procurement, construction, completion stages. Therefore, different approaches and tasks can arise during implementation. The principal 5D BIM use cases will remain under any circumstance quantity take-off and cost estimation.

3.1 Implementation of the BIM methodology

The application of the BIM methodology can be considered on different levels, such as implementation in the whole AEC industry [13], country-related adoption [14]–[17], development in a single firm [18], [19], and project-based implementation [20], [21]. Some authors emphasise that large construction companies operating on a design-and-build basis have already adopted BIM [22]. These contractors can usually manage their entire internal processes independently of other construction stakeholders, thus giving them an advantage over SMEs. As clients require BIM services more consistently, SMEs should advance the internal digitalisation process to remain competitive in the market.

Therefore, the guidelines for BIM implementation at the operational level are needed. They can include aspects such as the BIM adoption scope, discipline, model analysis and data transfer methods, interoperability, and staff responsibilities [13]. Feasible results with BIM in SMEs can be obtained if the internal BIM model structure, standard information linked to model objects, critical information workflows as well as collaboration with external stakeholders are defined [3], [23]. Company-related BEP and standards must elaborate on issues like “scope of implementation, infrastructure, contracts, implementation process” [24].

At the same time, BEP for enterprise-related implementation should prioritise the program–process–person perspective [25].

3.1.1 Factors influencing BIM implementation

According to the BIM implementation level examined, the critical success factors may be different. For BIM adoption in the national AEC industry, e.g., the factors like development standards, acceptance by professionals, educational offers, or pilot projects [6] are significant. For BIM implementation in construction projects highlight Darwish et al. fifteen factors as the most critical ones [21]. The first factors influencing BIM implementation are coordination between all project parties as well as training and development. Collaboration between AEC stakeholders, availability of competencies and experiences, BIM adoption strategy influence the BIM implementation in a project-related manner [21]. According to the research by Arayici et al., construction stakeholders, who have been implementing BIM, emphasise the importance of staff training on new processes as well as on new software and technology, effectiveness and establishment of the new process and workflow, understanding of BIM, a realization of competitive advantages, purchase of software and technology [8].

3.1.2 Challenges of BIM implementation

The main obstacles to BIM adoption may be divided into four large general groups, e.g., mentality, knowledge and skill, software, and hardware [3], or categorized into five limitation categories, such as technological, legal, personnel, cost, and management barriers [26]. However, scientific research papers on the hindrances to BIM adoption are primarily related to a specific country. The main barriers for the Chinese AEC industry formulated by Wu et al. [15] are senior management support and financial investments. New Zealand struggles with such challenges as lack of expertise, client demand, and collaboration, shortage of pilot projects and guidelines, cultural resistance, legal and technical problems [17]. In Iraq, the implementation barriers are divided into legal, organisational, and technological [16]. In comparison, the construction industry in Latin America is faced with change resistance, bureaucracy, lack of regulations, human resources, and administrative initiative, etc. [18]. Especially for SMEs, six challenges are formulated in previous research: limited resources, collaboration, lack of BIM knowledge, legal disputes, unsteadiness in regulations, adherence to the advantages of SMEs, data protection, and responsibility for the information [27].

3.2 Cost planning in the AEC projects

Cost planning and control are basic tasks in AEC projects. While planning is primarily concerned with realistic cost determination, cost control is of great importance during execution. According to the planning status, there are different levels of information in the cost calculation. In the German AEC industry, e.g., architects, planners, and project developers usually use the DIN 276 (German Institute for Standardisation e.V.) standard to determine project costs [28], [29]. Construction companies that take over the execution services must consider the following partial costs in their cost estimation: (1) material costs; (2) labour costs; (3) expenses of the construction tools; (4) indirect costs [30]. The costing of the construction company encompasses the following phases: pre-calculation (subdivided into quotation costing, contract costing, work execution estimate) and post-calculation [30].

Regardless of the project phase and construction participants, the quantity take-off is a foundation for the cost estimation. In the concept project phase quantity estimation provides the base for budget planning. In the tendering phase it supports the estimation of the project costs and the first statement on the execution time. Before the execution phase, it facilitates the calculation of the construction works and planning of construction activities. During the construction phase, actual quantity take-off is used for cost control [31]. The accuracy and level of granularity of quantity surveying vary throughout the project.

3.2.1 Traditional cost estimation

Some challenges in traditional cost estimation impact the project cost: problems with sharing information and lack of communication as well as collaboration between different project stakeholders, e.g., design and construction teams, regarding cost activities [32]. The total expenses of a project are often unknown until after the conceptual design stage, and in some circumstances, even after finishing the construction stage [33], [34].

The quantity survey is therefore error-prone in the conventional calculation method and requires an investment of time. Manual calculation on CAD documents, that only store limited information, is a basement of the traditional quantity take-off. This approach not only requires a high level of human resources but also causes deviations quite easily [35].

3.2.2 5D BIM methodology

During BIM-based cost planning, the so-called design-to-cost method is applied. The accuracy of cost planning, i.e., the progress from a cost estimate to a cost calculation, is largely dependent on the project development. With the advancing project phase, more information is added to the model. This results in a continuous refinement process, which leads to a final cost calculation. In the early stages of planning, a cost estimate is prepared using the *deductive costing* methodology, which operates *top-down*. This means that a general budget is derived based on real costs from past projects, considering decisive factors such as the construction method. Such an approach utilises cost elements [28] and is comparable to cost estimation with the help of the construction cost index [36]. Later, more detailed cost calculations are carried out *inductively* and thus *bottom-up*. The individual costs of the components are multiplied by the expected quantity to be executed. Unit prices are determined based on a price database. Quantity formulas are used to perform a component-based quantity take-off founded on the model property records.

To be able to guarantee cost safety within a project, cost planning must keep pace with the constantly progressing project planning. Model-based quantity take-off is the only method that makes progressive cost planning possible. However, model-supported quantity estimation does not produce all the required data to create a cost estimation and bill of quantities [31]. Model-based cost estimation combines quantities gained from models and additional information such as cost elements, unit prices, construction price index, specification of services, and other cost-related information. Moreover, the geometrical information of the BIM model must be transferred into geometrical information that conforms to the estimation procedure [37]. Additionally, from the perspective of a construction company, the model-based quantities are translated into quantities of materials, equipment, manpower, etc., and afterwards into costs [38].

In the scientific literature, are some frameworks for 5D BIM process integration that encompass the whole cost management process [39], focus on the tendering phase [40], try to facilitate costing in contractor-led projects [33], present general workflows of 5D BIM process by different stakeholders [7], provide guidance in BIM implementation for construction companies [24]. However, the research papers notice that it is challenging to overcome the 5D BIM implementation obstacles, e.g., the lack of a framework for information exchange and compatibility between software applications, cost data sharing, and poor model data quality [31], [33].

3.3 Management theories

BIM methodology brings the changes into the business process of the construction sector, which can be conducted on two levels: sector-wide and individual organisations [41]. Thereby, a simply methodical implementation of BIM does not cover all organisational areas and does not sufficiently lead the individual actors into the change. BIM adoption requires an entrepreneurial vision, which is developed and communicated by the executive staff. Consequently, strategy, structure, and management processes are essential aspects of BIM implementation, in addition to technology, individual skills and roles. BIM methodology concerns the management of information, such as information on building products, materials, structures, customer requirements, information on technical equipment, general requirements for planned installations, etc. Communication and collaboration between stakeholders are defined as keys to accomplishing smooth information management [24].

BIM is suggested as social-environmental, technological, organisational, and process–product innovation in the AEC industry [19]. This digital innovation requires changes in individual roles, needs time, appeals resources, acquisition of knowledge, financial investment, etc. For a successful implementation of the BIM methodology, the organisational change is required. Organisational change process needs a development strategy, the involvement and engagement of all stakeholders as well as management support [20]. The role of the manager is frequently placed very high, as one of the most important obstacles must be overcome, namely resistance to change. In this context, it is significant that the entire benefit of the BIM methodology for the enterprise is explained to each participant. According to Weise [42], change management with its variables (organisation and process orientation, resources, technology, strategy, culture, individual) plays a decisive role in the enterprise-related BIM adoption.

4 INVESTIGATION OF IMPLEMENTATION PROJECTS

5D BIM implementation projects selected for the investigation were carried out by different AEC stakeholders: architecture, planning, and engineering offices, construction companies, project development offices. The design and construction activities of parties involved in the study are multifaceted. The chosen companies are engaged in structural engineering, design and construction of industrial buildings, plant construction, bridge building, construction of residential and public facilities, infrastructure works, civil and underground engineering, refurbishment, building services and technical equipment, building automation as well as system buildings and finishing works. Among executing construction companies, there are some large companies that initially want to adopt 5D BIM in one department for the time being. As a local office does not exceed the size of an SME, it can be taken into consideration. 24 company-related implementation projects are summarised according to the roles of stakeholders and fields of activities in Fig. 2.

The 5D BIM implementation projects have the following limitations: (1) in some projects, only Phase 1 Concept 5D process (see Section 2, Fig. 1) was contracted, therefore the further development could not be evaluated; (2) some planning offices need much time to adopt 3D modelling and master the modelling tools for the design planning because of staff shortage before moving to 5D BIM; (3) 5D BIM implementation depends on the contracts with the client. As all projects are carried out in Germany, the Official Scale of Fees for Services by Architects and Engineers [29] and special features of cost planning (see Section 3.2) had to be observed. The most common BIM use cases are cost estimation in the design and tendering phases for planning offices, as well as quotation costing for construction companies.

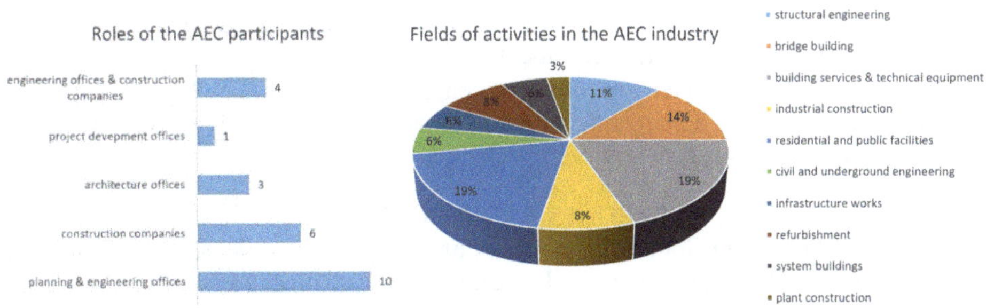

(a) (b)

Figure 2: Implementation projects. (a) Roles of the participants; and (b) Fields of activities.

The analysis of the 5D BIM adoption has shown that

- Communication within a company must be seriously reconsidered with the implementation of the BIM methodology. BIM model authors and estimators sometimes have heard each other's problems for the first time during the workshop.
- The size of the company has less influence on the adoption of the BIM methodology. The larger companies have more employees involved, but the same problems.
- If there was no responsible person for the enterprise-related BIM implementation, the digital transformation in the company would drift off.
- Clients' demand for BIM services is important for most participants.
- 5D BIM implementation is an iterative process [3] demanding constant harmonization between BIM model creation and refinement of the costing database. LOD of a designed model should be sufficient to provide accurate quantities.
- Many participants still see BIM only as a tool and not as a methodology and therefore see no need to change their old processes.
- Software selection often happens before the BIM use cases are clearly defined.
- There was no BIM implementation plan as well as a clear strategy in most projects.
- Many participants saw the implementation of the BIM methodology as a management decision but not as an advantage for their work practices.
- The companies were not interested in a model-based exchange with external parties.
- In the preliminary discussions it had to be explained that a feasible enterprise-related system for model-based cost planning could not be built in five days of workshop.
- Fields of activities influence the level of automation for model-based cost planning. For instance, more quantity take-offs can be automatically extracted from BIM model in building construction than in industrial construction. In the cost planning for the modular buildings allows almost the entire process to be automated.
- Construction companies without a modelling department are struggling to implement 5D BIM, while the model data from external architecture, planning, or engineering offices have not met their requirements.
- The previous estimating database is incorrect or out of data must be completely revised before making the switch to 5D BIM [43].
- Companies should plan more time for the adopting and restructuring of processes.
- Interoperability between the software applications is the major topic for the technological implementing and must be discussed in the planning of IT landscape.

- Different project participants have different views on cost estimation and, accordingly, on quantity take-off from the BIM model.
- National peculiarities, such as strong fragmentation of the German AEC sector and use of HOAI, complicate the implementation process. Due to the BIM methodology, a shift in the service phases is also observed in German offices.

5 RESULTS AND DISCUSSIONS

The following requirements for the development of a sustainable implementation concept are defined based on the literature review and investigation of practical implementation projects.

1. Concept must cover management, process, and technology issues [44], namely address strategic, operational, tactical, and technical questions [9].
2. Survey of current situation of BIM in an organisation, as well as effective external factors influencing BIM implementation, belongs to the preliminary adopting stage.
3. Establishing BIM is part of the company's strategic tasks. Therefore, a responsible BIM manager must be appointed in the company at the very beginning.
4. BIM should be regarded as a corporate standard that includes developing the company-related guidelines and BIM execution plan for projects.
5. Clear identification of BIM benefits, goals, and use cases for the company.
6. BIM implementation plan is not only "nice to have", it is to be created for a company.
7. All employees must be involved in the digital transformation.
8. Collaborative work in changing old processes and workflows.
9. BIM education and training: general in the first adopting stage and later role-specific.
10. Sequence of decisions: first strategy, then software selection.
11. Development of 5D BIM model content and information workflows as iterative approach, permanent coordination between the BIM model authors and estimators.
12. Selection of a pilot project and, if necessary, coordination with the client.

The concept for a smooth implementation of 5D BIM designed is presented in Fig. 3. To show the order and hierarchy of the main activities, the concept framework was created without assigning the roles. Depending on whether this model concept is realised for architecture, planning, and engineering offices or construction companies, the roles and responsibilities are respected regarding organisation size, internal and external relationship between the stakeholders. The information exchange must surely happen in the implementation phase and afterwards among BIM manager, BIM model authors, and estimators. BIM education and training is essential for the BIM adoption process.

6 CONCLUSION AND FUTURE RESEARCH

The developed concept provides a fundamental framework for AEC enterprises to make informed decisions in adopting 5D BIM in the body of the organisation structure. This concept framework should prepare a smooth, individually adaptable implementation by different stakeholders of the AEC industry. Hence, this article presents a significant part of knowledge about the 5D BIM implementation in the enterprise-manner. The research tries to overcome the most important limitations during 5D BIM adopting, as lack of skills and resistance to change, and represents a comprehensive approach. In comparison with other scientific papers [38], [40], [44], [45], this article encompasses the BIM implementation strategies and 5D BIM adoption methods as well as points the view of the external observer on the issue.

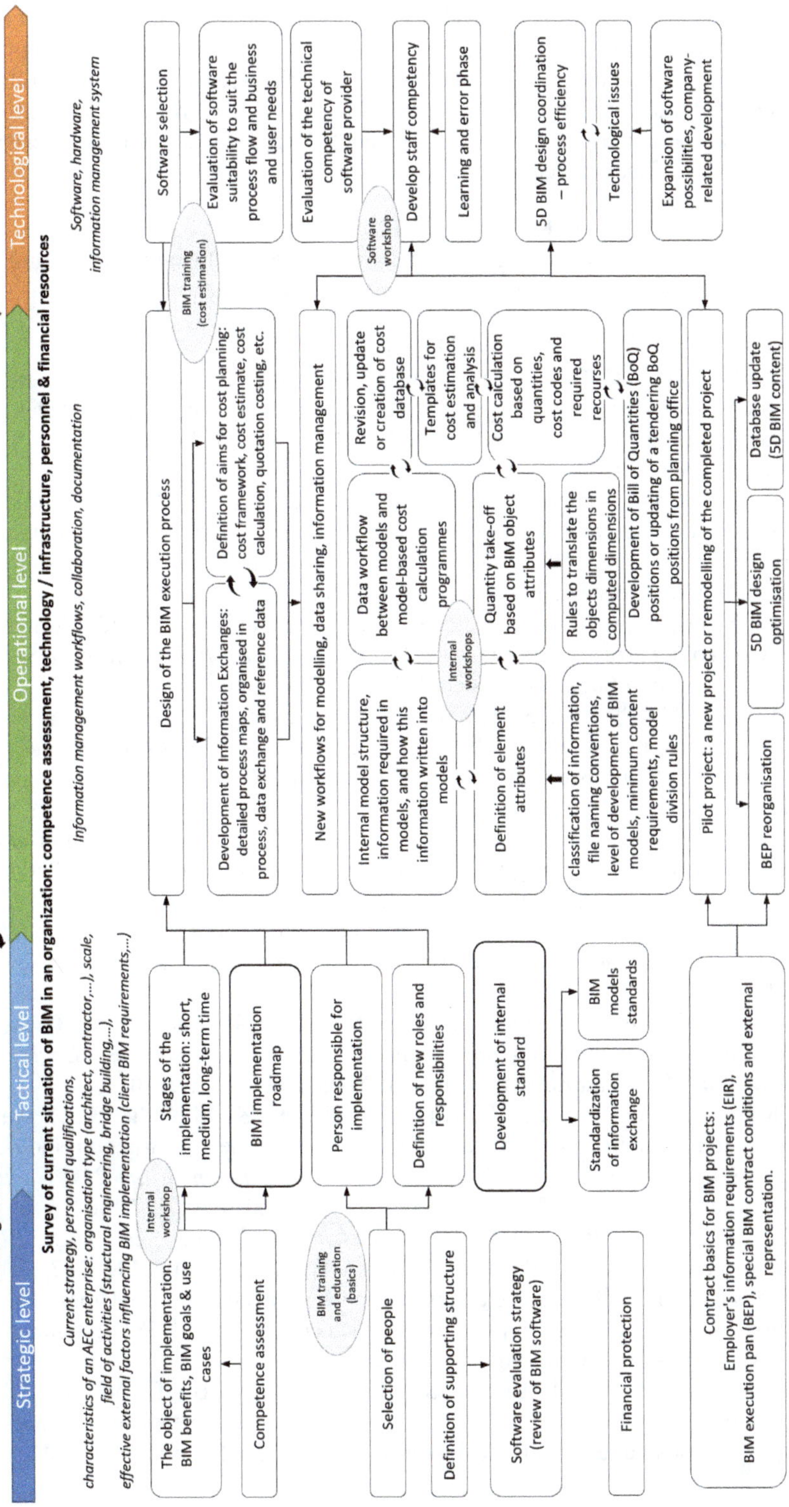

Figure 3: Concept framework for a sustainable 5D BIM implementation approach.

The future research can deal with the development of information exchanges for 5D BIM on the operational level such as requirements for detailed process maps. Although the implementation projects from the practical part of the research were carried out in Germany, results can be adopted internationally. In the forthcoming research, more attention should be paid to the national particularities and the more detailed procedures should be designed separately for planning offices and construction companies.

ACKNOWLEDGEMENTS

The authors would like to acknowledge the support of the 5D Institut GmbH and the contribution of the project data and sharing of the practical experience.

REFERENCES

[1] Deutsche Telekom, A.G. (eds), Der digitale Status quo im deutschen Baugewerbe, p. 14. https://www.digitalisierungsindex.de/wp-content/uploads/2021/01/Telekom_Digitalisierungsindex_2020_BRANCHENBERICHT_BAUGEWERBE.pdf. Accessed on: 1 Sep. 2020.

[2] PricewaterhouseCoopers (eds), Digitalisierung der Bauindustrie 2020. PwC-Studie zur Digitalisierung der Baubranche unter Berücksichtigung der Corona-Pandemie und des Infrastrukturausbau, p. 19. https://www.pwc.de/de/digitale-transformation/pwc-studie-digitalisierung-der-bauindustrie-2020.pdf. Accessed on: 1 Sep. 2020.

[3] Cesnik, J., Zibert, M., Lah, M. & Matic S., Required model content and information workflows enabling proficient BIM usage. *Proceedings of the IOP Conference Series: Materials Science and Engineering*, **603**, 032074, 2019. DOI: 10.1088/1757-899X/603/3/032074.

[4] Egger, M., Hausknecht, K., Liebich, T. & Przybylo, J., *BIM-Leitfaden für Deutschland*, Bundesministerium für Verkehr, Bau und Stadtentwicklung: Berlin, p. 18, 2013.

[5] Daniotti, B., Gianinetto, M. & Della Torre, S. (eds), *Digital Transformation of the Design, Construction and Management Processes of the Built Environment. Research for Development*, Springer: Cham, p. 103, 2020.

[6] Borrmann, A., König, M., Koch, C. & Beetz, J. (eds), *Building Information Modeling - Technologische Grundlagen und industrielle Praxis*, ed. J. Beetz, Springer Vieweg: Wiesbaden, 2015.

[7] Eastman, C.M., Teicholz, P., Sacks, R. & Liston, K., *BIM Handbook: A Guide to Building Information Modeling for Owners, Managers, Architects, Engineers, Contractors, and Fabricators*, John Wiley and Sons: Hoboken, NJ, 2011.

[8] Marshall-Ponting, A., Arayici, Y., Khosrowshahi, F. & Mihindu, S., Towards implementation of building information modelling in the construction industry. *Proceedings of 5th International Conference on Construction in the 21st Century (CITC-V)*. Istanbul, Turkey, May 20–22, 2009. DOI:10.13140/2.1.3776.6080.

[9] Baldwin, M., *Der BIM-Manager: praktische Anleitung für das BIM-Projektmanagement*, Beuth Verlag GmbH: Berlin, Wien, Zürich, pp. 112–137, 2018.

[10] Number of small and medium-sized enterprises (SMEs) in the European Union (EU27) from 2008 to 2021, by size. https://www.statista.com/statistics/878412/number-of-smes-in-europe-by-size/. Accessed on: 1 Sep. 2020.

[11] Policy Learning Platform SME Competitiveness (eds), Policy Brief, The SME policy of the European Union, p. 1. https://www.interregeurope.eu/fileadmin/user_upload/plp_uploads/policy_briefs/TO3_policy_brief_2__SME_policy_final_V23_01_2017_Siriuss_LS_clean.pdf. Accessed on: 1 Sep. 2020.

[12] Institut für freie Berufe Nürnberg (IFB) (eds), Umfrage zur Lage der Ingenieure und Architekten im Auftrag des AHO, der Bundesingenieurkammer, des VBI und des BDB – Index 2014, wirtschaftliche Strukturdaten, pp. 18–19, 2015.
[13] German Institute for Standardisation (DIN) (eds), *VDI-Richtlinie 2552-1, BIM-Grundlagen*, Beuth publishing DIN, p. 8, 2021.
[14] Wu, P., Jin, R., Xu, Y., Lin, F., Dong, Y. & Pan, Z., The analysis of barriers to BIM implementation for industrialized building construction: A China study. *Journal of Civil Engineering and Management*, **27**(1), pp. 1–13. DOI: 10.3846/jcem.2021.14105.
[15] Shibani, A., Awwad, A.K., Ghostin, M., Siddiqui, K. & Farji, O., Adopting building information modelling in small and medium enterprises of Iraq's construction industry. *Proceedings of the International Conference on Industrial Engineering and Operations Management*, Dubai, UAE, pp. 457–470, 2020.
[16] Doan, D.T., GhaffarianHoseini, A., Naismith, N., Ghaffarianhoseini, A., Zhang, T. & Tookey, J., Examining critical perspectives on building information modelling (BIM) adoption in New Zealand. *Smart and Sustainable Built Environment,* 2020. DOI: 10.1108/SASBE-04-2020-0043.
[17] Pinilla, C.A.V. & Riaño, L.N.A., Architectural model of implementation of building information modeling – BIM in the Colombian Construction Industry. *Proceedings of the International Conference on Industrial Engineering and Operations Management*, Toronto, Canada, pp. 142–152, 2019.
[18] Arayici, Y., Coates, S., Koskela, L.J., Kagioglou, M., Usher, C. & O'Reilly, K., BIM adoption and implementation for architectural practices. *Structural Survey*, **29**(1), pp. 7–25. DOI: 10.1108/02630801111118377.
[19] Villena, F., García-Segura, T. & Pellicer, E., Drivers of innovation using BIM in architecture, engineering, and construction firms. *Proceedings of the Construction Research Congress,* 2020. DOI: 10.1061/9780784482889.023.
[20] Bråthen, K. & Moum, A., Involvement matters: BIM implementation at project level in the AEC industry. *WIT Transactions on the Built Environment,* vol. 149, WIT Press: Southampton and Boston, pp. 157–168, 2015. DOI: 10.2495/BIM150141.
[21] Darwish, A.M., Tantawy, M.M. & Elbeltagi, E., Critical success factors for BIM implementation in construction projects. *Saudi Journal of Civil Engineering,* **4**(9), pp. 180–191, 2020. DOI: 10.36348/sjce.2020.**4**i09.006.
[22] Davies, R. et al., BIM in Europe: Innovation networks in the construction sectors of Sweden, France and the UK. *Proceedings of the 31st Annual ARCOM Conference*, eds A. Raiden, & E. Aboagye-Nimo, pp. 1135–1144, 2015.
[23] Abuelmaatti, A. & Ahmed, V., Collaborative technologies for small and medium-sized architecture, engineering and construction enterprises: Implementation survey. *Journal of Information Technology in Construction (ITcon),* **19**, pp. 210–224, 2014.
[24] Gercek, B., Tokdemir, O.B., Ilal, M.E. & Gunaydin, H.M., BIM execution process of construction companies for building projects. *Resilient Structures and Sustainable Construction*, eds E. Pellicer, J.N. Adam, V. Yepes, A. Singh & S. Yazdani, ISEC Press, 2017.
[25] Hjelseth, E., BIM understanding and activities. *WIT Transactions on the Built Environment,* vol. 169, WIT Press: Southampton and Boston, pp. 3–14, 2017. DOI: 10.2495/BIM170011.
[26] Enshassi, M.A., Hallaq, K.A.A. & Tayeh, B.A., Limitation factors of building information modeling (BIM) implementation. *The Open Construction and Building Technology Journal,* **13**, pp. 189–196, 2019. DOI: 10.2174/1874836801913010189.

[27] Li, P., Zheng, S., Si, H. & Xu, K., Critical challenges for BIM adoption in small and medium-sized enterprises: Evidence from China. *Advances in Civil Engineering*, **2019**, 9482350, 2019. DOI: 10.1155/2019/9482350.

[28] German Institute for Standardisation e.V. (DIN) (eds), *DIN 276:2018-12, Kosten im Bauwesen*. Beuth Publishing DIN, 2018.

[29] HOAI, *Official Scale of Fees for Services by Architects and Engineers (in Germany)*. Duesseldorf: Werner Verlag, 2013.

[30] Bargstädt, H.-J. & Steinmetzger, R., Grundlagen des Baubetriebswesens. *Schriften der Professur Baubetrieb und Bauverfahren*, pp. 220–222, 2016.

[31] Monteiro, A. & Martins, J.P., A survey on modeling guidelines for quantity takeoff-oriented BIM-based design. *Automation in Construction*, **35**, pp. 238–253, 2013. DOI: 10.1016/j.autcon.2013.05.005.

[32] Huang, X., Application of BIM big data in construction engineering cost. *Journal of Physics: Conference Series*, **1865**(3), 32016, 2021. DOI: 10.1016/j.aei.2016.07.001.

[33] Moses, T.G., *The Development of 5D BIM Framework to Facilitate Costing in Contractor-Led Projects*, Post-doctoral qualification, University of Wolverhampton: Wolverhampton, 2018.

[34] Ashworth, A. & Perera, S., *Cost Studies of Buildings*, 6th ed., Routledge: Florence, pp. 121, 530, 2015.

[35] Rui, Y., Review of building information modeling (BIM) application in construction industry. *International Journal of Innovative Technology and Exploring Engineering*, **8**(6C2), pp. 83–87, 2019.

[36] BKI Handbuch Kostenplanung im Hochbau, Rudolf Müller, 2021.

[37] Ma, Z., Liu, Z. & Wei, Z., Formalized representation of specifications for construction cost estimation by using ontology. *Computer-aided Civil and Infrastructure Engineering*, **31**(1), pp. 4–17, 2015. DOI: 10.1111/mice.12175.

[38] Mirarchi, C., Trebbi, C., Lupica Spagnolo, S., Daniotti, B., Pavan, A. & Tripodi, D., BIM methodology and tools implementation for construction companies (GreenBIM Project). *Digital Transformation of the Design, Construction and Management Processes of the Built Environment. Research for Development*, eds B. Daniotti, M. Gianinetto & S. Della Torre, Springer: Cham, pp. 201–208, 2020. DOI: 10.1007/978-3-030-33570-0_18.

[39] Li, H., Yuan, Z., Han, J. & Li, Z., Whole process cost management system based on BIM 5D. *DEStech Transactions on Engineering and Technology Research*, ICAEN, 2019.

[40] Leicht, D., Castro-Fresno, D., Dìaz, J. & Baier, C., Multidimensional construction planning and Agile organized project execution—The 5D-PROMPT method. *Sustainability*, **12**(16), p. 6340, 2020. DOI: 10.3390/su12166340.

[41] Smith, D.K. & Tardif, M., *Building Information Modeling: A Strategic Implementation Guide for Architects, Engineers, Constructors, and Real Estate Asset Managers*, John Wiley & Sons Inc., 2009.

[42] Weise, F., Change management im Kontext des Building Information Modelings in Organisationen der Bauwirtschaft – Eine Analyse Erfolgskritischer Faktoren in BIM-Anwendungsfällen. *Kompetenzen für die digitale Transformation 2020*, eds L. Lehmann, D. Engelhardt & W. Wilke, Springer Vieweg: Berlin, Heidelberg. DOI: 10.1007/978-3-662-62866-9_5.

[43] Sattineni, A. & Macdonald, J.A., 5D BIM: A case study of an implementation strategy in the construction industry. *Proceedings of the 31st International Symposium on Automation and Robotics in Construction and Mining,* pp. 361–367, 2014. DOI: 10.22260/ISARC2014/0048.

[44] Moses, T., Heesom, D. & Oloke, D., Implementing 5D BIM on construction projects: Contractor perspectives from the UK construction sector. *Journal of Engineering, Design and Technology*, **18**(6), pp. 1867–1888. DOI: 10.1108/JEDT-01-2020-0007.

[45] Viscuso, S., Talamo, C., Zanelli, A. & Arlati, E., BIM management guidelines of the construction process for general contractors. *Digital Transformation of the Design, Construction and Management Processes of the Built Environment. Research for Development*, eds B. Daniotti, M. Gianinetto & S. Della Torre, Springer: Cham, pp. 189–199, 2020. DOI: 10.1007/978-3-030-33570-0_17.

IMPLEMENTATION OF AUTODESK REVIT® IN DESIGN OF WATER TREATMENT PLANTS

HOSSAM KAMAL[1] & MOHAMED BAHGAT[2]
[1]Advridis Ltd., UK
[2]International Environmental Technology INTECH-HA, Egypt

ABSTRACT

Design of water treatment plant is one of the challenging engineering design sectors. This engineering multi-disciplinary field implies the interaction of different scientific efforts, besides the management and economic aspects. The traditional 2D design workflow compromises many risks of time and effort waste and miss-estimated quantities, time, and cost. Although, BIM techniques present an efficient 6D solution for many construction projects, their applications in such sophisticated electromechanical-based projects deserve a lot of efforts to be adapted. The present work is the result of collaborative work between the BIM teams in the two companies to adapt Autodesk AEC® package to the design of both wastewater and potable water treatment plant. The two teams carried out five successive projects to establish a standard workflow for the implementation of BIM techniques, based on Autodesk Revit®, in direct filtration, extended aeration, and sequential batch reactor Egyptian treatment plants. The results of the implementation of BIM techniques in water treatment plant design show the reduction of design time up to 67%, the reduction of man power up to 75%, increased accuracy of BOQ lists up to 94%, and transfer of project planning procedure to semi-automated schedules.

Keywords: water treatment plant design, BIM techniques, Autodesk Revit®.

1 INTRODUCTION

The design and construction of water treatment plants represent an interesting challenge of the engineering design entities. In April 2020, the International Water Association (IWA) defines the major trends impacting the water industry in the next decade as [1]:

1. Acting on the mounting effects of extreme weather events;
2. Protecting agricultural production;
3. Reusing wastewater to support a circular economy;
4. A customer-led revolution;
5. Smart and intelligent network technologies.

All these trends implies the utilization of enhanced optimised design concepts and effective smart tools.

BIM helps project stakeholders to visualize all the project components in a simulated environment. The obtained model is a digital representation of the building that helps project stakeholders to extract needed information and take decisions during the project planning, design, construction, and operation processes [2]. The implementation of BIM in utility design requires the creation of new workflow that facilitates the work and exploits the new software tools. Dodge Data and Analytics illustrated the importance and impacts of implementation of BIM in the "Business Value of BIM for Water project" report [3]. In 2017, the project of "Advance Works for Shek Wu Hui Sewage Treatment Works – Further Expansion Phase 1A and Sewerage Works at Ping Che Road" was awarded the Autodesk Hong Kong BIM Awards [4].

The present work is the fruit of cooperation between Adviridis Ltd. (UK) and International Environmental Technology INTECH-HA (Egypt) to found a BIM unit and to

www.witpress.com, ISSN 1743-3509 (on-line)
doi:10.2495/BIM210081

build the team and capabilities in 3D modelling and design process. INTECH-HA is one of the pioneer EPC companies in Egypt and MENA region in water treatment. They have elaborated more than 200 projects using the traditional 2D CAD design. On the other hand, Adviridis Ltd. is specialized in smart solutions for environmental design. This task was carried out in the framework of complete business process reengineering of ITECH-HA. This reengineering process includes BIM unit, procurement database creation, and workflow management recreation. In this paper, we shed lights on the BIM system transformation. After recruiting the starting BIM team of three engineers, the BIM unit creation project starts with a pilot project of a wastewater treatment plant of capacity 20,000 m^3/day designed on the advanced sequential batch reactor (ASBR) technology. Then, the team is extended to six BIM engineers and two other projects were carried out. Finally, the BIM unit of eight engineers carried out three simultaneous projects with three different design technologies.

The main outcomes of this cooperation are:

1. the reduction of design time up to 67%;
2. the reduction of man power up to 75%; and
3. increased accuracy of BOQ lists up to 94%.

From our point of view, the present work represent novelty in three folds:

1. The adoption of BIM procedures in one of the developing countries "Egypt" in different infrastructure projects.
2. The implementation of BIM workflow as a part business process reengineering procedure
3. The adaptation of Autodesk Revit® software in electromechanical-based projects.

The present paper starts by discussing the starting challenges. Then, the implemented workflow is introduced. Finally, the outcomes and the impacts on the business process are depicted. Generally, the present work shows the outcomes of the first phase of business process reengineering in engineering design. The future phases will widen the implementation of BIM in the project management procedures in the same company.

2 CHALLENGES OF CREATION OF BIM THINKING

To start a new business depending on a new technology is a hard job. However, it is an easy job compared to introducing the same new technology to a business entity that have already been working for more than two decades using traditional technologies. At the planning phase of this transformation project, the challenges were classified in three axes:

1. Personal workflow management
2. Technology adaptation

2.1 Challenges of workflow management

The first step in dealing with the axis of human resources is to evaluate present resources and their management protocol. The actual case was that the company consists of four sectors; engineering, projects, installations, and procurement. The main concern was the engineering sector. This sector was consisting of a general director, two directors of mechanical and electrical design, five electrical designers, eleven mechanical designers, and seven draftsmen. The structure design was considered as a part of project sector. The instrumentation and control design was considered as a part of installation sector. The workflow management starts from a process design producing process flow diagram (PFD).

Then, the mechanical designer proposes the general arrangement and preliminary piping and instrumentation diagram (P&ID). Then, both the mechanical and the electrical designer start working on detail design. After that, the civil guides are produced and are sent to the structure designer. In all these processes, the responsibility of the designer is to sketch her/his design and send it to a draftsman. The draftsman draws the 2D plans and sections and print them to be reviewed by the designer.

Within that environment of work, some challenges were sported:

- the designer spent about 55% of job time in reviewing the printed drawings.
- the senior designers had lost their ability to work on CAD software or to follow-up the capabilities of new releases.
- the draftsmen took about 18 months of successive work to understand the specific types of machines and terminology (decanter, scrapper, fine screen, etc.).
- the structure design was lagged after the mechanical design and should verified from the mechanical engineer manually.
- the electrical engineer should recalculate (or redesign) the after revision of mechanical and structure.
- the designs loop was iterated at least three times to arrive to the final design submitted for approval.
- Both engineers and draftsmen were adapting the 2D thinking in building the plans, projections, and sections. While 3D prospective was built only if a presentation was required for special purpose.
- The BOQs were counted manually.
- The cost and time of construction were calculated on spreadsheets depending on manual data entry.
- The psychological impact of introducing new technology with new workflow was huge on both engineers and draftsmen specially with the fear of losing their jobs or authorities during hard economic environment.

2.2 Challenges of technology adaptation

The water treatment processes depend on wide variety of equipment. Some of these equipment are on-the-shelf ones like pumps, motors, generators, … etc. However, the core equipment of the treatment processes are not common and they are made unit-by-unit according to the plant design parameters. This implies that the item model library is continuously revised and updated in each project. The preliminary evaluation considered that about 40–25% of equipment models should be rebuilt in each project. Besides, the item that has a certain function has different designs according to supplier. This implies that in each new project the same item may be built at least three times and the dimensions and arrangement are modified accordingly. Fig. 1 shows the clarifier bridge from two suppliers for the same plant. One of the major challenges of water treatment plant design arrangement is the limited available land relative to the required capacity. For example, the WWTPs of Port Fouad and Dar Elsalam have the same capacity of 20,000 m^3/day and the same technology of extended aeration, but the available area of Port Fouad WWTP is 98.788 m^2 while that of Dar Elsalam is 145.512 m^2. This implies changing of 45% of the used equipment to achieve the required output.

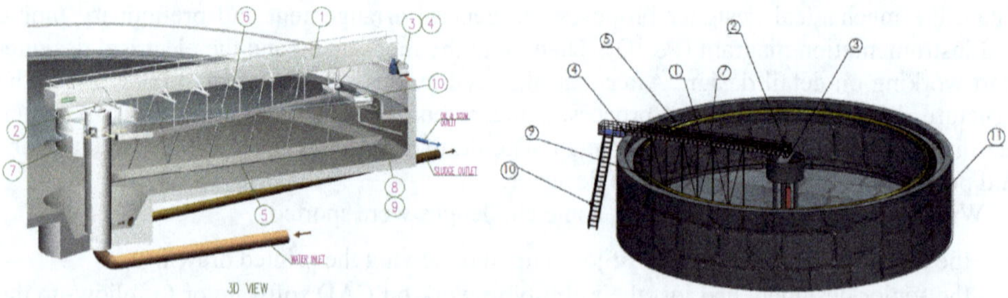

Figure 1: Different structure of scrapper of primary sedimentation tank.

3 THE RE-ENGINEERED BIM WORKFLOW

The transformation project starts by the analysis of current situation and building the issue register to quantify the business process waste types and sources. Root-cause analysis with Pareto chart are built and they lead the team to achieve a project of business process reengineering of the engineering sector. This project establishes the new system of building information modeling (BIM) in the International Environmental Technologies Company (INTECH). This system is realized by constructing a new unit; named Technical Support Unit (TSU). TSU includes two sections:

1. Information management section (IMS); and
2. Modeling and drafting section (MDS).

The project starts by specifying requirements and concluded by achieving a pilot project under the new system. The business process concept of work is to create a dynamic teamwork that can support both engineering and tendering processes.
The project objectives are set as:

1. Coordinate the RFI, RFQ, and RFP (in collaboration with Tender Department):
    - Simplify the RFI process with a standardized workflow that maintains transparency for the entire team.
    - Efficiently create, coordinate, and manage all submittals in a single document library. Initiate requests, manage reviews and approvals, and sync critical information for easy access.
2. Project document management
    - Access to drawings and documents: Ensure all project teams have access to the information that they need on whatever device they are working on. Powerful search capabilities, a single viewer for plans, models and photos helps to keep the whole team on the same page.
    - Precise construction document control: Standardize approval workflows, control access to information, and maintain version history.
    - Centralized document management: Bring drawings, models, specifications, RFIs, photos, and markups together in a single document management platform.
3. Coordinate the work attribution to human resources (in collaboration with Engineering Department)
4. Perform quality assurance and quality control of engineering design processes and activities.

5. Review of constructability and detect clashing and errors of design in real time.
6. Manage the urgent changes of design occurred during construction process.
7. Construction data and analytics
8. Perform the closing procedure of projects including documentation, archiving, and experience transfer.
9. Organize the paper-form document archiving system.
10. Construct and maintain an efficient electronic archiving system.
11. Drafting and drawing of project plans.
12. Building of component libraries to standardize and facilitate work.
13. Training/continuous training of workforce on new systems.

Fig. 2 depicts the structure of the created TSU and the assigned objectives to each position of the unit.

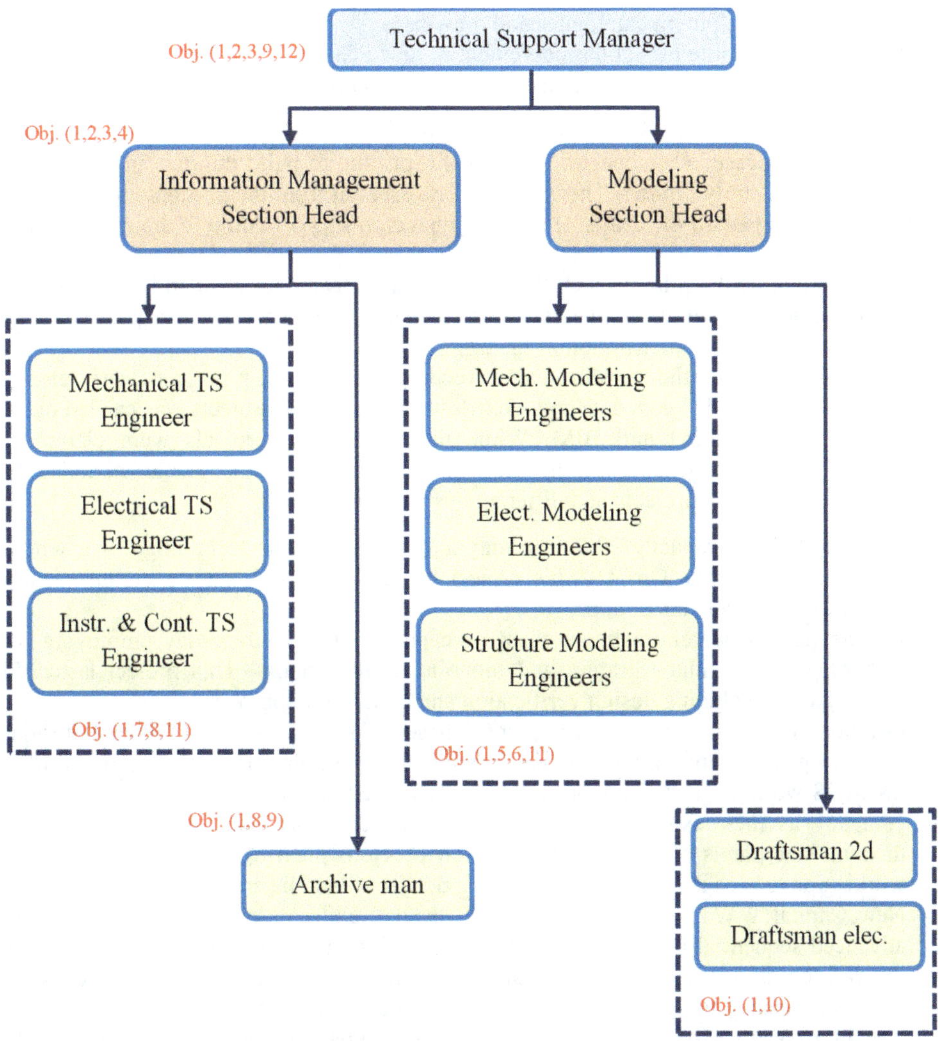

Figure 2: The structure of the created technical support unit.

The pilot project was the Tuna Elgabal wastewater treatment plant (WWTP) having a capacity of 20,000 m^3/day and employing the technology of extended aeration of Bioworks GmbH. The plant was designed using the traditional methods of 2D modelling. The design activities lasted 89 working days (including design revisions and consultant approvals) by two mechanical engineers, one electrical engineer, one control engineer, one instrumentation engineer. The plant included 17 types of main equipment (fine screens, coarse screens, mixers, sludge mixers, aeration tanks, electrical generators, etc.) and 14 types of supporting equipment (valves, sensors, cables, administration building, etc.). After recruiting three BIM engineers, the pilot project started on parallel tracks:

1. Training of two mechanical engineers and one electrical engineer on the BIM concepts and Autodesk Revit® software.
2. Training of three draftsmen on the Autodesk Revit® software.
3. Training of newly recruited BIM engineers on the water treatment concepts.
4. Collection of equipment catalogues and drawings.
5. Upgrade of available PCs and procurement of new PCs.
6. Purchase of software packages and infrastructure components.

The pilot project lasted 67 working days as the design had been already verified. The three draftsmen succeeded to deal with BIM concepts and to build the item models on the Autodesk Revit® family builder. This success was essential from the psychological point of view as it encouraged all the draftsmen and engineers to support and join the new business process.

After completing the pilot project, the transformation team was built and the training of the rest of junior and senior engineers and draftsmen was carried out during 6 weeks on three groups to avoid the interruption of the original workflow.

In order to evaluate the transformation process, four other new projects were selected to be designed using BIM concepts and workflow. One of these projects was carried out in parallel tracks; traditional and BIM. While the other three projects were elaborated completely in BIM unit. Fig. 3 illustrates some examples of the plants modelled by Autodesk Revit®. These projects had different design technologies:

1. Tahta WWTP of capacity 20,000 m^3/day and employing the technology of extended aeration of Bioworks GmbH. This project lasted 43 working days including design verification and consultant approvals.
2. Assuite potable water treatment plant of capacity of 40,000 m^3/day employing the technology of lamella filtration and monolithic filter floors. This project lasted 51 working days including design verification and consultant approvals.
3. October potable water treatment plant (phase 3/4) of capacity of 400,000 m^3/day employing the technology of lamella filtration and monolithic filter floors. This project lasted 68 working days. This project was considered as a milestone for INTECH-HA company as their experience in potable water treatment plants was limited to the medium size plants of 40,000 m^3/day. Their previous plant of medium size took 109 working days because of the overwhelming details of lamella and filter nozzles.
4. New Capital WWTP of capacity 250,000 m^3/day and employing the technology of advanced sequential batch reactor (ASBR). The design activities of this project lasted 46 working days despite its huge challenges. This project was a challenge because of the large capacity of the plant and because that many technologies were employed like two different types of odour control systems, two types of chlorination systems, GRP covers for main process tanks, and many others.

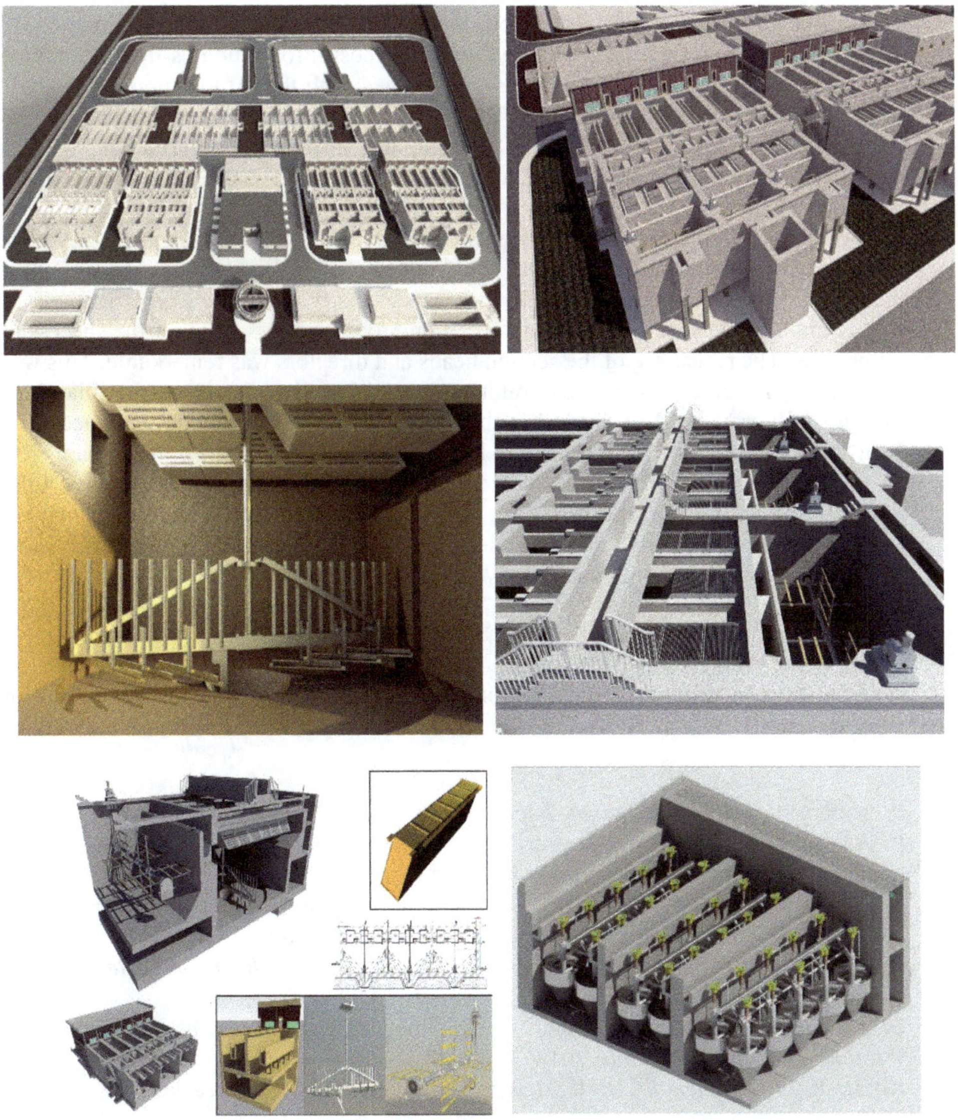

Figure 3: Some examples of the plants modelled by Autodesk Revit®.

4 THE BIM TRANSFORMATION EXPERIENCE

After completing the business process reengineering process of BIM transformation, we can conclude our experience in the following points:

1. The results of the implementation of BIM techniques in water treatment plant design show the reduction of design time up to 67%, the reduction of man power up to 65%.
2. One of the major contributions of BIM system is automated and accurate elaboration of bill of quantities (BOQ). The accuracy of BOQ lists is increased up to 94%. In traditional 2D design there is no way but counting items manually. The BIM-based workflow eliminates the job of counting engineer.

3. BIM is known as 6D system as it incorporates cost, time, and as built information. In our present case we succeeded to fully benefit of cost information as the information extracted from Autodesk Revit® software is merged with that of the supplier database and the cost analysis is elaborated.
4. On the other hand, the utilization of time and installation information will be considered in the future phase as this job needs the inclusion of the other sectors of the company.
5. The Autodesk Revit® software is excellent in architecture-based design. However, in the electromechanical-based designs it could not be used as stand-alone. The workflow should include the Autodesk Civil 3D® software specially for piping isometric.
6. The psychological aspects of the transformation process are essential for the success of the process. The resistance of the section heads and directors was remarkable. This was eased after the achievement of the pilot project and realization of the fact that BIM concepts will promote their work. For the draftsmen, the BIM training was successful for five out of seven and we managed to keep the two failed draftsmen in the team.
7. As the Egyptian market is not fully accustomed to BIM concepts, we didn't fully eliminate the traditional workflow. We keep the two draftsmen who cannot comply with BIM concepts to elaborate the electrical single line diagrams, the mechanical P&ID, and the digitization of paper drawings.

ACKNOWLEDGEMENTS

By the end of this work, we'd like to present our deep thanks and gratitude to the teams of Adviridis Ltd. (UK) and International Environmental Technology INTECH-HA (Egypt) for their time and efforts.

REFERENCES

[1] Wennberg, C., *5 Major Trends Impacting the Water Industry in the Next Decade*, IWA, 22 Apr. 2020. iwa-network.org/five-major-challenges-and-emerging-trends-impacting-the-water-industry-in-the-next-decade/.
[2] Azhar, S., Khalfan, M. & Maqsood, T., Building information modelling (BIM): Now and beyond. *Construction Economics and Building*, **12**(4), pp. 15–28, 2012.
[3] Jones, S.A. (ed.), *SmartMarket Report*: *Business Value of BIM for Water Project*, Dodge Data & Analytics, 2018.
[4] Söbke, H., Theiler, M., Tauscher, E. & Smarsly, K., *BIM-based Description of Wastewater Treatment Plants*, 2018.
[5] Riffat, R., *Fundamentals of Wastewater Treatment and Engineering*, CRC Press, 2013.

SECTION 3
BIM AND AUTOMATION IN CONSTRUCTION

VISUAL EDUCATION ON CONSTRUCTION SITES USING BIM

SEBASTIAN SEIß & BILAL SHAH
Bauhaus-University Weimar Chair Construction Management, Germany

ABSTRACT

The advancement in building information modelling (BIM) technology has largely benefited architecture, engineering, and construction (AEC) stakeholders. Automation and visualization offered by different BIM tools are regularly used by architects, engineers, and planners to make the planning and design process more efficient and accurate. However, the potential of BIM for helping construction site workers is not fully realized; the existing methods for using intelligent 3D BIM models to instruct site workers are limited and can potentially develop. A BIM-based 4D visualization is a practical use case for showing the sequence of site tasks to workers and enables them to identify their scheduled tasks. By now, 4D BIM does not support workers in the process of site implementation and work safety. The use of videos for instruction purposes has been an effective practice in other sectors, but it has not been used in AEC for facilitating site works. Thus, the research develops a method by which video tutorials are connected to task objects in a 4D model. The aim is to access instructional videos following the scheduled tasks in the 4D model to construction workers. The proposed method allows site workers to visualize a sequence of site tasks on a 4D model and to understand construction tasks by an automized displaying of linked instruction videos. The procedure is evaluated on a case study and a survey of workers and site managers. Based on the analysis of survey results, factors such as future relevance, practicality, and usability of the proposed method are clarified, and future works are recommended.

Keywords: visual education, BIM, 4D planning, health and safety, construction operations.

1 INTRODUCTION

In a construction project, transferring information about the intended product to the lowest level is of utmost importance. The on-site workers do not understand the technical terms, which are generally used on construction sites by engineers and site managers. It is tough to communicate to site workers about performing a particular construction task with expected accuracy. The communication barriers, e.g., language and the inability of site workers to understand verbal instruction due to the complexity of the task, have been a long-stayed problem in the construction industry. This communication problem results in faulty and incomplete work. Due to inadequate advancement in exploring ways to instruct low-skilled workers, there has been a heavy reliance on verbal instructional methods on site.

The advancements in the method of BIM have allowed us to explore visualization in the construction industry in the most productive way. The progress in BIM has benefited the stakeholders involved at the highest level, but it has not been used to help the site workers significantly. The data-enriched 3D BIM models can be used as an instructional tool for all levels, i.e., Engineers, Architects, site workers, trainees, and owners. The intelligent 3D Models resulting from the BIM process allow us to link extra information with the model objects as a kind of information repository. The information can be accessed throughout the lifecycle of the building for different purposes. The interoperability property of the object-oriented models enables the exchange of project information among participants.

The use of videos and visualization for instructional and safety purposes has been a massive success in different sectors. One of the main reasons for videos being an effective medium of instruction is the self-learning ability. Also, instructional video clips help to

WIT Transactions on The Built Environment, Vol 205, © 2021 WIT Press
www.witpress.com, ISSN 1743-3509 (on-line)
doi:10.2495/BIM210091

build confidence for executing a particular task by visualizing each detail of its execution process. The development of such a visual instructional model for low-skilled workers can prove to be an effective tool in communicating construction tasks.

The current usage of construction videos for instructional tasks will be explored before the concept development. The research focused on an automatized linking of instruction videos to the 3D Model objects and the 4D schedule of a BIM. The studies proved to be an important development in visual education for site workers and paved the way for this research. It is vital for avoiding miscommunication to develop an approach that focuses on construction activities and instruction for the steps involved in site and work safety tasks. The use of 4D BIM provides an effective tool to develop a visual environment in which instructional video tutorials can be linked to the scheduled tasks of the 4D model. In this way, site workers can visualize the sequence of construction activities through 4D simulation and get the necessary work instructions to execute the tasks. This study aims to develop a method for linking instructional and safety videos with the 4D model to guide site workers for their scheduled tasks and educate them about the sequence of activities by 4D simulation.

2 BACKGROUND ON VISUAL EDUCATION

Visualization is defined as the learning tool used to convey, portray, educate and store data in all practical fields, from architecture to chemistry, meteorology to radiology, and finance to history. If the visualization is planned and designed appropriately, it can enhance human information processing [1].

Multimedia-based learning is certainly more effective than typical verbal instructions, as it provides an interactive, self-paced learning environment and appeals to different learning types. The study of Issa states, "Multimedia presentation enhances people's ability to visualize; it is the impetus to learning that takes instruction from the highly subjective means of verbal communication and develops mental images that help many people understand new information more clearly." Additionally, this learning method helps develop confidence in learners; it provides the opportunity to learn and improve themselves instead of being trained, which increases worker's interest in getting more involved in tasks [2].

Many researchers studied visual education of different forms to educate site workers with difficulties, e.g., language barriers, low experience, and low skill in understanding verbal instructions. Visual education by videos or pictures has also been effective for instructing the assembly and installation of various home appliances. Some of the examples of the use of visual education as an effective medium in different research are discussed as follow. The differences between assembly and construction task instructions were identified by Mourgues and Fischer [3]. These differences influence the attributes gained from the above-stated principles. The main differences are the user knowledge about the task, operation environment, and complexity of tasks. The study identified problems with the existing verbal instructional method on the construction site and defined a template with the work structure and works instructions for cast-in-place concrete labourers. The template used can be observed in Fig. 1.

The template was validated in the field and found to be effective compared to the traditional verbal method. Labourers and site personal preferred the instruction based on this template as it reduced rework and questions. The template contributed significantly to the cast-in-place construction domain and provided a strong foundation for developing new methods for visual site instruction [3].

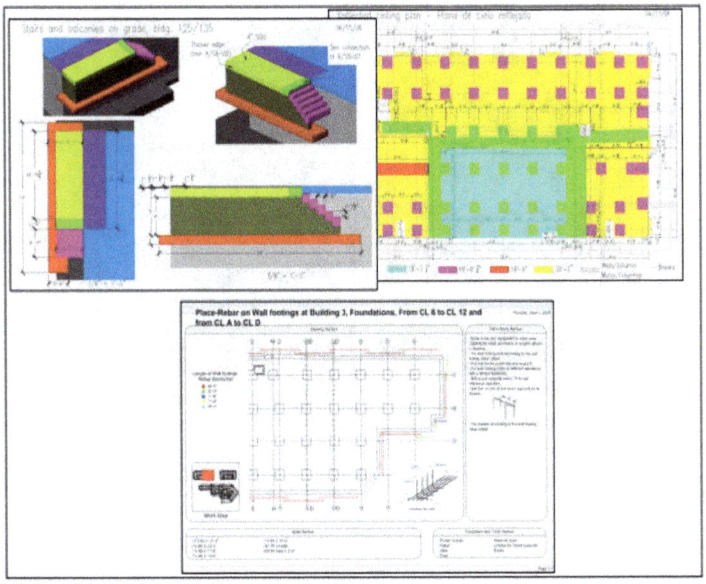

Figure 1: The template used for cast-in-place concrete labourers [3].

One of the advantages of using 3D visualization is the training of workers with limited education. As most site workers might be less educated, it is difficult for them to follow written instructions. The other problem could be language, meaning construction workers from different regions find it difficult to follow complex tasks, as they are unfamiliar with the local language. 3D visualization is an effective training tool that can address these problems and communicate tasks to workers better.

To confirm the usability of 3D visualization, a training module was modelled by Clevenger and Del Puerto, which focused on the relationship of assembling construction elements such as walls and safety provisions. The wall assembly model was initially developed to identify issues during the design phase and prevent them from becoming a problem. Once the model was prepared, it was realized that it could serve additional purposes and could be used as content for a training module. (see Fig. 2). The safety provisions recommended by Occupational Safety and Health Administration (OSHA) is associated with every construction stage [4].

Video tutorials are used for on-site safety instructions in animated videos to create awareness in site workers regarding site safety. The United States Department of Labourer covers different construction features through video tutorials. It creates awareness among workers about different unwanted situations during the execution process of various tasks [5].

The verbal instructional method used on site is observed to have limitations in instructing site workers because of their limited knowledge and language barriers. The study introduces a video-based instructional approach in which video with systematic instructions for a construction task is considered as an alternative instructional method. The use of video tutorials encourages self-learning and ensures efficient transfer of knowledge. Video tutorials have played an important role in assisting users in learning software tools

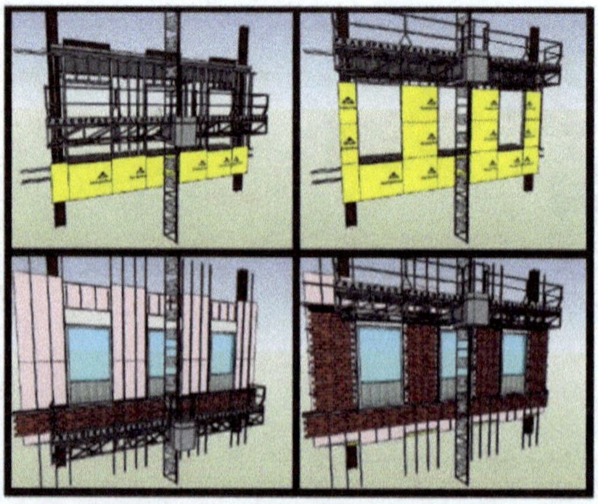

Figure 2: 3D-visualization of a wall describing construction sequence and safety controls [4].

related to the construction industry and creating site safety awareness. A general approach for developing task-related videos is presented by Nasir and Bargstädt, which comprises four phases; needs recognition, content design, making instructional content, and assessment. BIM 3D models are the primary source for creating a virtual environment of a video tutorial; therefore, videos developed by this approach are called BIM-based video tutorials for construction tasks. It is expected that videos developed through this process would help simplify the construction tasks for site workers [6], [7].

3 METHODOLOGY

To implement the idea of providing scheduled-based instructional content to the site worker, combining the 3D model with a schedule to execute activities should be helpful. The combination of the 3D model with time information is referred to as a 4D model. The other main aspect of this research is discovering how external instructional content such as videos related to scheduled tasks and safety are connected to 4D-model information. In this way, the proposed method would help site workers understand the sequence of tasks and methods to perform the scheduled task with the required safety.

The research approach carried out for this study is design-based, proposed by Verschuren and Hartog [8]. The stages mainly comprised the following:

1. Needs identification and making assumptions for the proposed model
2. Development of a concept in a way that satisfies the needs
3. Creation and implementation of a test version
4. Evaluation

This study focuses on creating an alternative instruction method by identifying its requirements and defining assumptions for the proposed new method. The requirements are based on a literature review, and an iterative coordination process with Hochtief end users. Based on assumptions and requirements, the concept is developed, after which a test version of the proposed concept is created and implemented on the target users. The test version embodies the characteristics of the expected final product. Finally, it is evaluated by the end-users to find out if it satisfies the requirements and assumptions. The validation is done by a 4D model combined with an instructional video on target users. For measuring the satisfaction level, staff working at different positions are interviewed. The results are structured using a standardized questionnaire and analyzed to identify the problem stage.

4 SYSTEM DEVELOPMENT FOR VISUAL EDUCATION USING BIM

The focus of the study is to provide schedule-based instructional content; therefore, DesiteMD is used. DesiteMD enables the connection of 3D model objects to time information like Navisworks and provides the user with a full Application Programming Interface (API) access. Once 3D model objects are connected to the activity schedule, the model is referred to as the 4D model, which shows construction sequencing of different activities through 4D simulation. The other part of the study is the connection of instructional and safety videos to 4D model information. It is achieved by using the API of DesiteMD; the API offers functions that link instruction videos with a 4D model and automatically display the videos to the 4D schedule sequence.

As the target user of the system is a site worker, the system should allow easy access to the intended content. The vital aspect of this study is to develop a simple user interface. The webform module in DesiteMD API receives signals from the UI (user interface) of DesiteMD; it establishes relationships with scheduled tasks for this study [9].

The concept of the system is illustrated in Fig. 3. The system is based on a two-layer architecture of user interface on the top and the software system on the bottom. The visual education concept is based on a 4D BIM and links videos to the provided tasks in the model. First, the user will run the 4D sequence of the model by the user interface. During the animation, DesiteMD searches for links between the building object, schedule tasks, and video links. The database contains the videos as links based on Uniform Resource Locator (URL) of common online video platforms in relation to standardized tasks. If the model contains an object and the object is related to one or multiple tasks of a schedule, the program will reason the related video link. The video will be then automatically displayed in the user interface. The videos are implemented by iFrame in DesiteMD and can be provided by each video platform like YouTube or Vimeo. It becomes clear that the tasks used in the schedule have to be according to the task stored in the database to ensure a link between the task and the provided video.

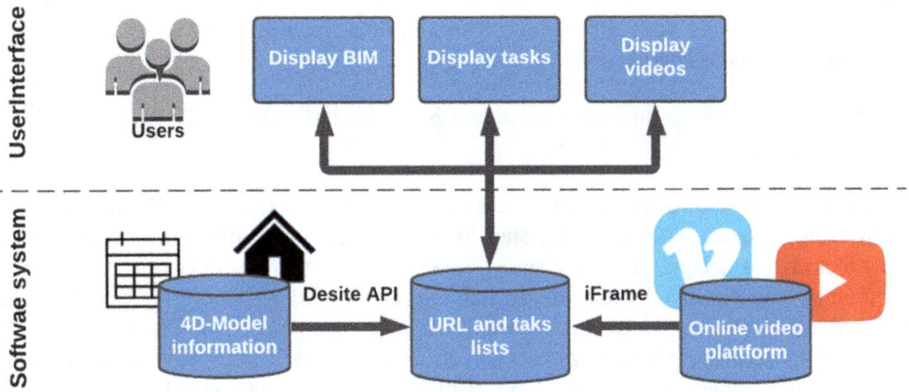

Figure 3: Concept for assigning videos to 4D model information.

4.1 User interface and access to instructional videos

An overview of the user interface is given in Fig. 4. On the top of the software application is the usual menu of the software. The Screen is divided into two main areas. On the left side, information about the schedule and the videos according to the task of the 4D model are illustrated. On the right side, the BIM is visualized according to the colour related tasks of the schedule. In this example, the concreting process of a column is visualized in the video window on the left side, including the scheduled task with a start, end date and duration. The column to be concrete is visualized in the centre of the 4D-BIM in yellow.

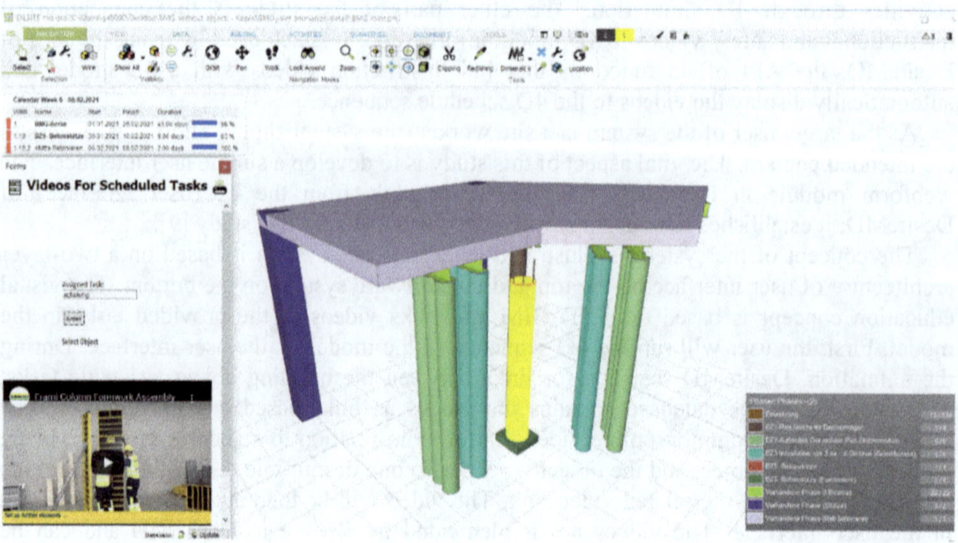

Figure 4: The user interface of DesiteMD.

The user can individualize the user interface to add more information according to the task needs, like concrete class or used formwork. The software DesiteMD enables the construction workers to watch videos on multiple mobile devices accessible to site supervisors and construction workers.

4.2 Automatic access of instructional videos based on a 4D BIM for visual education on site

As discussed earlier, the system can access all the videos related to the 4D object. It is used for watching videos that display the start to finish process of constructing an object, e.g. if the column is selected, it will display videos related to reinforcement, formwork and concreting. The displayed videos depend on the represented activity of the schedule and the selected object by the user. For example, if the user selects an identified 3D object representing activity, e.g. column formwork, and all task videos related to the column formwork will appear. The user will select the appropriate video, which resonates with the scheduled task to watch it (see Fig. 5).

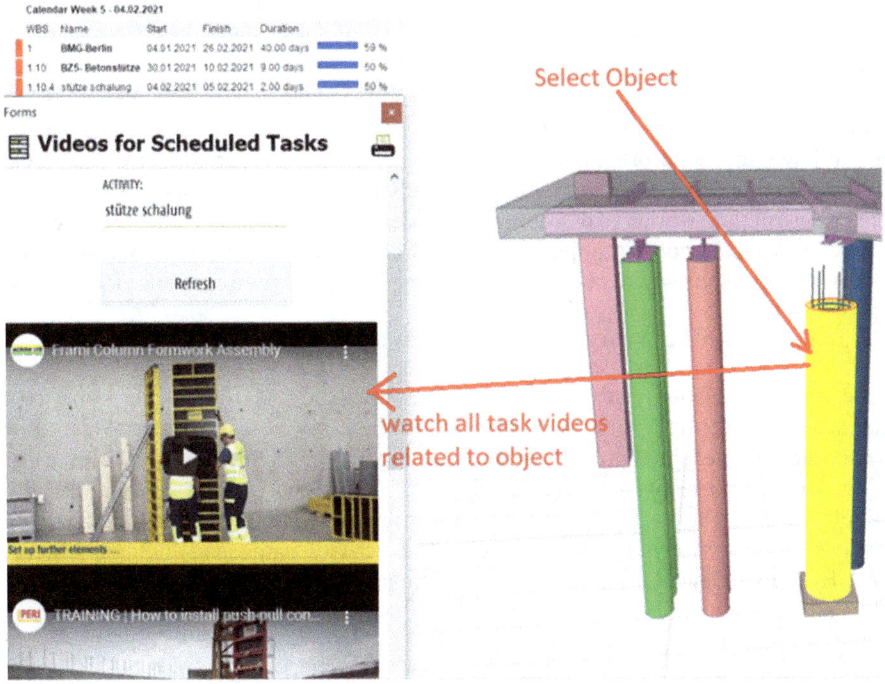

Figure 5: Access all task video connected to 4D activity and the building object.

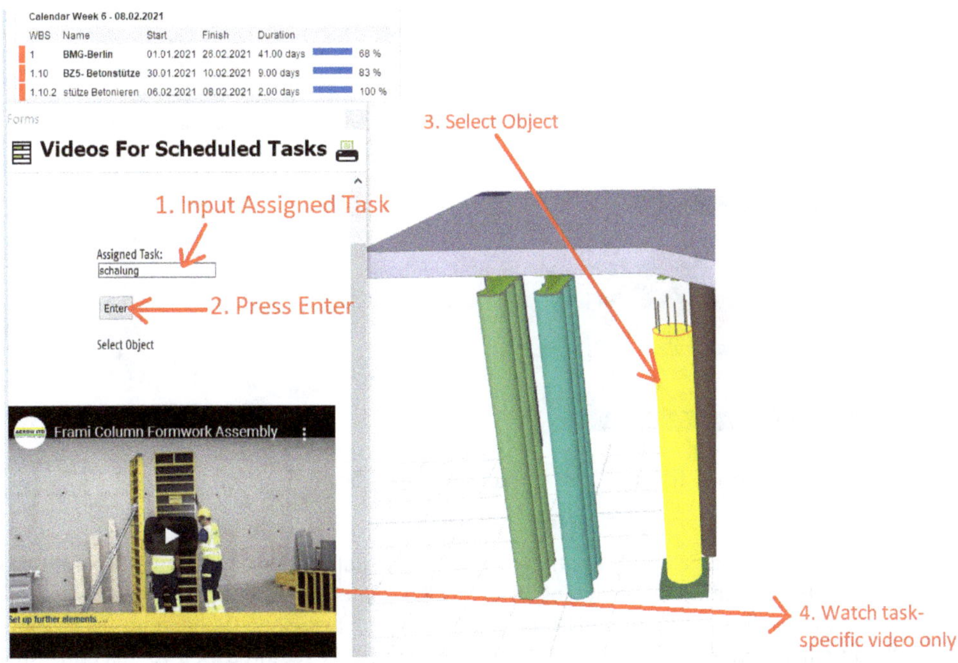

Figure 6: Process for watching task-specific video through 4D-model.

To access only task-specific videos from objects connected to multiple activities, additional control called "Assigned Task" is added to the user interface of the web service module.

It should be noted that the colour scheme is used for identifying a task-specific object in the same way as used for displaying all videos linked with task objects. The selection of objects follows the same process as mentioned in Fig. 6. Apart from showing videos for activities not connected to objects, the other advantage of using this method is that it enables construction workers to watch only videos assigned, especially their scheduled activity. The selected column object in Fig. 7 is connected to more than one activity, but due to user input "Formwork" in the field "Assigned Task," only formwork video appears.

4.3 Warning for activities with special safety instructions

The use of 4D simulation can highlight activities involved in different stages that require special safety instructions. After viewing the 4D simulation, access to safety video follows the procedure defined in the previous section. Fig. 7 shows the appearance of a safety warning for current activity in 4D animation. The object of the current activity is highlighted with red colour by default. The 4D animation stops on the activity "Ring beam concreting," and a safety warning appears. The activity name and building object are noticed. By clicking the "OK" button, the 4D simulation will continue.

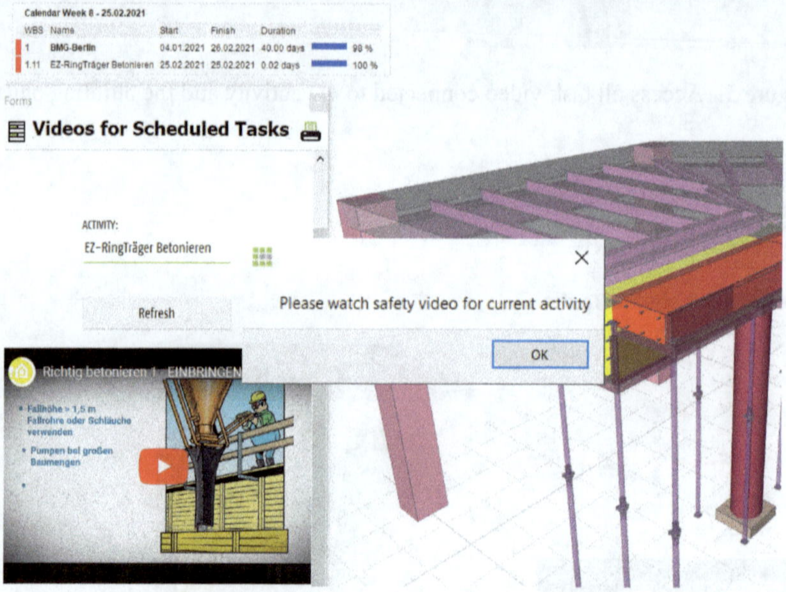

Figure 7: Safety video for activity "Ringbeam Concreting".

As a warning had appeared on activity "Beam Concreting" during 4D animation, task object is selected, resulting in displaying safety video related to concrete.

To watch the safety video in a 4D model that displays task-specific content, user input, e.g., "Safety instructions" in the field "Assigned Task," is required before selecting the object.

Although 4D simulation is used to get aware of activities that require special safety precautions, its start/pause-control also facilitates a user to watch an instructional video related to the required task at any point. It provides a possibility to the user for viewing construction sequence and video content simultaneously.

5 SURVEY ON THE PROPOSED SYSTEM

The questionnaire was sent to 110 employees of the construction company Hochtief. Out of 110 persons, 39 had participated. The questionnaire contained the link for video, which displayed 4D animation and methods for accessing instructional videos through 4D model related to different construction activities. The video was edited so it explained each step by using text and mock characters.

The survey investigated that 87% of the participants believed that videos related to scheduled tasks with the BIM model could help site workers execute tasks effectively. Especially benefits in the visualization *"1 picture is worth 1000 words!"* and the missing need for language skills are highly positive rated by the participants.

The participants realize the potential for using such 4D animation and video connection for solving problems. One of the significant aspects noticed in the answers is the emphasis on the need for visual-based instructions compared to verbal or written. It is clear from the answers that communication problems greatly influence the level of understanding for complex tasks; therefore, using such a visual environment can help in better interpretation of tasks. The answers reflect on the importance of using proper up-to-date videos related to more complex tasks. It is essential to provide access and a clear method to site workers to use such 4D models and watch video material through it.

Furthermore, 65% of the participants thought that the proposed instructional method is relevant and usable in the future; around 35% of the participants expressed either doubt or considered it irrelevant. Therefore, the process is deemed suitable for future tasks because it can make challenging tasks easily understandable; if the proposed method is used to its full potential, it can limit human error by bridging the communication gap.

It is concluded from the recommendations that the involved site workers, i.e., site managers, labourers need to be educated regarding the potential benefits of this method and the procedure to use it. Implementing a practical task can be a useful practice because the proposed method will only convince site workers if they see the practical outcome of the method to understand the task in a better way. The potential users of this method are site workers; therefore, the 4D model should be accessible to them, and the method to access content should be simple and clearly explained. The recommendations also reflected the possible use of instructional mediums other than videos, e.g., detailed drawings, technical data, etc. The connection of instructional videos with 4D model objects is considered a useful practice. Still, the videos should focus on more complex site tasks rather than simple formwork, as shown in the survey video.

6 CONCLUSION AND FUTURE WORK

The proposed system enables construction workers and site managers to access the right instructional and safety video content at the right time by linking video content through an algorithm and database. The participants, especially the site manager, realize the potential to use such 4D animation and video connection to solve daily construction site problems. One of the major aspects noticed in the answers is the need for visual-based instructions compared to verbal or written. It is clear from the answers that communication problems greatly influence the level of understanding for complex tasks; therefore, using such a visual environment can help in better interpretation of tasks. It can also be derived that the approach is used for increased coordination among planners and contractors regarding challenging tasks, e.g., heavy reinforcement. The animation displaying a correct work sequence, e.g., reinforcement, can help less experienced engineers implement the reinforcement plan correctly. The answers reflect on the importance of using proper up-to-date videos related to more complex tasks. It is essential to provide access and a clear method to site workers to use such 4D models and watch video material through it.

It is concluded that the involved site workers, i.e., site managers, labourers need to be educated regarding the potential benefits of this method and the procedure to use it. Therefore, the 4D model and videos should be accessible by mobile devices. The recommendations also reflect the possible use of instructional mediums other than videos, e.g., detailed drawings, technical data, etc. The connection of instructional videos with 4D model objects is considered a helpful practice. Further, the development of a video library focusing on more challenging site tasks is necessary. The contained videos can therefore be real or animated. As videos are assigned to the activities based on their name, it is essential to create a BIM standard for naming construction activities. In this way, the method will be generalized and usable for all the companies working with BIM. In further research, the script should be enveloped to add a rating option to the user interface so that workers can value the instructional video per its helpfulness.

ACKNOWLEDGEMENT

We want to thank the company Hochtief for their support in processing this work and participating in the survey.

REFERENCES

[1] Tversky, B., *Spatial Schemas in Depictions*, 2001.

[2] Issa, R.R.A., Cox, R.F. & Killingsworth, C.F., Impact of multimedia-based instruction on learning and retention. *Journal of Computing in Civil Engineering*, **13**(4), pp. 281–290, 1999.

[3] Mourgues, C. & Fischer, M., A work instruction template for cast-in-place concrete construction laborers. Center for Integrated Facility Engineering, Stanford University. Working Paper 109, 2008.

[4] Clevenger, C. & Del Puerto, C.L., Using 3D visualization to train Hispanic construction workers. *Associated Schools of Construction 2010*, 2011.

[5] Nasir, A.R., *A Digital Task Instruction Model for Low-skilled Construction Workforce*, Bauhaus-Universitätsverlag Weimar im Jonas Verlag für Kunst und Literatur GmbH, 2017.

[6] Nasir, A.R. & Bargstädt, H.J., An approach to develop video tutorials for construction tasks. *Procedia Engineering*, **196**, pp. 1088–1097, 2017.

[7] Bargstädt, H., Nasir, A.R. & Ignatova, E., Can BIM support better working conditions for low-skilled labor. *CONVR* 2014, p. 44, 2014.

[8] Verschuren, P. & Hartog, R., Evaluation in design-oriented research. *Quality and Quantity: International Journal of Methodology*, **39**(2005) pp. 6–39, 2005. DOI: 10.1007/s11135-005-3150-6.

[9] DesiteMD-API, DesiteMD (API). API (Tutorial).

OPPORTUNITIES AND CHALLENGES IN THE IMPLEMENTATION OF BUILDING INFORMATION MODELLING FOR PREFABRICATION OF HEATING, VENTILATION AND AIR CONDITIONING SYSTEMS IN SMALL AND MEDIUM-SIZED CONTRACTING COMPANIES IN GERMANY: A CASE STUDY

NIJANTHAN MOHAN, ROLF GROSS, KARSTEN MENZEL & FABIAN THEIS
Aachen University of Applied Sciences, Dresden University of Technology

ABSTRACT

Even though BIM (Building Information Modelling) is successfully implemented in most of the world, it is still in the early stages in Germany, since the stakeholders are sceptical of its reliability and efficiency. The purpose of this paper is to analyse the opportunities and obstacles to implementing BIM for prefabrication. Among all other advantages of BIM, prefabrication is chosen for this paper because it plays a vital role in creating an impact on the time and cost factors of a construction project. The project stakeholders and participants can explicitly observe the positive impact of prefabrication, which enables the breakthrough of the scepticism factor among the small-scale construction companies. The analysis consists of the development of a process workflow for implementing prefabrication in building construction followed by a practical approach, which was executed with two case studies. It was planned in such a way that, the first case study gives a first-hand experience for the workers at the site on the BIM model so that they can make much use of the created BIM model, which is a better representation compared to the traditional 2D plan. The main aim of the first case study is to create a belief in the implementation of BIM Models, which was succeeded by the execution of offshore prefabrication in the second case study. Based on the case studies, the time analysis was made and it is inferred that the implementation of BIM for prefabrication can reduce construction time, ensures minimal wastes, better accuracy, less problem-solving at the construction site. It was observed that this process requires more planning time, better communication between different disciplines, which was the major obstacle for successful implementation. This paper was carried out from the perspective of small and medium-sized mechanical contracting companies for the private building sector in Germany.

Keywords: building information modelling, HVAC, prefabrication, construction, small and medium scaled companies.

1 INTRODUCTION

The construction industry is witnessing a paradigm shift that tends to achieve higher productivity, efficiency, quality, sustainability and infrastructure value while reducing the life cycle costs and time of a project [1]. The evolution of computers and information technology lead to a digital revolution in the AEC industry, which in turn gave rise to the BIM technique. BIM is a process that generates and manages digital representations of buildings, in terms of both physical and functional characteristics [2]. Over the past 20 years, the BIM technique has been successfully adopted in different parts of the world and has been beneficial for all the stakeholders involved in a project. On December 15, 2015, the Federal Ministry of Transport and Digital Infrastructure (FMTI) in cooperation with Planen-Bauen 4.0 GmbH proposed a three-step plan (Fig. 1) for implementation of BIM for road and rail projects over the period from 2015–2020 [3]. Even though Germany is in the pilot phase of the implementation process, many companies are still hesitant about exploring the capabilities of BIM.

WIT Transactions on The Built Environment, Vol 205, © 2021 WIT Press
www.witpress.com, ISSN 1743-3509 (on-line)
doi:10.2495/BIM210101

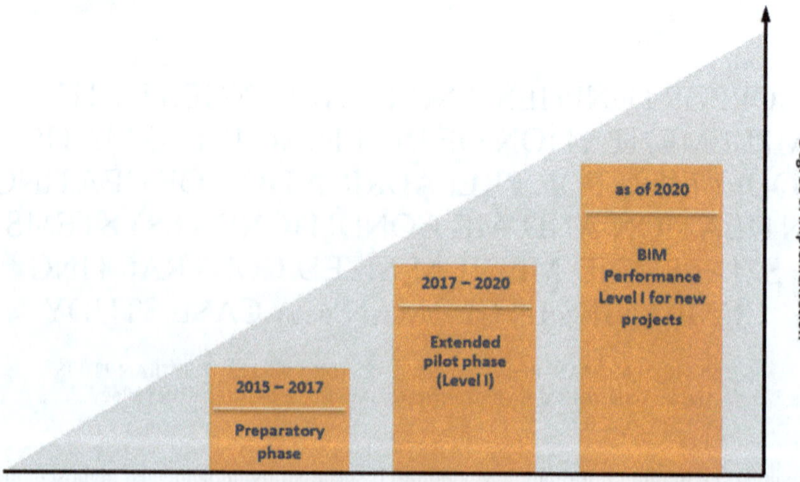

Figure 1: Phases in the implementation of BIM in Germany [6].

The most common misconception among the stakeholders is limiting BIM to 3D modelling software. BIM is not only a software but a holistic process that integrates process, people and the software platform establishing a well-coordinated workflow among parties involved in a project. Another common misconception about BIM is the assumption that this concept is suitable only for larger firms and projects that are more significant. The BIM concept is equally beneficial irrespective of the size of the project or the company, as in either case the financial costs occurred is the same, albeit on different scales [4].

2 PROBLEM STATEMENT

The major problem in the HVAC construction industry is the ineffective communication and coordination between different disciplines involved in a project which results due to the stocking of information among the stakeholders. Therefore, most of the decisions are made on the construction site under the pressure of project deadlines which results in a lot of rework and manual creation of additional documents.

The traditional two-dimensional (2D) plans are individual plans that do not have digital coordination (information about pumps, valves etc.) with the drawings of the other disciplines. Therefore, a change made in one of the plans should be incorporated in others manually thereby increasing the chances of errors occurring in the documentation. 2D plans have a limited understanding among the installers at the construction site, which marks the necessity for a better representation of the models. The consequence of a manually developed bill of materials results in an incorrect order for components thereby delaying the project delivery.

The current trends in the potential and status of implementation of digitalisation processes of the construction companies in Germany are illustrated in Fig. 2. From the statistics, it is evident that the contracting companies have a huge potential to use various technologies for digitalisation in construction but the implementing methods are still lacking. This is mainly due to the high risks concerning the huge investments (human resource, software and hardware) and demanding effective communication and coordination process with BIM. The above-stated reasons result in reluctance with the small and medium scale contracting companies in adapting to digitalisation/BIM process for construction.

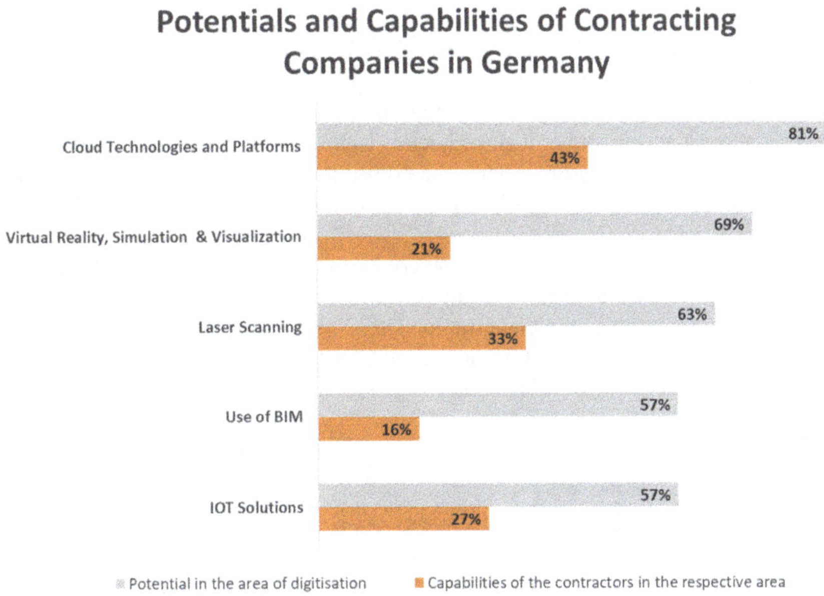

Figure 2: Potential and skills of digitalisation of construction companies in Germany. *(Source: Digitalisierung der Bauindustrie, PwC-Studie, December 2020.)*

The BIM workflow is not a concrete guideline but rather tailor-made for an organisation according to its capabilities and requirements. At the world stage, the BIM concept was proposed as a solution to enhance communication and coordination in a construction project. It guides an organisation to develop a standardised workflow to enrich the process and procedures administered at the firm. BIM emphasises Virtual Design and Construction (VDC), i.e., a 3D model as a digital twin of the project is developed which enables the user to resolve any discrepancies before going into construction. The 3D model also provides better representation for the installers and facilitates the prefabrication of the components. Model-based material lists ensure optimal ordering of materials and reduce wastage at the construction site. Overall, the implementation of a BIM workflow aids an organisation to produce detailed, high-quality plans and documentation for a project, while also optimising the time taken for construction.

In this context, the prefabrication process (an advantage of BIM) was selected for the study to depict the opportunities and challenges as it is a more hands-on process and has a direct impact on the workflow at the construction site.

3 BIM FOR PREFABRICATION

Without BIM, designing prefabricated Mechanical Electrical and Plumbing (MEP) system components becomes an uphill struggle, and contractors risk creating parts and pieces that may not be compatible with each other as planned once installed on the construction site.

However, using highly detailed and coordinated BIM 3D models of various architectural and MEP systems, each trade contractor can be confident that their system will fit and be installed according to the planned design. According to the above-mentioned, the following process flow for the design and construction of HVAC systems with the prefabrication process was developed.

Fig. 3(a) depicts the design process of a construction company (contractors) in Germany. The execution drawings or 3D model are received from the engineering planning office and the manufacturer-specific information (geometry and technical specification) of the components are added and thereby the installation drawings are developed by the contracting companies. A coordination and collision check is to be carried out with other trades to avoid discrepancies during construction. Using the coordinated model, assembly/fabrication part drawings are developed which is to be followed by the execution of material procurement and fabrication process.

The installation process of a construction company (contractors) in Germany is depicted in Fig. 3(b). The procurement/fabrication process is followed by the transportation and logistics of the components at the construction site. The components are then installed at the planned positions and an inspection is carried out. Once all the parameters are checked, the system is commissioned and simultaneously the as-built documentation is prepared.

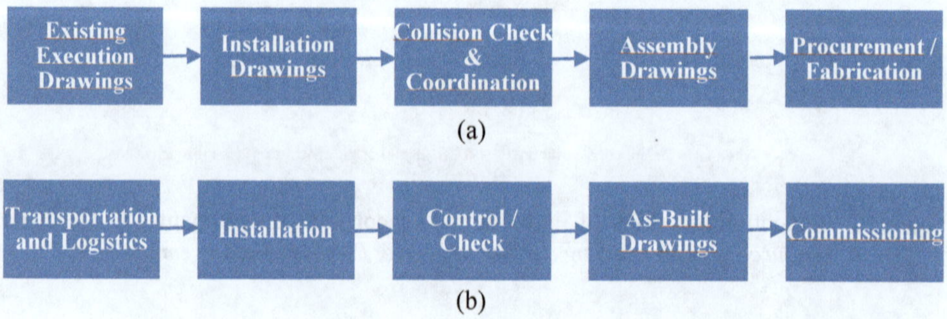

Figure 3: Proposed process flow diagram for a mechanical contracting company with prefabrication method (a) Design process; and (b) Installation process.

The case study on the opportunities and obstacles of implementing prefabrication process for the construction of HVAC system (mechanical rooms) was carried out in two phases.

3.1 Case study 1: On-site fabrication process with BIM models

The case study aimed to introduce and integrate the use of BIM models at construction sites as an advancement from traditional 2D drawings. Instead of the conventional installation at the respective rooms, a fabrication tent was constructed at the site and the welders fabricated the installation parts using drawings from the BIM model. This provided an opportunity for the installers to cross check with the actual installation position and get adapted to the use of 3D models at construction sites. Fig. 4 depicts the execution of case study 1 at a construction site.

Fig. 4(a) illustrates the BIM model of an HVAC technical room (with two buffer tanks and heat exchangers for space heating & cooling purposes), which was developed with a level of detail 500 using Autodesk Revit Software. The components used in the model, such as valves, buffer storage and pumps, were obtained directly from the manufacturer and are the digital replica of the procured components for the project. The HVAC BIM Model was coordinated with other disciplines like electrical and plumbing, and the final collision-free drawings and model was provided to the installers at the construction site.

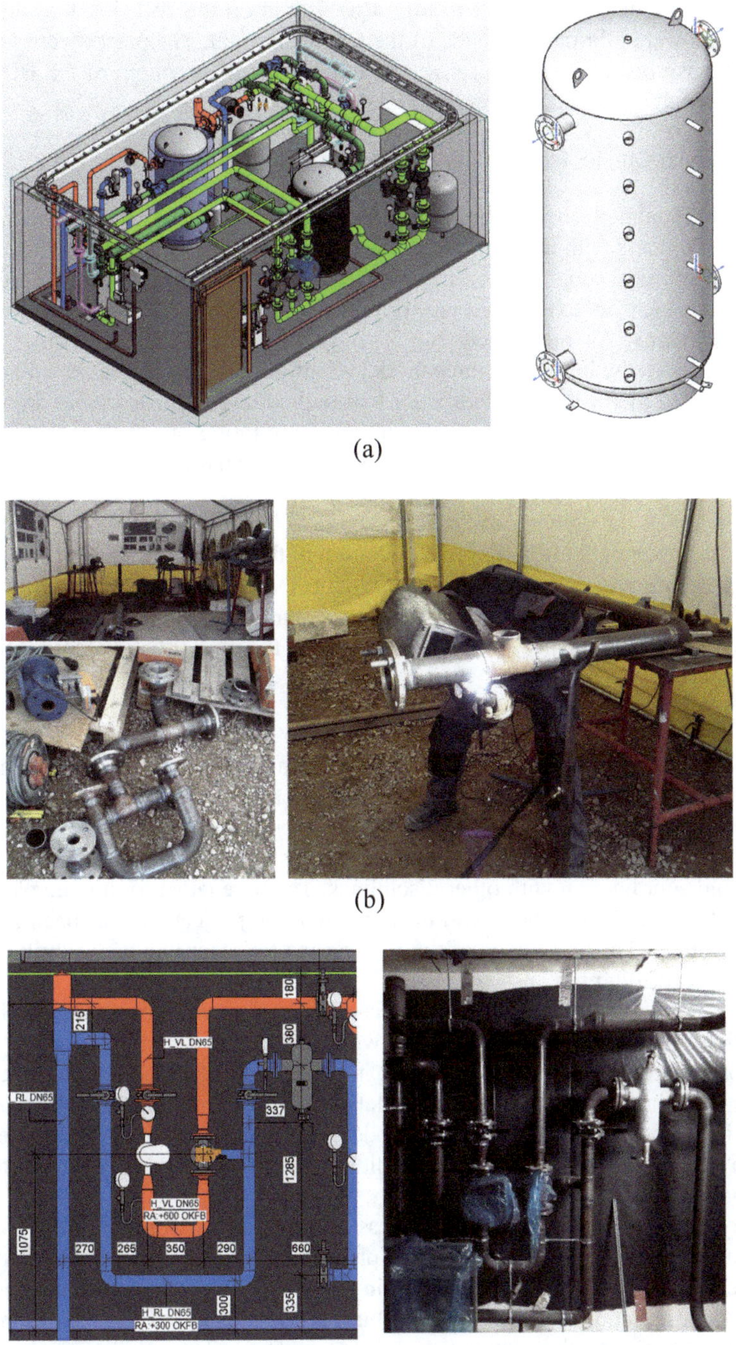

Figure 4: Execution of case study 1 – On-site prefabrication. (a) BIM model with manufacturer-specific components; (b) Fabrication tent; and (c) Comparison between BIM model and as-built.

The fabrication tent is depicted in Fig. 4(b). Based on the collision-free drawings, the assembly parts were fabricated outside of the technical room. The pieces were assembled at the planned positions in the technical room, and a comparison photo of the BIM model to that of the as-built condition is shown in Fig. 4(c).

3.2 Case study 2: On-site fabrication process with BIM model

The case study aimed to examine the changes and challenges in implementing the prefabrication process with BIM models using external service providers. In this method, the model of the technical room was split into different assembly part drawings (Fig. 5(a)) and was made available to an external service provider who delivers the finished fitting pieces of the HVAC systems. As mentioned in Fig. 3(a), to avoid onsite difficulties during the installation process, the individual trade model was thoroughly coordinated with other trades and was approved by the general contractor before the draft of the assembly drawings. Each dimension, such as pipe diameters, length and pressure rating of the flanges, were depicted extensively in the assembly drawings to avoid discrepancies during fabrication and installation.

Fig. 5(b) depicts the delivery of fabricated parts to the warehouse and the sorting process of the individual components before transportation to the construction site.

The pipe fitting pieces were then transported to the construction site, similar to HVAC machines like chillers, boilers, etc. The parts were marked with an identification system for easy sorting at the project site. The final assembly of the pipe fitting pieces is illustrated in Fig. 5(c).

4 RESULT ANALYSIS

Using the above-mentioned case studies, a comparison between the conventional construction process to that of the process with prefabrication using BIM models have been made for time parameter and are illustrated in the graphs shown in Fig. 6.

The first column depicts the time taken for the design of the technical room. It involves modelling and coordination with other disciplines. The time taken for on-site planning (Fig. 6(a))/for prefabrication (Fig. 6(b)) and installation at the project site has been illustrated in the second column. The percentage of time consumed for recording the as-built condition is shown in the third column.

From the graphs, it is inferred that the percentage of time taken for design is higher in the prefabrication process because of the more work required for creating assembly drawings and for coordination. The higher design time, in turn, is compensated by the reduction in on-site planning and installation as the BIM model is a digital twin of the technical room, which was extensively coordinated. As the BIM model is a replica of the real-time project site, the time taken for recording the as-built condition is considerably lesser than that of the conventional process.

The above graphs illustrate that the process using prefabrication with BIM models takes about twice the amount of time in the design phase that includes the creation of construction drawings, coordination and collision checks in comparison with the traditional method. The time taken for installation and creation of as-built documents in the prefabrication process is less in comparison with the conventional process. The values of the graphs have been collected with various interactions with the installers at the project site.

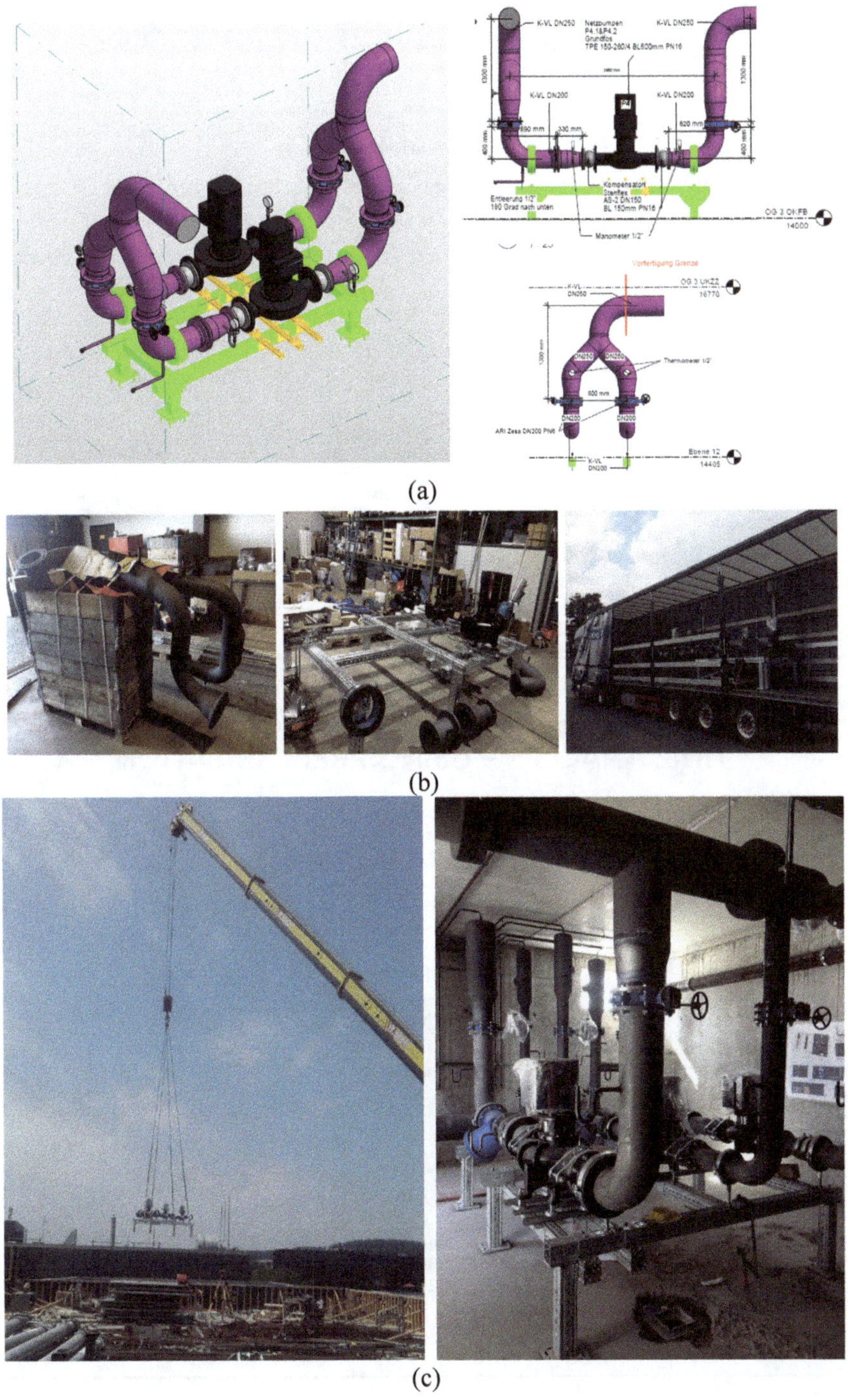

Figure 5: Execution of case study 2 – Offshore prefabrication. (a) Assembly drawing with manufacturer-specific components; (b) Prefabricated fitting pieces; and (c) Transportation and installation.

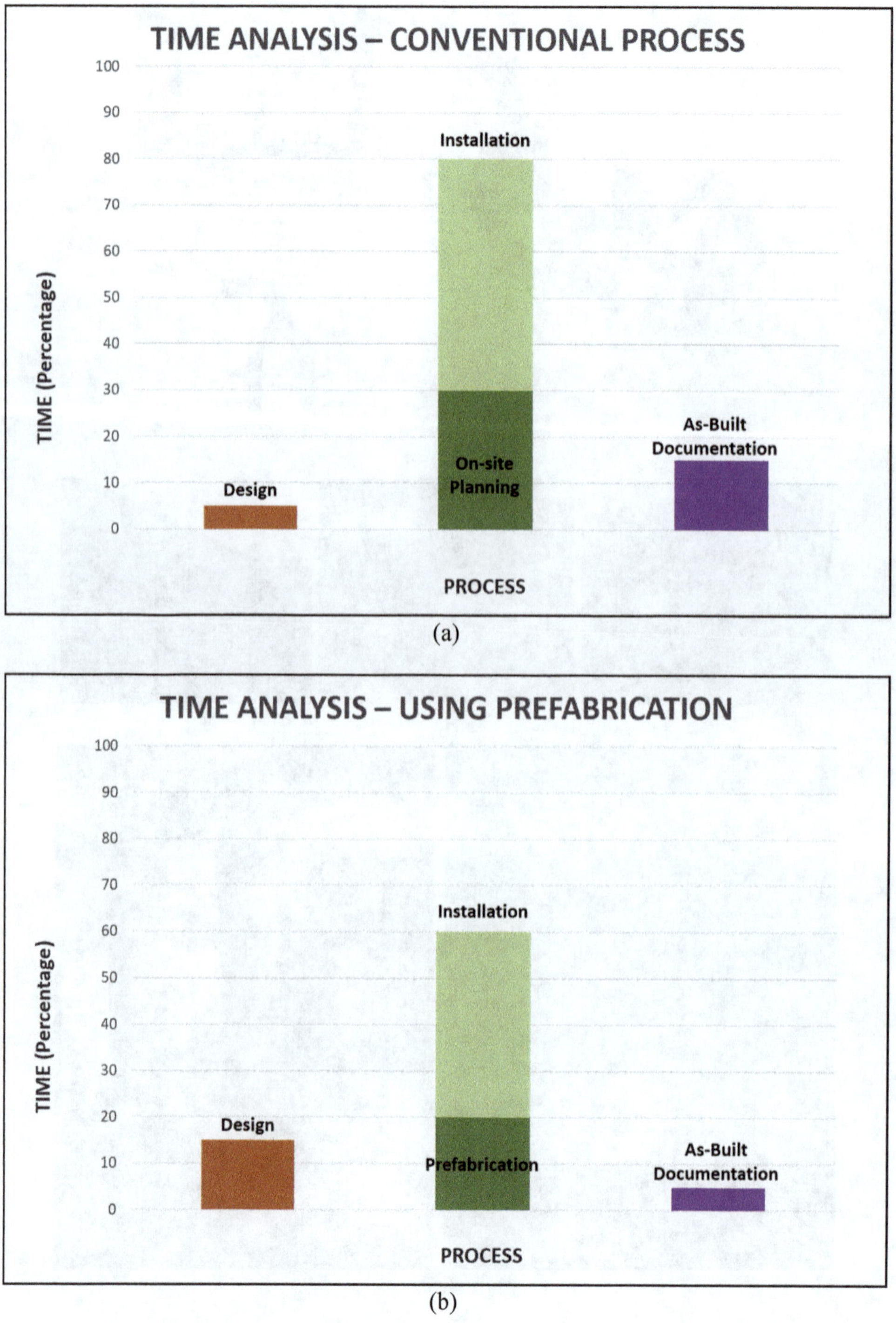

Figure 6: Time analysis. (a) Conventional process; and (b) Using prefabrication process with BIM models.

4.1 SWOT analysis

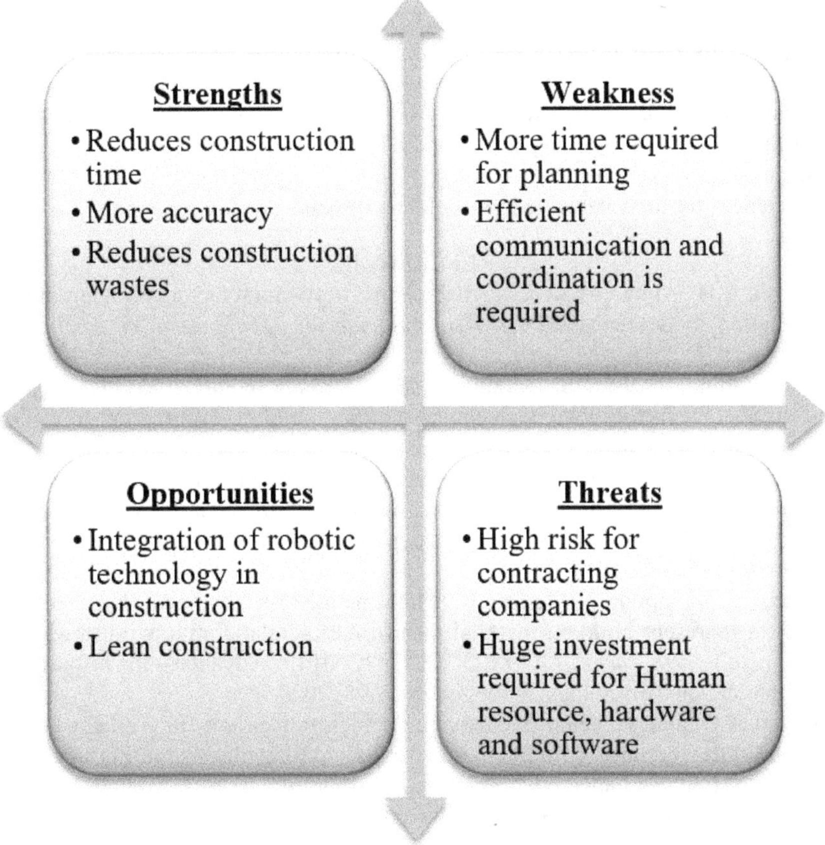

Figure 7: SWOT matrix – BIM for prefabrication of HVAC systems.

5 CONCLUSION

BIM is neither just technology nor just software. BIM is a best-practice process for the AEC industry, and like any other concepts such as PRojects IN Control Environments (PRINCE2), Agile development etc., BIM has its impact on project management, project coordination and procurement processes in an organisation.

Even though the prefabrication process requires more time in the design phase, it reduces the time taken for installation and creation of as-built documents that in turn results in an overall reduction of about 20% of the time taken for the construction process. The reduction in the time consumption reflects directly with the costs incurred. The contracting company can reduce their costs up to 15% (considering the loss in the profit margin for the materials such as pipes and pipe fittings – which are supplied by the external service providers), thereby benefitting the contractors.

Implementation of BIM for prefabrication of HVAC systems may seem an ambitious endeavour because of its diverse components, extensive information data and high capital.

However, when done right, it will open a new frontier for the profile and capabilities of an organisation.

ACKNOWLEDGEMENTS

We would like to express our sincere thanks to the management and the employees of Wilhelm Theis GmbH, a mechanical contracting company in Germany for supporting us in the implementation of the research work. Furthermore, we would like to thank the research groups at the Institute of Building Informatics, Dresden University of Technology and the Department of Civil Engineering – Smart Building Engineering, Aachen University of Applied Sciences for their valuable inputs on this research work.

REFERENCES

[1] Wadhah, A.H., Abbas, M.A. & Nagham, N.A., Motivation factors for adopting building information modeling (BIM) in Iraq. *Engineering, Technology & Applied Science Research*, p. 1, 2018.

[2] How Building Information Modeling (BIM) Has Revolutionized the Construction Industry. https://www.viatechnik.com/blog/building-information-modeling-bim-revolutionized-construction-industry/. Accessed on: 31 May 2021.

[3] Bastian, F., BIM – Status of implementation in Germany. *Annual Conference of the European Society of Law*, University of Fribourg, Switzerland, pp. 4–9, 2017.

[4] David, P., Paul, S. & Stefan, M., *Building Information Modelling for Dummies*, John Wiley, pp. 345–350, 2016.

[5] Valluru, P., Karlapudi, J., Menzel, K., Mätäsniemi, T. & Shemeika, J., A semantic data model to represent building material data in AEC collaborative workflows. *Boosting Collaborative Networks 4.0.* PRO-VE 2020. IFIP Advances in Information and Communication Technology. Springer, vol. 598, 2020.

[6] Stufenplan Digitales Planen und Bauen, Bundesministerium für Verkehr und digitale Infrastruktur, Germany, p. 5. https://www.bmvi.de/SharedDocs/DE/Publikationen/DG/stufenplan-digitales-bauen.pdf?__blob=publicationFile. Accessed on: 30 May 2021.

[7] https://theis-gmbh.de/. Accessed on: 30 May 2021.

BIMPO: A GENERATIVE PARAMETRIC TECHNIQUE FOR BUILDING ENVELOPE DESIGN

AIDA GAMAL FATHY, ZEYAD EL-SAYAD, DINA SAADALLAH & ALI FOUAD BAKR
Department of Architecture, Faculty of Engineering, Alexandria University, Egypt

ABSTRACT

Parametric design tools are digital tools used by the architect, for the purpose of creativity. BIM platforms tend to simplify the work of designers and engineers, if their products are within standards. Within the digital construction field, creativity and information modelling are frequently in contrast. An interaction between parametric design and BIM platforms is necessary to avoid the influence of predefined parametric elements on the design concept. Parallel origami is suggested as a generative tool to produce complex shapes. This research proposes a new tool for Revit architecture software (BIM tool), based on the BIM-Parametric-Origami (BimPO) integration framework. This tool directly generates optimized alternatives of complex or non-conventional solar devices, for specific target performances of building envelopes. The BimPO script provides more possibilities for an increasing number of form generation. Results will prove BimPO's ability to efficiently achieve numerous Origamic Solar Device (OSD) alternatives, for improved energy efficiency. Moreover, selection criteria are proposed, which could indicate optimal OSD configurations, with a balance of daylight and thermal performance (TH), based on daylight simulation.

Keywords: BIM, parametric design, programming, visual programming, origami, generative design, form generation, solar devices, building envelope.

1 INTRODUCTION

Building Information Modelling (BIM) is an intelligent 3D model-based process, which improves the output and management performance of projects. Adaptation of BIM is steadily gaining popularity in the construction industry from 13% in 2011 to 74% in 2018. Users have been quite aware of BIM, but its adoption still varies from one country to the next (Denmark 74%, Canada 67%, UK 48%, Japan 46%, and the Czech Republic 25%) [1]. Moreover, the usage of BIM is growing and within five years, it is predicted that BIM usage in the design professions in most countries will reach more than 80%. These facts suggest that the use of BIM is turning into the norm within the construction field [2].

Improvements in the technology of building construction have accelerated within the latest few years and developments in BIM have been essential to fulfil the more recent requirements. with the rise in complexity that has come with contemporary architecture, whether it is to do with the complexity of form, the need for better performance, or facility management, there is a greater need for tools to help the process of form generation. As the number of digital tools in use increases, there is a greater need for interoperability between different applications. BIM tools allow the integration between various disciplines, by having a semantically rich model to-and-from which data is transferred [3].

These issues in the early phase of design are counteracted by the use of conceptual modelling tools, such as in Revit. These tools often create more general and simpler simulations, which yield an approximation of the outcome of the process. This includes the distribution of solar energy in a building throughout the year or an approximation of the most optimal window glazing. VPL (Visual Programming Language) software has been the most recent addition to modelling tools. VPL is a third-party tool for modelling software. It creates the possibility of taking modelling to the next level. Using VPL, designers can use a lot of scripting to create more complex models. This is usually not possible with the

WIT Transactions on The Built Environment, Vol 205, © 2021 WIT Press
www.witpress.com, ISSN 1743-3509 (on-line)
doi:10.2495/BIM210111

existing BIM tools. It also gives more flexibility to other actors in the project, by letting them set more parameters, in order to optimize the model even further.

2 BUILDING INFORMATION MODELING (BIM)

To help architects cope with the complexity of adaptive building requirements, processes and tools with abilities to generate the design of dynamic envelopes, which serve as shading devices, have become available. Nowadays, accurate information about all physical and functional features of a single building can be organised in a BIM.

BIM is one of the most important techniques in the architecture, engineering, construction, and operations (AECO) industry. A BIM is an emerging tool/methodology/paradigm/process of virtual design and construction, involving the generation and management of digital representations of the physical and functional characteristics of a facility. It creates and uses the coordinated, consistent, computable information of the 3D models of the project's components, interconnected with the holistic information conceived as a source of shared knowledge. It aims to support decision-making. All of this allows for better analysis and control of a facility from its earliest conceptual stages, through design and construction, its operational life, and eventual demolition than manual processes. When accomplished, these computer-generated-semantic-3D models include accurate geometry and data required to support the construction, fabrication, and procurement activities through which the building is realized [4]–[9]. Today, a lot of tools exist to enable BIMs to evaluate global energy needs and increase the energy efficiency of buildings [10].

3 PARAMETRIC DESIGN (PD)

Parametric design is integrated within BIM to overcome the gap in research related to its inability to generate different forms. The Oxford Dictionary defines a parameter as "a numerical or other measurable factor contributing to a set that determines a system or adjusts the status of its operation," and the word parametric as "relating to or expressed in terms of a parameter or parameters." Based on the fundamentals of the literature, we can integrate PD in a design process, based on algorithmic thinking [11], which uses parameters and rules to constrain them [12], [13].

PD also relates to the BIM model as the latter uses PD's concepts of associative geometry and topological relationships [14], [15], which establish dependencies among different design elements.

3.1 Parametric modelling and visual programming/scripting interfaces

Parametric modelling and visual programming/scripting interface parametric modelling are also an important part of the BIM technology. They allow for the exploration of real-time design options and the management of the relationships among building elements. Parametric modelling makes parameters available for processing by computational algorithms and mathematical formulas, before being used or assessed.

Visual programming is a concept that provides designers with the means for constructing programmatic relationships utilizing graphical user interfaces, rather than writing codes from scratch. It allows building designers to easily experiment with various design alternatives, without professional knowledge of programming or scripting. There are various visual programming tools in BIM and CAD tools, such as Bentley Generative Components and Grasshopper for McNeel Rhinoceros. Architects are able to make different relationships by linking pre-packaged nodes together to make a custom algorithm [10].

The Autodesk Revit API (Application Programming Interface) helps designers add features to the software and create custom tools and plugins. Dynamo is a Revit plugin that utilizes the Revit API. It is built using the Windows Presentation Framework. As a parametric modelling engine, Dynamo is inspired by other tools to extend Revit's parametric modelling abilities, by adding an associativity, which is not available in the off-the-shelf application. This includes driving parameters based on external inputs, such as solar geometry. Dynamo helps users automate processes and computational design, through a node-based Visual Programming Interface [16].

4 ORIGAMI (O) AS A FORM-FINDING TOOL

Origami is the conventional Japanese craft of paper folding. The Japanese word "origami" is a combination of two Japanese words "ori" meaning to fold and "kami" meaning paper [17].

Origami, from an ancient Japanese craft to today's recently recognized "engineering discipline", offers many opportunities to engineers and designers to recognize mathematical relations, unprecedented forms, and static and dynamic structures [18].

4.1 Reason to choose origami as an application for parametric modelling

Origami has endless possibilities. However, the software and applications available for origami design are limited in terms of design freedom [18].

4.2 Origami-based design

In later years, building envelope explorations were carried out using inspiring new ideas from origami. Origami suggests a set of paper-folding methods, which can be tested with parametric modelling software. Origami is based on the relationships, as well as the shape, between different points and surfaces, which can be defined by rules. It can be turned into parametric models [19].

5 BIMPO

BimPo is a new approach integrating BIM, parametric design, and origami (see Fig. 1). This approach aims to overcome the previously mentioned gap, by generating various combinations that could improve daylight performance, as well as any other objective set by the designer. Hence, using this approach, a wide range of optimized solutions is reached.

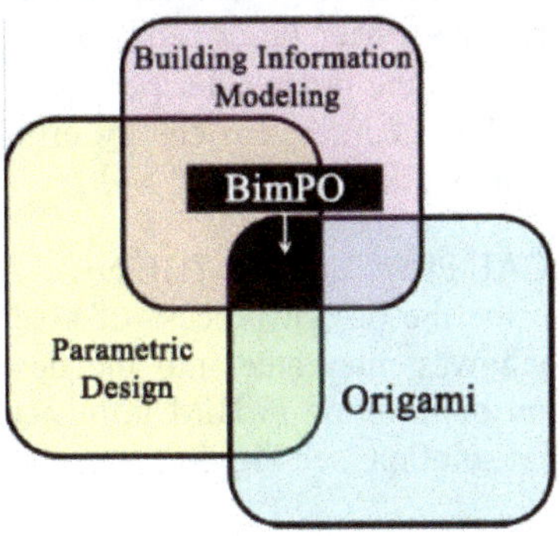

Figure 1: BimPO is a new approach to integrate BIM, parametric modelling, and origami.

6 METHODOLOGY

The methodology used in this research is employed to identify, explain, and test a suggested, new approach. A research's aims and objectives are represented by input 1 (in1). By integrating existing related work (in2) with new knowledge gained through literature reviews and data collection (in3), the methodology can then generate the new approach (output or (out)), after passing through an iterative, three-stage theory-building process. This process includes three repetitive stages – description, scripting, and testing (see Fig. 2):

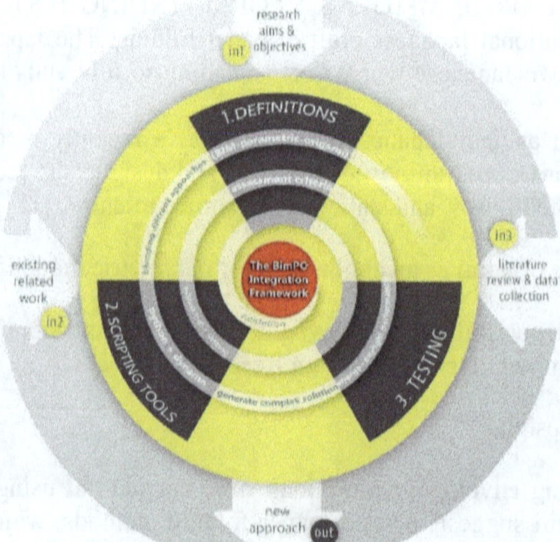

Figure 2: Methodology of the research.

- First, the Definition Stage identifies the benefits of blending the three different approaches: BIM, parametric modelling, and origami.
- Second, the Scripting Stage builds upon descriptions to infer the new theory, based on the integration of the three different approaches to generate complex dynamic components. In essence, the scripting stage develops a testable theoretical proposition, which clarifies what has previously been described.
- Third, the Testing Stage inspects the proposed script for validity, tests concepts or the relationships between them for accuracy, for energy efficiency (EE) and daylighting strategies, to develop the framework or theory.

7 THEORETICAL PROCESS OF THE NEW APPROACH

In this paper, the authors describe the design process of a set of OSDs developed using BimPO, and how this approach was integrated into the design workflow. The authors demonstrate how the integration of scripting in BIM influences the whole design process and also the selection of the final solution (see Fig. 3).

8 CASE STUDY

The goal of the project is to use the BimPO approach to generate alternatives for OSDs, based on the path of the sun in Alexandria, Egypt. For this project, the authors used:

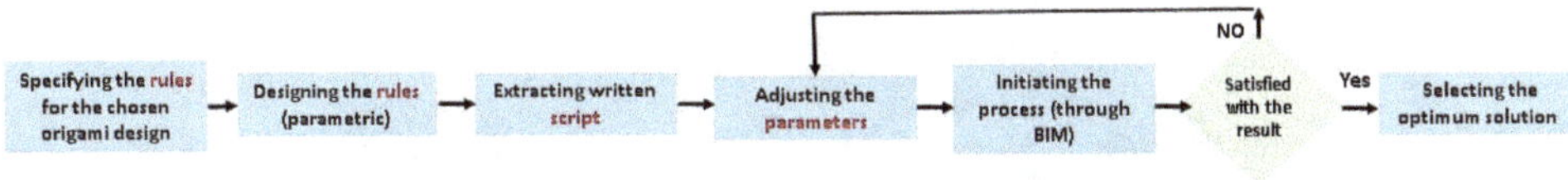

Figure 3: Theoretical process of the new approach.

- Revit architecture version 2021
- Dynamo version 2.1
- Refinery

8.1 First stage

In this stage, the authors identified the equation that controls the OSD (origamic model), by using the adaptive component in Revit 2021 (BIM) (see eqn (1) below):

$$\text{base} = \text{sqrt}((\text{width factor}^2) - (\textit{opening factor}^2)). \tag{1}$$

In this project, two OSD parameters are controlled in Dynamo.

1. Opening factor
2. Width factor (B_factor) (see Fig. 4).

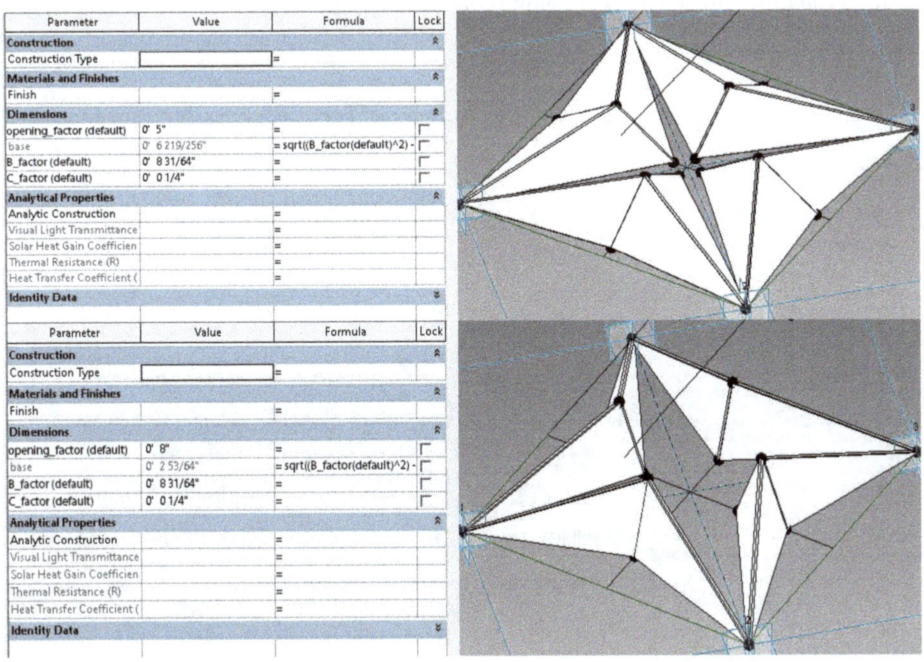

Figure 4: OSD parameters defined by the adaptive component. (Software: Revit 2021.)

A simple cylinder is created for the mass model and the adaptive components (OSD) are distributed on the surface of the cylinder (see Fig. 5).

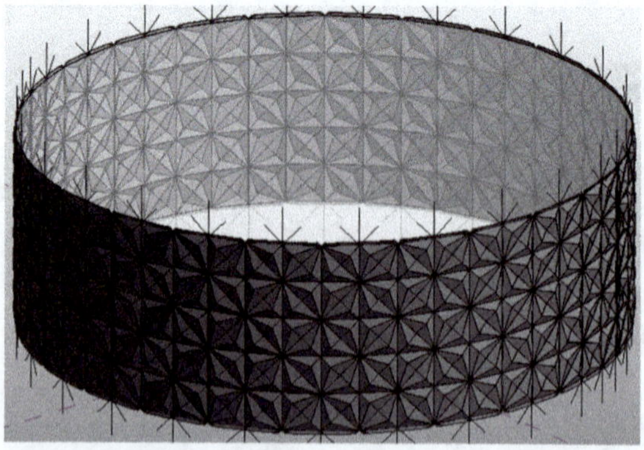

Figure 5: The adaptive components (OSD), surrounding the surface of a simple cylinder. (Software: Revit 2021.)

8.2 Second stage (scripting)

1. Exporting every single OSD coordinate, while utilizing exported data nodes in Dynamo.
2. Opening a new file and loading the adaptive component onto the file.
3. Importing the edited excel file using Dynamo (see Fig. 6).

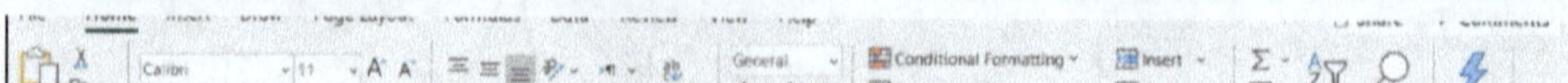

Figure 6: Graph showing steps from Dynamo for importing excel file to get the locations of origamic shading. (Software: Excel 2020 and Dynamo 2.1.)

4. Getting access to the parameters of the OSD by name (see Fig. 7).

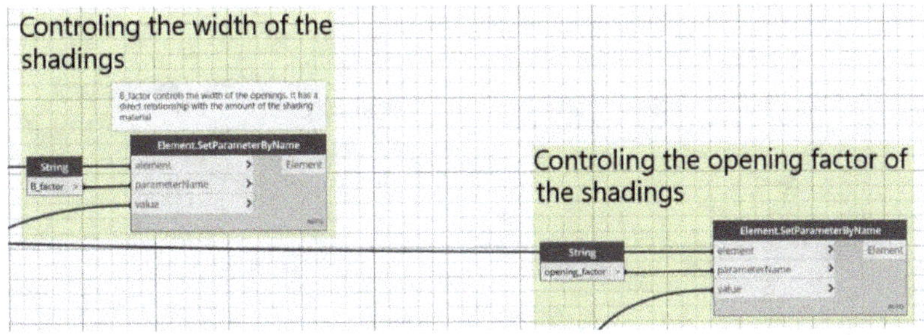

Figure 7: Graph showing steps from Dynamo to get access to the parameters of OSD by name. (Software: Dynamo 2.1.)

To reach the parametric values for the opening factor, the following procedure was followed:

1. Finding the normal values for each OSD (see Fig. 8).

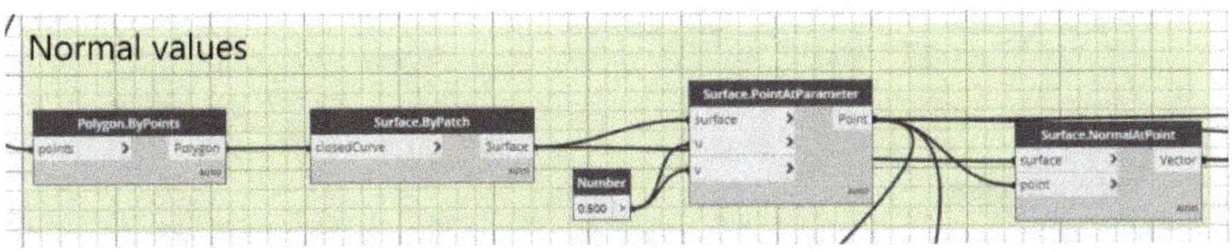

Figure 8: Graph showing steps from Dynamo to reach normal values for each OSD. (Software: Dynamo 2.1.)

2. Using three variables (x, y, z) to create the position of the dynamic vector of the sun. In this procedure, the application of the sun's path in Revit was replaced by the vector of the sun present in Dynamo. This was done with information on the sun's position for Alexandria, Egypt in Revit. Those variables are used as input to generate alternatives with the help of Refinery optimization (see Fig. 9).

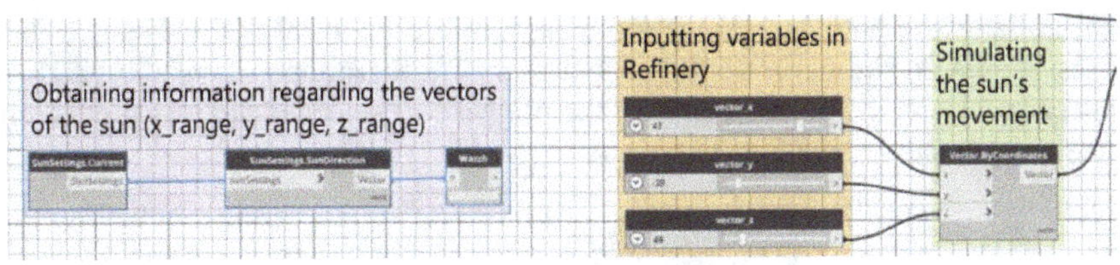

Figure 9: Graph showing steps from Dynamo for the input of three variables (x, y, z) into Refinery to generate the position of the dynamic vector of the sun. (Software: Dynamo 2.1.)

3. Calculating the angle between the normal values of OSD and the position of the sun's vector through the dot product and remapping the values of these angles for use as values for the opening factor (see Fig. 10).

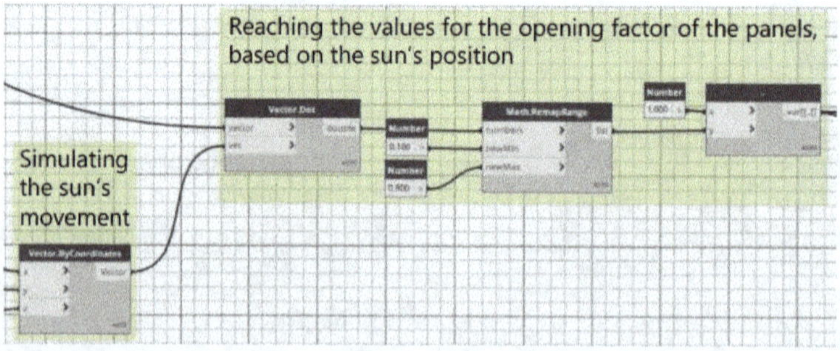

Figure 10: Graph showing steps from Dynamo to obtain the value of the opening factor of the OSD based on the sun's position. (Software: Dynamo 2.1.)

The same values were utilized to choose the colors of the OSD, based on the angle between the vector marking the sun's position and its normal values (see Fig. 11).

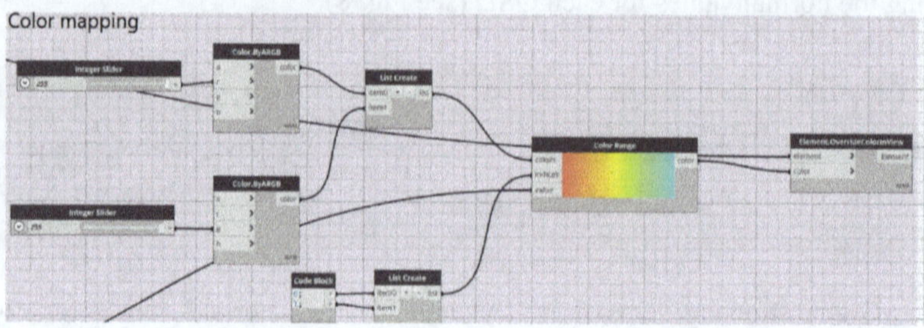

Figure 11: Graph showing steps from Dynamo to color the OSD. (Software: Dynamo 2.1.)

To obtain the parametric values for the width factor (B_factor), the following procedure is done:

1. Creating a mass model in Revit. This mass is the neighbouring building of the project (see Fig. 12).

Then, utilizing the boundary of the model as a reference to reach the distances between the closest points on the curve (building boundary) to the midpoints of the OSD. In this case, the boundary on the opposite side of the project was selected and was plotted on level 0, as the reference curve (see Fig. 12).

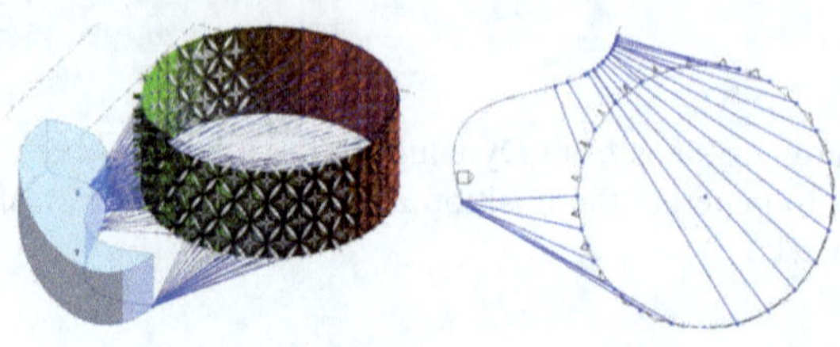

Figure 12: Obtaining the parametric values for the width factor utilizing the boundary of the model as a reference. (Software: Revit 2021.)

2. Finding the distance between the midpoints of the OSD and the closest point to them on the chosen curve (see Fig. 13), then utilizing the remapped values to control the width factor of the OSD. This is done in order to get different values for the distance for later use in Refinery. The location on the curve changes through rotation. Three variables are used as input (see Fig. 14):

    - X and y: to change the origin of the rotation
    - Angle: to change the angle of the rotation

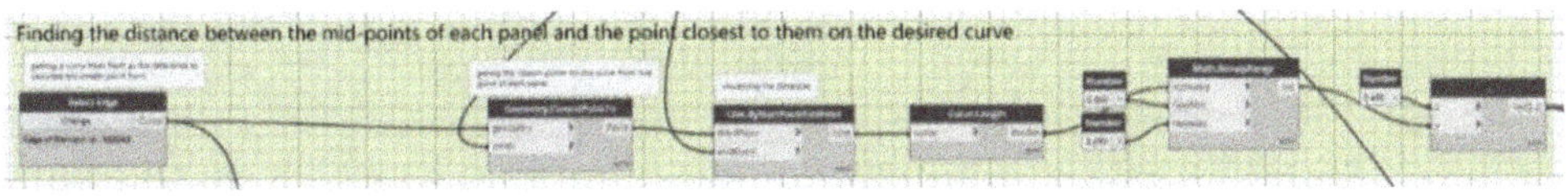

Figure 13: Graph from Dynamo showing steps to find the distance between OSD points and the curve. (Software: Dynamo 2.1.)

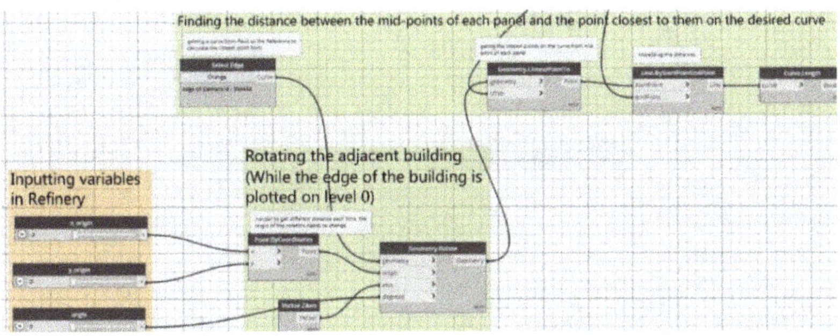

Figure 14: Graph explaining the use of the three variables as input to change the angle of rotation. (Software: Dynamo 2.1.)

8.3 Third stage

The script is exported to Refinery. There are two objectives for the use of Refinery to generate different alternatives (see Fig. 15). They are:

1. Maximizing the value of the summation of the values of the opening factor
2. Minimizing the value of the summation of the values of the width factor

The generation process is created with a population size of 48 and leads to 20 alternatives.

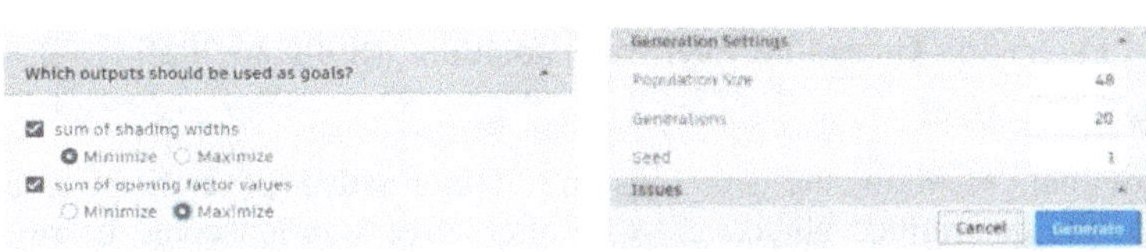

Figure 15: Customization options for the generation of alternative solutions. (Software: Refinery.)

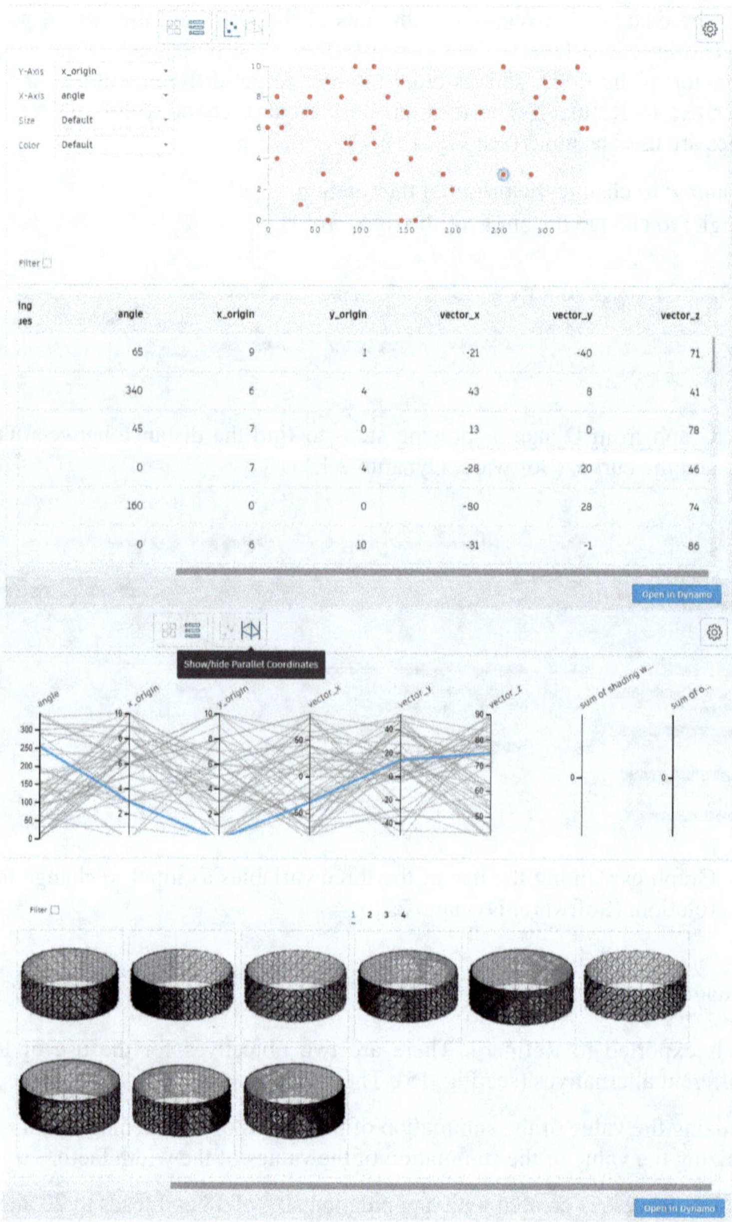

Figure 16: Generating alternative solutions. (Software: Refinery.)

These alternatives show the solutions in 20 alternative generations (see Fig. 16), according to the determined objectives. Any alternative solution could be selected to be displayed as a Revit project and its corresponding variables could be extracted. One solution is selected in Refinery (see Fig. 17). Architects can test their solutions based on many concepts (energy efficiency, daylight strategy, etc.).

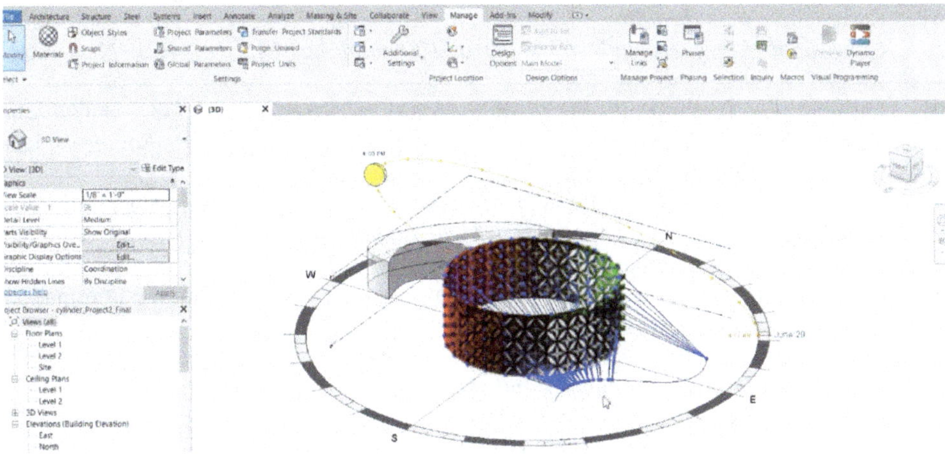

Figure 17: The final selected solution. (Software: Revit 2021.)

9 CONCLUSIONS

It is very important to provide more opportunities for form generation. Currently, there is gap between new trends and the different types of software that architects use. Basically, there is an absence of basic programming knowledge, which in turn makes an architect implement his/her creative ideas no matter the difficulties.

This is the purpose for the use of BimPO, a new method used to develop the BIM platform, using VPL and its capacity to integrate BIM, parametric design, and origami.
This exploration was accomplished in order to establish early design processes, which are effective as alternative solutions to isolate design problems associated with complex BIM façades design, responding to environmental changes.

Nowadays, the potential for the use of technology is impressive. Studies to be pursued in this field should now put more focus on this method. It is important to understand that new technologies require conducting new research and applying new frameworks. The legislative, professional, and economic apparatus is currently not prepared to work with this philosophy. However, it is absolutely necessary to catch up with the times.

The study regarding the method performed in this research could also be applied in other areas of design. It represents a line of thought to share with different disciplines, to facilitate the integration of technology.

ACKNOWLEDGEMENTS

The authors would like to acknowledge the effort made by Professor Ali F. Bakr, as well as all members of the research team to make the case study possible.

REFERENCES

[1] Malleson, A.K., NBS International BIM Report 2018, NBS: United Kingdom, 2018.

[2] Malleson, P., Kato, H., Popíšilová, B., Watson, D. & Friborg, G., NBS International BIM Report 2016, NBS: United Kingdom, 2018.

[3] Finances, Best 20 Building Information Modeling (BIM) Software in 2018, 2018. https://financesonline.com/building-information-modeling/.

[4] Krygiel, E. & Nies, B., *Green BIM: Successful Sustainable Design with Building Information Modeling*, John Wiley & Sons: New Jersey, 2008.

[5] Kymmell, W., *Building Information Modeling: Planning and Managing Construction Projects with 4D CAD and Simulations*, McGraw-Hill: New York, 2008.
[6] Eastman, C., Teicholz, P., Sacks, R. & Sacks, K., *BIM Handbook – A Guide to Building Information Modeling for Owners, Managers, Designers, Engineers, and Contractors*. John Wiley & Sons, Inc.: New Jersey, 2011.
[7] Matthew, G., Jason, G., Melissa, T., Seokho, C. & Fiona, C., Building information modelling: An international survey. *World Building Congress, 2013*, pp. 5–9 May 2013, Brisbane, Australia, 2013.
[8] Ladenhauf, D. et al., Computational geometry in the context of building information. *Energy and Buildings*, **115**, pp. 78–84, 2015.
[9] Agugiaro, G., Energy planning tools and city GML-based 3D virtual city models: Experiences from Tento (Italy). *Applied Geomatics*, **8**(1), pp. 41–56, 2016.
[10] Kim, H. & Yan, W., Parametric BIM-based energy simulation for buildings with complex kinetic facades. *Proceedings of the 33rd eCAADe Conference,* Vienna, Austria, 2015.
[11] Jabi, W., Soe, S., Theobald, P., Aish, R. & Lannon, S., Enhancing parametric design through non-manifold topology. *Design Studies,* pp. 1–19, 2017.
[12] Barrios, C., Parametric affordances: What, when, how. *Proceedings of the ACADIA Regional Conference,* 2011.
[13] Gero, J., An empirical foundation for design patterns in parametric design. *Proceedings of the 20th International Conference of the Association for Computer-Aided Architectural Design Research in Asia*, 2015.
[14] Gerber, D. & Pantazis, E., A multi-agent system for facade design: A design methodology for design exploration, analysis and simulated robotic fabrication. *Proceedings of the 36th Annual Conference of the Association for Computer Aided Design in Architecture*, 2016.
[15] Oxman, R., Thinking difference: Theories and models of parametric design thinking. *Design Studies*, 2017.
[16] Kensek, K.M., Integration of environmental sensors with BIM: Case studies using Arduino, Dynamo, and the Revit API. *Informes de la Construcción,* **66**(536), pp. 31–39, 2014. DOI: 10.3989/ic.13.151.
[17] Gavy, *Origami Inspired Architecture and Interior Design – What Creativity Does*, 2012.
[18] Tantawy, D., Origamic architectural form design system. *International Journal of Sciences: Basic and Applied Research*, **21**(2), pp. 66–86, 2015.
[19] Gao, W. & Ramani, K., *Kaleidogami™: Multi-Primitive Reconfigurable Artistic Structures*. School of Mechanical Engineering School, Electrical and Computer Engineering, Purdue University.

APPLICATION OF GENERATIVE DESIGN FOR STRUCTURAL OPTIMIZATION AT THE CONCEPTUAL DESIGN PHASE

ANDERS SAGVÅG BIRKEMO[1,2] & SAMINDI M. K. SAMARAKOON[1]
[1]Department of Mechanical and Structural Engineering and Materials Science, University of Stavanger, Norway
[2]Multiconsult Norge AS, Norway

ABSTRACT

Optimization of structural components leads to higher performing designs, by reducing unnecessary cost and increasing the efficient use of material. The structural optimization process is a rigorous iterative time-consuming process. Developments within algorithmic aided design (AAD) tools have made it easier for structural engineers to achieve automation of optimization routines. Generative design (GD) provides optimization routines which can aid in the pursuit of optimal solutions, given a set of criteria. The use of GD in combination with finite element analysis (FEA), to optimize structural design, has yet to see noticeable adoption within the structural engineering community. This paper discusses two illustrative cases to demonstrate how GD tools have been applied to optimize structures at the conceptual phase. The results have been validated using FEM-design software. A workflow has been proposed, based on the lessons learnt from the cases. This workflow inherits characteristics that, when followed, yielded consistent and reliable results.
Keywords: algorithmic aided design, generative design, structural optimization, workflow.

1 INTRODUCTION

With the advent of new and more powerful technologies, designers and architects are empowered to realize increasingly complex structures [1]. Structural engineers are often tasked with taking such concepts from an idea to a product which is buildable and structurally sound, while adhering to the initial design. For structural engineers, the final delivery will often be in the form of a highly detailed and information-rich 3D model (model-based delivery), traditional 2D drawings, or a combination of the two. The purpose of such deliverables is to serve as the reference for the builders at the building site. Naturally, these references need to be of high quality and free from errors, while precisely mimicking the physical and functional characteristics of the building. To achieve this, all the aforementioned deliveries are regularly generated in some form of BIM software such as Tekla or Revit [2]. Such software offers easy collaboration between project team members and enables thorough quality control procedures. Consequently, to keep up in an environment where the bar is raised continuously, proficiency in such tools is fast becoming a necessity [3]. As a result, efficient modelling methodologies such as parametric design have an increasing presence in the typical structural engineering workflow. Furthermore, as this methodology is seeing more use, new and inventive workflows are proving to alter the ways structural engineers work on a daily basis. Moreover, structural engineers regularly use Finite Element Analysis (FEA) tools to evaluate the structural integrity of structures. To do so, analytical models must be generated. In much the same manner as BIM models, such analytical models can be made parametrically. In addition, by having an algorithmically defined structure, additional benefits are enabled. Since the position of all elements is governed by the logic described by the script, optimization algorithms can be deployed which in turn can manipulate the geometry to achieve certain objectives. Structural engineers aim to optimize structures in terms of the amount of

WIT Transactions on The Built Environment, Vol 205, © 2021 WIT Press
www.witpress.com, ISSN 1743-3509 (on-line)
doi:10.2495/BIM210121

material needed while maintaining the structural integrity of the building. To analyze and design structures, structural engineers commonly use FEA software. In other words, for the optimization algorithm to properly assess each iteration of the structure, FEM analyses need to be incorporated into the automated workflow. With the use of Grasshopper plug-in Karamba3D, such capabilities are enabled. With the power of Grasshopper, Karamba3D and optimization algorithms such as simulated annealing and evolutionary solver, structural engineers can let the algorithms turn out solutions in the conceptual design phase. Furthermore, being able to deploy algorithms which can aid in the finding of optimal solutions can provide significant yields, in terms of both the overall price of the structure and the time savings related to the conceptual design phase. Furthermore, having integrated structural analysis within the parametric modelling routine will enable high levels of interoperability between the Building Information Modelling (BIM) model and the analysis model.

Through case studies, this paper aims to illustrate workflows structural engineers might adopt in the future. Although the applications shown are limited, the fundamental purpose is to highlight the possibilities that exist in the employment of optimization algorithms in structural design. Finally, a general workflow will be suggested that describes how to attack a problem, with the aforementioned algorithms.

As this paper mainly conducts research through case studies, qualitative research will be the overarching method; this also falls in line with the use of surveys to obtain feedback from industry professionals on the different topics. However, to evaluate the accuracy of the case studies, quantitative methods are needed to compare the numerical results from various sources. In other words, the quantitative research method will be used to validate the outcome of certain cases, to be able to qualitatively evaluate the cases as a whole.

2 GENERATIVE DESIGN VERSUS TRADITIONAL METHODS

Traditionally, the optimizing of structures is done iteratively, with a more trial-and-error approach. Obviously, through previous projects, an experienced engineer will have found solutions to problems that might resemble the task at hand. Their previous experience will then serve as a foundation for how to proceed in the pursuit of solving the said task. This approach is tried and proven and, with the collective experience of the project team, will reliably produce adequate results. However, past solutions do not necessarily mean that it is the most optimal way of solving the particular problem. In addition, it is next to impossible to rule out biases based on past experiences. Consequently, the exploration of other and possibly superior alternatives might be neglected, on the sole base that a deemed preferred solution has already been used. Generative design refers to the process of deploying optimization algorithms that aid in the search for optimal solutions, given a set of inputs and constraints, as shown in Fig. 1. Then, the optimization algorithms explore the possible solution space through the iteration of generational genome populations. A common method used in generative design is evolutionary solvers. Various generative design tools have been developed, as discussed in the following section, and the general approach can be seen in Fig. 1.

To implement generative design approach, there are variety of software combinations can be used. Moreover, Rhinoceros: 3D-CAD software developed by McNeel is mostly used in the industry and has long been a favourite among designers. In later years, with the inclusion of visual programming capabilities provided by the imbedded application Grasshopper, Rhinoceros has seen noticeable adoption among architects and civil engineers who seek to enhance their efficiency in modelling BIM models. In addition, FEA software

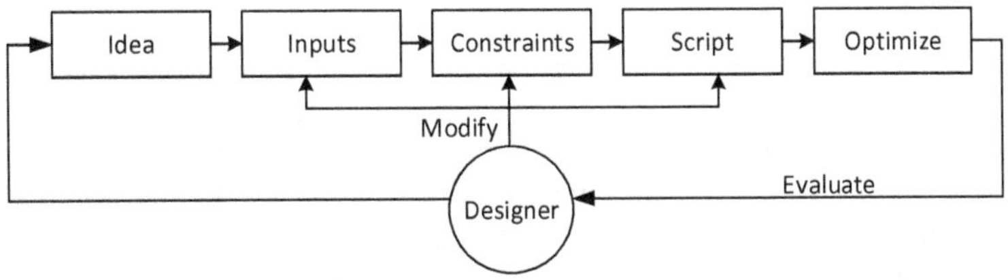

Figure 1: Generative design approach.

providers have also seen the potential and are now providing plug-ins that enable bi-directional interoperability between their respective software and the modelling capabilities that Rhinoceros brings. With the use of Grasshopper, the user can, with a high degree of precision and efficiency, develop scripts that produce complex geometry that would be cumbersome to do manually. It could be argued that Grasshopper is the go-to tool in terms of Algorithmic Aided Design (ADD) and has by far the largest community support, compared to its competitors. Furthermore, Galapagos is an optimization routine, also developed by Rutten [4], which is embedded within Grasshopper. Galapagos enables the user to define a set of input parameters which the algorithm can manipulate. Furthermore, these parameters are usually inputs to a problem which needs optimization. For Galapagos to learn which combination of the parameters yields the best solution, a fitness criterion needs to be defined. The fitness criterion discusses the performance of that particular iteration, given the associated parameter values. Subsequently, when the algorithm is set up as described, the routine can be set in motion. Before starting, Galapagos asks whether the fitness score should be minimized or maximized. Finally, the user can choose from two different optimization techniques, evolutionary solver or simulated annealing. Evolutionary solver is by far the most common method and is also the one applied to the relevant cases in this paper.

Karamba3D is an FEA plug-in for Grasshopper [5] which enables scripting of structural analysis in the same manner as is regularly done with geometry within Rhinoceros. It is very "lightweight", providing fast solutions to structural analysis problems defined by geometry and information programmed in a Grasshopper script. This capability makes it a prime candidate to be paired with generative design routines like Galapagos. Exploiting this capability will be further studied in this paper. This study focuses on the application of generative design, using Karamba3D, Galapagos and Grasshopper for the optimization of structural members.

2.1 Collaboration of BIM models with digital tools at conceptual design phase

Conceptual design is the starting point of the structural design process, at which design requirements given in the relevant standards (i.e., strength, durability, robustness, and safety) and the client's demands are taken into account. During the modelling process at the conceptual stage, it is vital to have a fruitful collaboration with the architect, structural engineer, and client, to discuss all the various considerations, to achieve the optimal design. However, it has been challenging to achieve productive collaboration during the conventional structural design process because of the lack of information/data exchange and poor visualization of the end product. In this case, the use of BIM-based structural design enhances the collaboration among different participants in a project. Moreover, currently, BIM-based structural design has been widely-adopted in construction practice, to overcome such challenges and to achieve customer satisfaction. Furthermore, the use of BIM during structural design results in systematic modelling processes, a powerful interactive visualization platform, and standardized data exchange interfaces during the life cycle of the construction. However, the benefits of using generative design for structural optimization and integrating of BIM have not been well understood because of the lack of knowledge about workflows and the use of different digital tools. Particularly during the conceptual design phase, as given in Fig. 2, digital tools can be used to implement a generative design approach to achieve optimized structural solutions while collaborating with BIM models.

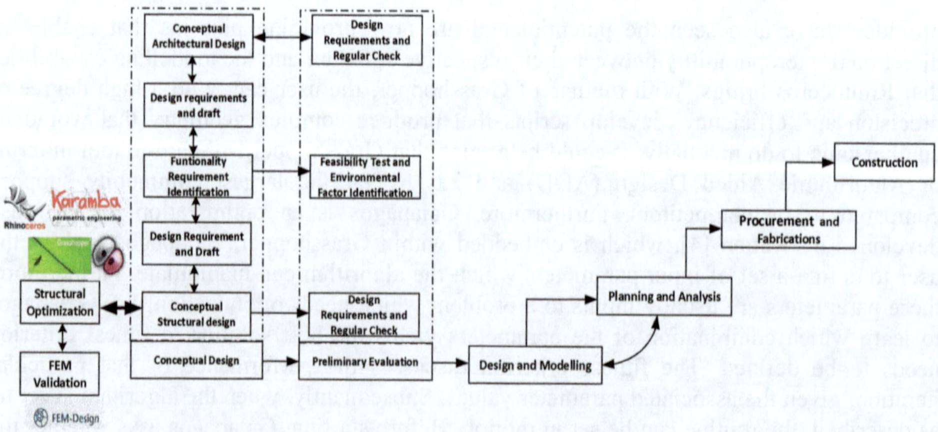

Figure 2: BIM-enabled structural design in construction. (Source: Modified from Chi et al. [6]).

3 ILLUSTRATIVE CASES TO IMPLEMENT GENERATIVE DESIGN FOR STRUCTURAL OPTIMIZATION

To study the potential of structural optimization algorithms, a set of cases has been constructed. These cases were developed with input and guidance from industry professionals and stem from real cases they have experienced in their work. To study these cases, a flowchart has been established, as given in Fig. 3. According to Fig. 3, case description gives a brief description of the problem at hand which includes an explanation of the structure and why the problem is suitable for optimization. Then, to being the optimization, it is necessary to idealize the case and set limitations. Commonly, structural engineers idealize the structure when performing the structural analysis; the case-specific idealizations that have been made, in addition to the limitations of the procedure, will be described during this stage. During this process, it is vital to set the optimization goal, based on the target outcome.

For each optimization problem, the algorithm needs a set of inputs at the beginning. Furthermore, these inputs are classified as either static or variable. Static inputs are parameters that remain unaltered through the optimization process, effectively functioning

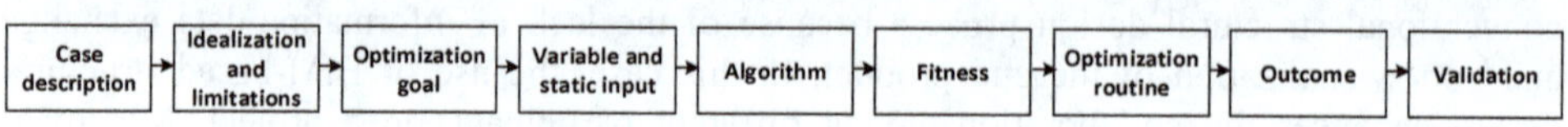

Figure 3: Flowchart for optimization of structural members.

as the constraints to the given problem. Variable inputs, however, are the parameters the optimization algorithm can alter in the pursuit of the most optimal solution. Whether the inputs are static or variable will depend on the optimization goal and are therefore case-specific. Moreover, algorithms can be used to produce an array of design options based on the requirements set at the beginning. The performance of the structure will be a sum of parameters deemed important for the value of the given iteration. Parameters such as passing the preliminary Serviceability Limit State (SLS) check, as well as the total weight of the structures, are examples of ways to measure the performance of the iteration. After the algorithm has run its course, the best solution, with its accompanying genes and fitness, is presented. To validate the result, stand-alone FEA software such as FEM-Design is used. The top solution proposed by the optimization algorithm will be exported through automated workflows enabled in Grasshopper, and more thorough analyses can be undertaken.

4 CASE 1: STRUCTURAL OPTIMIZATION OF 65-M STEEL TRUSS WITH RESPECT TO SELF-WEIGHT

The steel truss shown in Fig. 4(a) is 64.6 m in length, 4.8 m high, and there is 9.6 m distance between adjacent truss members. It has been targeted to optimize the weight of the steel truss using a generative design approach, as discussed in Fig. 3, and using the software given in Section 2.5. In this case, the goal of the optimization was to find the lightest steel truss by satisfying the constraint of 100 mm maximum displacement due to the applied loads. The load used in this analysis is weight of the roof of 1 kN/m^2 and snow load of 2.8 kN/m^2. Moreover, the yield strength of steel in the truss is 355 MPa (S355). During this analysis, one load combination was chosen. However, the analysis can be carried out for all the load combinations given in most of the standards, to obtain an optimized solution.

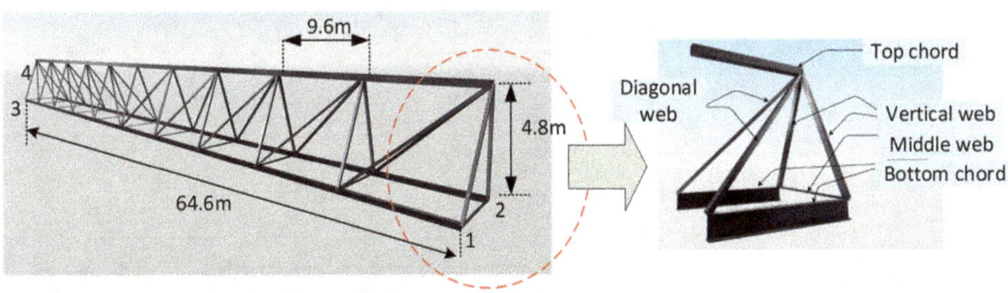

(a) (b)

Figure 4: (a) 3D view of the steel truss; and (b) Different chord in the truss.

Initially, length, height, distance between truss members and loads were assigned as static input to Grasshopper software. Although these static inputs remain static throughout the optimization, they could easily be changed to suit a new and different sized truss with alternative imposed loads. The steel truss was allowed to rotate major axis and supported over pin and roller supports. The steel truss was subjected to loading from both the roof and the snow that could accumulate on it.

It is vital to define variables at the beginning of the analysis in Grasshopper software. In this analysis, input variables, such as number of bays in truss (10–20), height (4.8 m–6.64 m) of the truss, and cross-truss members at top, bottom, middle, vertical and diagonal have been used. The algorithm got two genes to alter, in order to find the most optimal solution: the total number of truss bays and the height of the truss.

To optimize the structure, the algorithm could use a pool of different cross sections (see Fig. 4(b)), the diagonal web members, vertical web members, and middle web members, and the top chord could pick from a list of rectangular hollow sections (RHS), while the bottom chord could pick from a pool of HEB sections. The different structural members, as well as the span of possible cross sections have been given in Table 1.

Table 1: Variable input: cross sections.

Member	Span of values
Diagonal web	[RHS20 × 20 × 2, RHS400 × 400 × 16]
Vertical web	[RHS20 × 20 × 2, RHS400 × 400 × 16]
Middle web	[RHS20 × 20 × 2, RHS400 × 400 × 16]
Top chord	[RHS20 × 20 × 2, RHS400×400 × 16]
Bottom chord	[HEB100, HEB1000]

4.1 Algorithms for optimization of the weight of the steel truss

Based on the input variables summarized previously, a generative algorithm was constructed for optimization of the steel truss via the technical route of the random generation of parameters within the given range and selection by constrained conditions. The algorithm is based on Grasshopper to realize the automatic generation of different steel trusses based on the input variables. The framework of the algorithm is shown in Fig. 5.

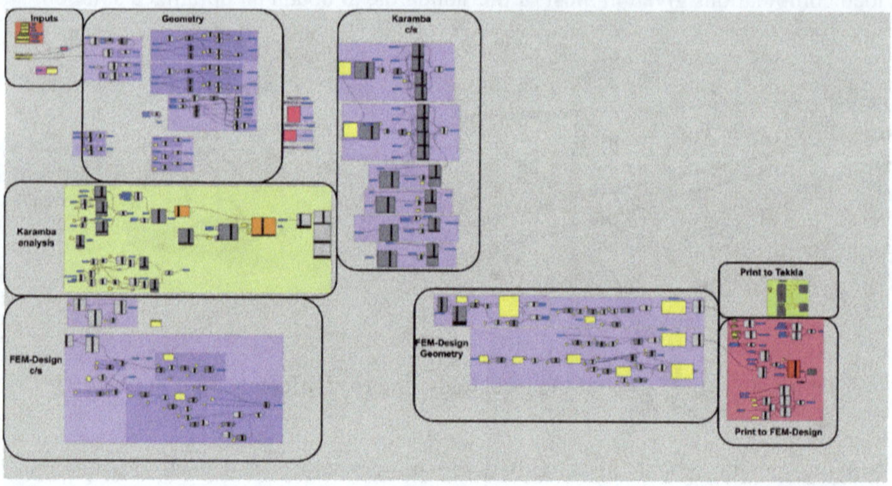

Figure 5: Case 1 – Framework of the algorithm constructed to optimize the weight of the steel truss.

According to Fig. 5, first the lines which will later represent truss members are drawn up, and all the top points where the diagonals intersect the top chord are gathered. Furthermore, all lines are then assigned a random cross section picked from a pool of predefined sections. Subsequently, self-weight and snow load are applied as point loads to the previously gathered top points of the truss. Depending on the number of truss bays, the magnitude of the loads will vary. The location of the support points is set, and supports are made with the appropriate degrees of freedom. Now Karamba3D has enough information to evaluate the structure. When it is finished, Karamba3D determines the best combination of cross section for each member to meet the constraint of maximum displacement while not exceeding the utilization of each member. When all members have been given a new cross section, the total mass of the structure is calculated. The total mass will be the fitness of that particular iteration and is sent back to Galapagos. This is repeated until it finds the lightest structure that still fulfils the requirements. Once Galapagos has found a suitable solution, it is sent to FEM-Design for validation. In addition, if needed, the structure can also be sent to Tekla to populate a BIM model.

4.1 Optimization routine using Galapagos

The optimization routine makes use of the evolutionary solver within Galapagos. The objective of this analysis was to obtain the lightest structure by satisfying the given constraints, i.e., the algorithm was set to minimize the fitness value. The total population of each generation was bound by a maximum of 50 genomes; this should be sufficient to properly explore the fitness landscape, while keeping the runtime of the optimization routine within reasonable limits. The initial boost was set to 4; i.e., generation 0 will have four times as many genomes as the following generations. This was done to thoroughly cover the fitness landscape such that the following generations did not colonize local optima instead of the best global one. With a total runtime of 15 minutes, the evolutionary solver converged towards a solution after 41 generations, yielding a total weight of 77516.56 kg. After running the optimization routine, the fittest genome was found to be when height was 6.64 m and number of bays in truss was 10.

4.2 Optimized cross sections for steel truss

The fittest genome and the steel truss including the applied load were found. Then, Karamba3D assigned the following cross sections to the different structural members in the steel truss, as shown in Fig. 6. The optimized cross sections obtained for different truss members are shown in Table 2 for half of the truss members; member ID is given based on Fig. 6.

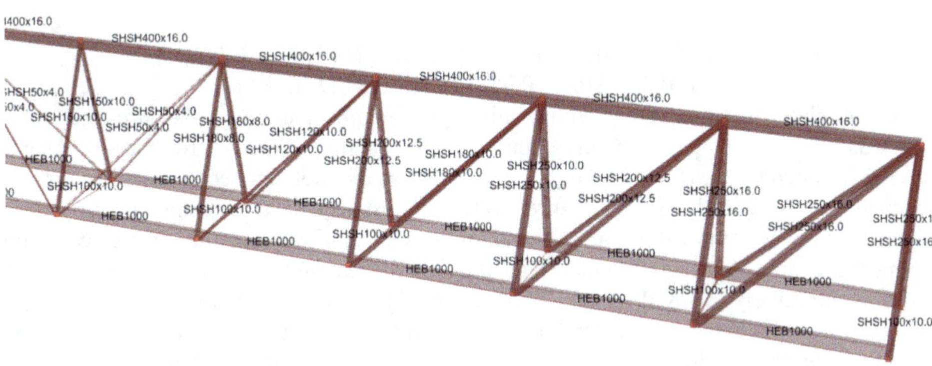

Figure 6: Case 1 – Optimized cross sections in half of the truss.

Table 2: Case 1 – Optimized truss members in half of the truss.

Member ID	Profile
Diagonal 1	SHS250 × 250 × 16
Diagonal 2	SHS200 × 200 × 12.5
Diagonal 3,4	SHS180 × 180 × 10
Diagonal 5	SHS50 × 50× 4
Vertical 1,2	SHS250 × 250 × 16
Vertical 3	SHS250 × 250 × 10
Vertical 4	SHS200 × 200 × 12.5
Vertical 5	SHS180 × 180 × 8
Middle	100 × 100 × 10
Top chord	400 × 400 × 16
Bottom chords	HEB1000

4.3 Validation of the results using FEM-design software

The optimized steel truss was further validated using FEM-Design software. To validate the results, the geometry, loads, cross sections and supports were exported to FEM-Design software. The process was automated through the FEM-Design plug-in for Grasshopper. The results from the SLS load combination yielded a maximum vertical displacement of 98 mm. The results from the ultimate load combination yielded a maximum utilization of 68% (see Fig. 7). After the structural integrity was validated within FEM-Design, the truss was printed to Tekla.

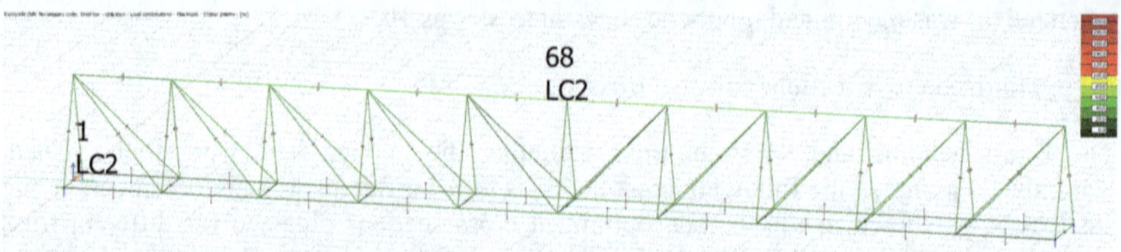

Figure 7: Case 1 – Validated result – Utilization.

5 CASE 2: OPTIMIZATION OF TENDON PROFILE FOR A THREE-SPAN POST-TENSIONED BEAM

To accommodate large open areas within buildings, columns need to be placed further apart to increase the span length. Consequently, long-spanning beams are needed. Using reinforced concrete beams for long span lengths is not an economical solution. Consequently, structural engineers often turn to post-tensioned systems to solve this problem. These systems offer the engineer options to achieve structural integrity while keeping the beams relatively slender. The process of designing post-tensioned beams is performed iteratively, with the main outcome being to determine the necessary cross-sectional area of the beam, the tendon profile, and, finally, the required tendon force. Initially, some assumptions regarding losses, trial cross section and tendon profile are made. Furthermore, the necessary tendon force is calculated to balance the loads vs. the applied forces. Finally, as the geometry of the tendon and the required force are known, the actual losses can be calculated. Compared to the initial assumptions, the engineer can now go back and re-iterate, to optimize the structure further. This process continues until the performance of the solution is deemed acceptable. The following case will present an automated approach to designing a post-tensioned system. The case draws inspiration from a real structure.

5.1 Optimization of tendon profile in pre-stressed concrete beam

A three-span pre-stressed concrete beam (i.e., post-tensioned) with parabolic tendon profile has been studied, as shown in Fig. 8. In reality, the distributed load acting on the beam by the post-tensioned tendon is not uniform, i.e., it varies with the slope. In other words, the idealization that has been carried out in this case is that the distributed loads are uniform within each parabola section. Furthermore, the optimization routine does not include losses in tendon force. In addition, the beam is only subjected to one load combination (SLS) during the optimization. For this case, the optimization goal was to find a tendon profile that needed the least amount of jacking force while still maintaining compression throughout the beam.

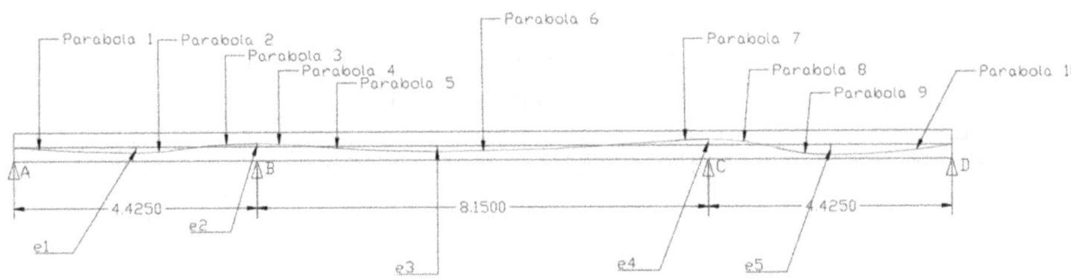

Figure 8: Case 2 – Post-tensioned beam (dimensions in meters).

In this study, span lengths of the post-tensioned beams have been selected as 4.42 m and 8.15 m, as shown in Fig. 8. The post-tensioned beam is supported over four columns (i.e., pinned support), and the parabolic tendon profile consists of ten different parabolas. Moreover, a T-shape beam cross section has been used in this study, as shown in Fig. 9.

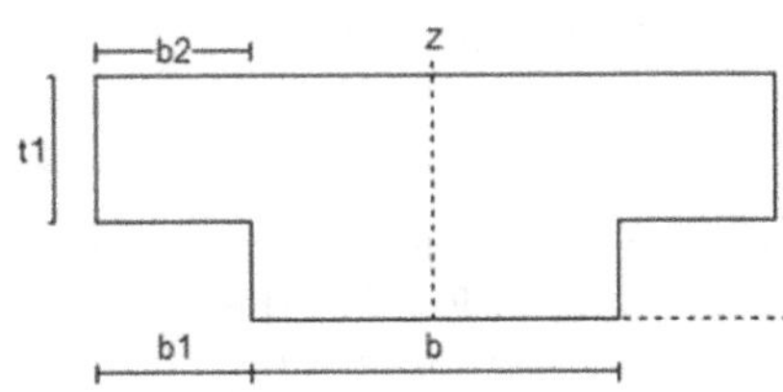

b	700	mm
b1	300	mm
b2	300	mm
h	500	mm
t1	300	mm
t2	0	mm
Zt	284	mm
Areal	5.30E+05	mm2
Iy	9.83E+09	mm4

Figure 9: Case 2 – Static inputs – Cross section of beam.

In addition to the self-weight of the beam, three concentrated loads were placed 3 m from “A” with a magnitude of 210 kN in “AB” span, at 8 m from “A” with a magnitude of 115 kN in “BC” span, and at 14.095 m from “A” with a magnitude of 115 kN in “BC” span. The characteristic compressive strength of concrete has been used as 35 MPa. When designing pre-stressed concrete structures, it is vital to control stresses at SLS. In this case, the post-tensioned beam has been designed in such a way to maintain zero tensile stresses across the cross section of the beam. During the analysis, the partial factors of safety for permanent load:1 and variable load:1 were used in the load combination in the optimization routine.

To be able to optimize the tendon profile, the algorithm can alter a set of variables. To define the shape of the tendon profile, the algorithm can alter the location of a set of points (i.e., e0, e1, e2, etc.) of interest. The variable inputs for the algorithm were set up in such a way that they could alter the locations of the given points of interest by a fraction of the dimensions of the given structure, e.g., span length and cross-sectional height. Doing so meant that it would be a simple task to optimize a different beam by altering the span length and cross section.

5.2 Algorithm for optimization of tendon profile

To optimize the tendon profile using the algorithm shown in Fig. 10, the practical tendon profile described in Gilbert [7] has been used. Gilbert [7] describes how it is possible to idealize the tendon profile as parabolas with a given radius of curvature. Subsequently, the radius of curvature, together with the applied tendon force is directly related to the equivalent uniformly distributed load acting on the beam from the tendon. The algorithm uses the equations making up the approach and makes it available to the optimization routine. Initially, after all variable and static inputs have been set, the algorithm is set in motion.

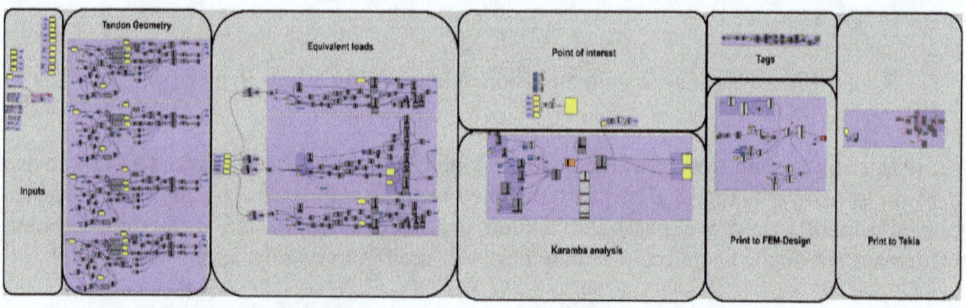

Figure 10: Case 2 – A framework algorithm.

5.2.1 Procedure for constructing algorithm

Gilbert [7] discussed a method to find a practical cable profile when pre-stressed concrete beams consist of different parabolic segments, as shown in Fig. 11. This method was used while developing the algorithm. In this method, initially, the distance (h_{i_1}) which inflection point "i_i" is offset "e_2" is calculated using eqn (1). Then, the algorithm has all the variables it needs to calculate the radius of curvature for the three parabolas, $r_1(parabola\ 1), r_2(parabola\ 2)\ and\ r_3(parabola\ 3)$ in span AB, using eqns (2)–(4) respectively.

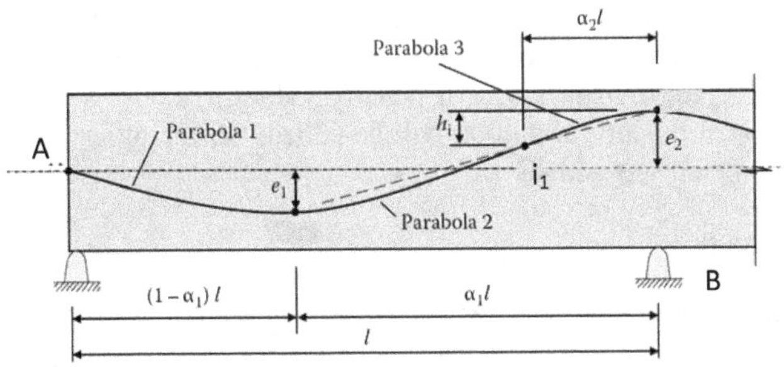

Figure 11: Case 2 – Illustration of tendon profile parabolas [7].

$$h_{i_1} = \frac{\alpha_2}{\alpha_1}(e_1 + e_2). \quad (1)$$

$$r_1 = \frac{l^2(1-\alpha_1)^2}{2e_1}. \quad (2)$$

$$r_2 = \frac{l^2(\alpha_1-\alpha_2)^2}{2(e_1+e_2-h_{i_1})}. \quad (3)$$

$$r_3 = \frac{\alpha_1\alpha_2 l^2}{2(e_1+e_2)}. \quad (4)$$

When the radius of curvature is known, the equivalent uniformly distributed loads can be calculated using eqn (5). Similar equations have been developed for Span "BC" and "CD", and the algorithm has been constructed.

$$q_i = Pk_{P_i}, \quad (5)$$

where P = Effective pressing force in post − tensioned tendon and $k_{P_i} = \frac{1}{r_i}$, i = 1,2,3.

Using the coordinates of the eccentricities and inflection points, together with the now known radius of curvature for each parabola, all parabolas can be drawn as curves in Rhino software. These curves will later be used to print the tendon shape to Tekla software.

5.2.2 Use of Karamba3D

Once the coordinates for each parabola and equivalent loads are calculated, it is necessary to develop the FEA model. Individual beam elements are modelled between each of the parabolas, ten in total. Furthermore, the equivalent loads and point loads are applied. The outcome of the analysis is the stress state at predefined points of interest. These stress states are finally used to evaluate the performance of the structure and enable Galapagos to find the most optimal solutions. Furthermore, the equivalent uniformly distributed loads are now made available to the FEA nodes in the Karamba3D plug-in. Together with the predefined beam elements and boundary conditions, the Karamba3D structural analysis node evaluates stresses in the cross section at 12 points of interest. The stress conditions, together with the applied tendon force, are fed to Galapagos as the fitness for the optimization.

5.3 Fitness

To measure the performance of each iteration, a fitness condition is constructed. In this case, the fitness conditions depend on two different inputs. The first input is the effective pre-stressing force in the post-tensioned tendon, the second is a sum of a condition-based equation determined by the stress condition at the 12 points of interest. For every point with stress above 0 MPa, i.e., tensile stress, a penalty value of 2,000 is added, e.g., if all 12 points are in tension, a penalty of 24,000 will be added. On the other hand, if all points are in compression, no penalty is added.

5.4 Optimization routine

When the variable inputs have been set up and a way to assess the performance of each iteration has been added, Galapagos has enough information to start the optimization routine. Galapagos has control of all the variable inputs and is given instant feedback on the performance through the fitness condition. Consequently, it gradually learns which combination of the variables yields the best performance and, after a set amount of time, presents the best solutions it could find. After a runtime of 45 minutes and 103 generations, Galapagos converged towards a solution only requiring 1,007 kN of force. The magnitude and position of each equivalent load are presented in Fig. 12.

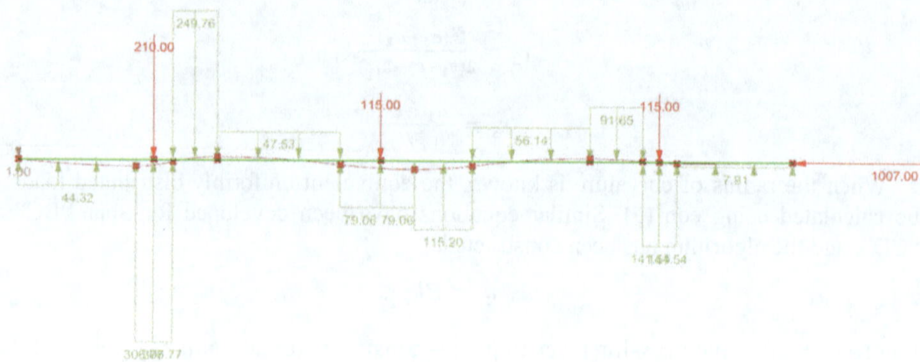

Figure 12: Case 2 – Illustration of loads ("red" color numbers are given in "kN", and "green" color numbers are given "kN/m").

5.5 Validation of the optimized tendon profile

To validate the result, the beam was also analyzed in FEM-Design software. The geometry of the structure, loads and load cases were sent automatically through the FEM-Design Grasshopper plug-in. However, it is not currently possible to model post-tensioned cables through the plug-in. The tendon profile and equivalent forces were therefore manually generated inside FEM-Design with the known locations for the points of interest. After running the analysis, the results given in Fig. 13 were found. According to Fig. 13, not only is the entirety of the beam in compression as intended, the minimum tensile stress is 0 MPa, indicating a well-optimized profile and tendon force. Finally, the geometry was automatically printed to Tekla to populate a BIM model.

6 DISCUSSION

In the pursuit of exploring the capabilities of Algorithmic Aided Design, this paper has presented two cases in which the generative design approach has been used for the

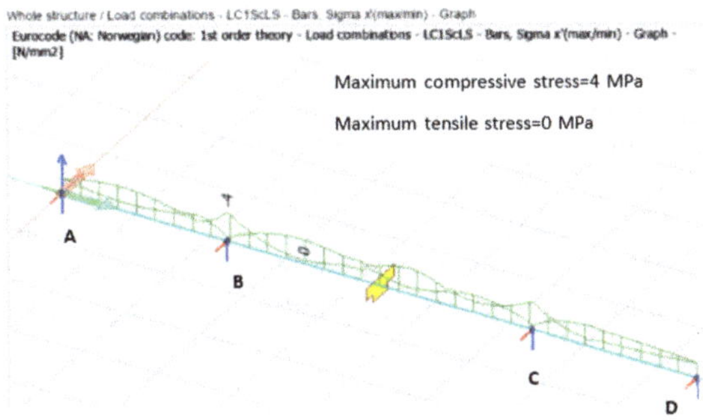

Figure 13: Case 2 – Validation of results.

optimization of structural members. A video demonstrating cases 1 and 2 can be seen at https://youtu.be/wIo5PulGBrM. After developing the scripts and algorithms and obtaining feedback from industrial professionals, the potential use of each case study in practice is discussed below.

6.1 Case 1 – Structural optimization of 65-m steel truss with respect to self-weight

The process of setting up this script provided great insight into the intricacies of using optimization routines. The script was not only set up to optimize the structural members, but it also had to find an efficient way to gather the optimized structure and export it to FEM-Design for validation. Lastly, by using Karamba3D to select the optimized cross sections, the number of free variables, i.e., choosing the profile of each structural member, was greatly reduced. Consequently, the runtime of the optimization routine was significantly accelerated.

6.2 Case 2 – Optimization of tendon profile for a multi-span post-tensioned beam

Developing the script to run the optimization algorithm was quite time-consuming; in fact, the traditional way would probably be faster. However, since the script is set up in such a way that it is easy to change the loadings, dimensions, and span lengths, in future there is a possibility to use the developed script with some modifications for a similar optimization. As a result, significantly less time is needed to set it up. Furthermore, the results show that there is a satisfying correlation between the stress evaluation in Karamba3D and that in FEM-Design.

6.3 Feedback and observations

A questionnaire was sent out to obtain feedback on structural engineers' (i.e., Multiconsult, Norway) point of view about the use of optimization routines. Most of the structural engineers gave positive feedback about the cases studies and considered Generative Design a good tool in the conceptual design phase or in cases where a reduction in material costs is a significant priority. However, some hurdles still stand in the way of using such algorithms, such as the complexity of the scripts and the time needed to define them. However, presently, the structural engineers lack knowledge and training on developing optimization routines. Therefore, proper training will be able to overcome such hurdles in the future.

Both cases 1 and 2 show that it is indeed possible to set up routines to find reasonably good solutions to a problem, given a set of variables and constraints. However, the discussion of these cases was not necessarily about solving the presented problems but on illustrating the degree of automation a structural engineer can enable with the technology currently available. With the ever-growing prevalence of Machine Learning and Artificial Intelligence, automation has and will affect all industries; it is therefore naïve to think that structural engineering will somehow escape this. Consequently, to remain competitive, similar workflows should be explored to harvest the benefits, once the technology has reached adequate maturity. The cases described here can be considered stepping stones to fully automized machine learning algorithms solving structural engineering problems autonomously.

6.4 Proposed workflow based on case 1 and case 2

A workflow has been proposed based on two case studies, as shown in Fig. 14. Initially, static inputs are defined based on the type of optimization problem. Then, variable inputs (genes) and the range of their values should be defined. The inputs define the structural problem which is fed into Karamba3D. Moreover, Karamba3D analyzes the structure and outputs performance parameters in real time. Together with the performance parameters, some of the variable inputs (genes) will often make up the fitness of a given genome. The fitness and variable inputs (genes) are made available to the optimization routine, Galapagos. The optimization routine is run with a given optimization goal, either maximize or minimize. After the optimization routine has finished, the top solution is sent to FEM-Design for validation. After the structure has been validated, calculation reports can easily be generated.

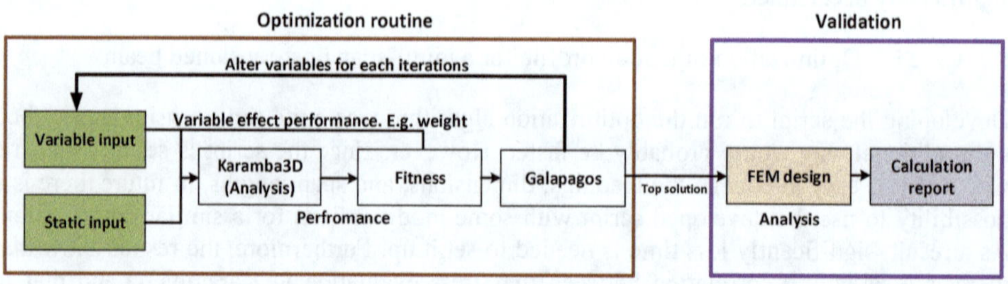

Figure 14: Proposed workflow for optimization of structural members using generative design approach.

7 CONCLUSION

The objective of this paper has been to explore cases, to obtain an understanding of how AAD is currently used, which possibilities it brings, and, lastly, to investigate what the future might hold. By having the analysis model defined parametrically, new and innovative technologies are enabled. Generative design, which is one of these, was explored in cases 1 and 2. Furthermore, these cases were used to verify whether structural optimization routines could produce reasonable solutions. As the verified results indicate, both cases highlight structures that satisfy the given criteria. Machine learning and artificial intelligence will affect all industries in the future; the cases described in this paper give a taste of how structural engineers might deploy computers to tackle the problems of tomorrow. Lastly, as an outcome of this paper, a proposed workflow has been presented. The workflow has been developed through trial and error, when working on the different cases, and inherited characteristics that, when followed, yielded consistent results. Consequently, this workflow can serve as a user guide for newcomers looking to make use of the technologies studied in this paper.

REFERENCES

[1] Banfi, F., Fai, S. & Brumana, R., BIM automation: Advanced modeling generative process for complex structures. *26th International CIPA Symposium on Digital Workflows for Heritage Conservation*, Copernicus GmbH, pp. 9–16, 2017.

[2] Kovacic, I. & Filzmoser, M., Designing and evaluation procedures for interdisciplinary BIM use – An explorative study. *Engineering Project Organization Conference*, Devil's Thumb Ranch: Colorado, pp. 29–31, 2014.

[3] Russell, D., Cho, Y.K. & Cylwik, E., Learning opportunities and career implications of experience with BIM/VDC. *Practice Periodical on Structural Design and Construction*, **19**(1), pp. 111–121, 2014.

[4] Rutten, D., Galapagos: On the logic and limitations of generic solvers. *Archit. Des.*, **83**(2), pp. 132–135, 2013.

[5] Preisinger, C., Linking structure and parametric geometry. *Architectural Design*, **83**(2), pp. 110–113, 2013.

[6] Chi, H.L., Wang, X. & Jiao, Y., BIM-enabled structural design: Impacts and future developments in structural modelling, analysis and optimisation processes. *Archives of Computational Methods in Engineering*, **22**, pp. 135–151, 2015.

[7] Gilbert, I.R., *Design of Prestressed Concrete to Eurocode 2*, 2nd ed., CRC Press, 2017.

SECTION 4
BIM AND CULTURAL HERITAGE

HBIM WORK METHODOLOGY APPLIED TO PREVENTIVE MAINTENANCE: A STATE-OF-THE-ART REVIEW

JORGE GARCÍA-VALLDECABRES[1], ANTONIO GALIANO-GARRIGÓS[2],
LUIS CORTES MESEGUER[3] & MARIA CONCEPCIÓN LÓPEZ GONZÁLEZ[1]
[1]Department of Architectural Graphic Expression, Universitat Politècnica València, Spain
[2]Department of Architectural Constructions, University of Alicante, Spain
[3]Department of Architectural Constructions, Universitat Politècnica València, Spain

ABSTRACT
The consolidation of BIM work environments and the efficient management of construction information are reaching previously forgotten facets of the buildings' life cycle. Preventive conservation and maintenance, which affect the building operation that accounts for 80% of CO_2 emissions, are being subject to an improvement in their procedures based on the digitization of information. These actions are reaching historical buildings where procedures are not only being developed to improve their maintenance, but also in areas such as historical documentation and cultural management. The generation of repositories with documentation of the buildings is making it possible to establish new relationships with society, improving the conservation of buildings and designing actions that avoid future risks. There are different types of maintenance: corrective, preventive and predictive and despite its advantages, the agents in charge of these buildings do not always perform it properly. Furthermore, dysfunctions and lack of coordination are detected as many agents act without having adequate information. To solve these problems, BIM work environments are positioned as a valid alternative since they provide a centralized and updated repository where the information is arranged so that it is available to the different agents when they need it. This information can be historical or technical, including digital models and evaluation of deterioration and intervention processes. The objective of this research is the development of an analysis of the degree of implementation of BIM strategies applied to preventive maintenance in historic buildings and the analysis of the investigations carried out. This work is part of the first results of the research project entitled "Analysis and Development of the integration of BHIM in GIS for the creation of a tourist planning protocol of the cultural heritage of a destination (HBIMSIG-TOURISM)" granted by the Ministry of Research of Spain under the 2020 call for R&D&I Projects oriented to the Challenges of the Society with the reference PID2020-119088RB-I00.
Keywords: preventive conservation, management, BIM, HBIM, heritage.

1 INTRODUCTION

The protection and conservation of cultural heritage has been established as one of the necessary actions to be developed by modern societies in the process to define a feeling of belonging [1]. Historical buildings, as elements of an urban fabric, are part of a mental map that people develop which allows them to orient themselves in the city and in the circuits that it has [2]. For that matter, the cultural activity in which heritage participates becomes part of the cultural structure of the city and helps to put it in value by improving the relationship between the physical environment and the human perception of spaces [3]. Moreover, historical heritage helps the balance between the activities that occur in public spaces by being able to generate poles of attraction based on the people who are attracted by the interest that the building generates. In many cases, these buildings contribute to generating a balance in the activities that take place in the urban fabric, demonstrating the quality that it has [4].

However, to have a building considered historical heritage participate in the city activity, it must be kept in an ideal state of maintenance, allowing it to develop the use for which it was designed or to have the necessary modifications to adapt to a new one. The Athens

WIT Transactions on The Built Environment, Vol 205, © 2021 WIT Press
www.witpress.com, ISSN 1743-3509 (on-line)
doi:10.2495/BIM210131

Charter of 1931 also states that there is no better tool for the conservation of heritage than affection and respect by people, in addition to recommending assigning useful uses for these buildings. More recently, in the 2000 Krakow Charter, the spirit of the Venice Charter is ratified and its first two objectives specify and highlight the importance of heritage maintenance as an elemental part of the process of preserving a property. The National Plans for the conservation of heritage usually recommend improving the training and information of the stakeholders participating in the process of exploitation and conservation of this type of buildings. At the same time, it is needed the improvement of the information available to potential visitors interested in knowing the characteristics of the historical heritage at different levels.

Developing preventive conservation plans (CP) is often considered a sign of maturity in a sustainable process of managing cultural heritage. The CP could be defined as a set of actions that aims to preserve heritage by ensuring its accessibility by future generations. The CP also defines actions which allow to perpetuate the useful life of this type of buildings in optimal conditions, as well as designing conservation–restoration actions and defining rules to allow them to be enjoyed by interested citizens without causing deterioration [5]. These PC plans must be flexible and sequential, enabling a strategy that requires a schedule with short, medium and long-term objectives [6]. In this sense, the National Preventive Conservation Plan in Spain [7] has been developed with the objective of developing research actions to identify and analyze risks, conservation methods and techniques. It must also identify criteria and work methods, coordination of the actions, the optimization of the resources and the way to access information, as well as the revision of collections and funds [8] (Fig. 1).

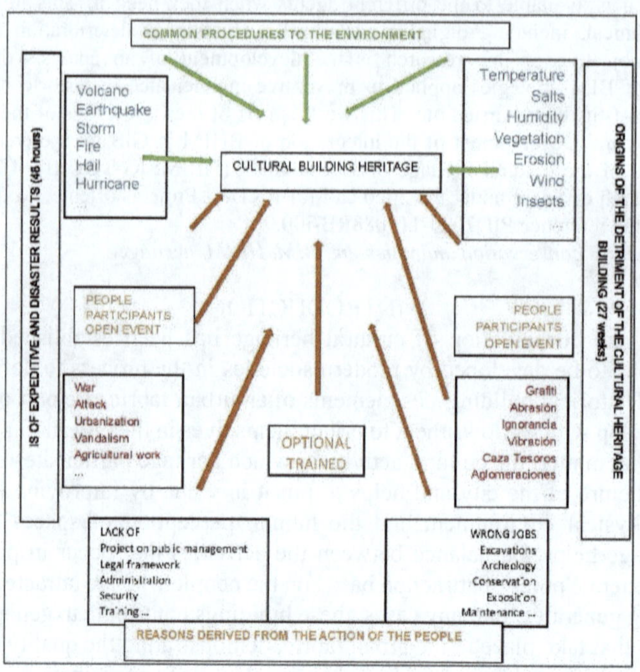

Figure 1: Causes of deterioration of cultural property. *(Source: Own interpretation of painting by Guichen [5].)*

Maintaining cultural heritage in adequate conditions allows improving its presence at the level of society and its inclusion in cultural circuits, generating conservation conflicts due to an excessive influx of visitors, even causing deterioration (Fig. 2). For this reason, it is necessary to include visitor management in preventive conservation plans and develop an adequate management of visits [9].

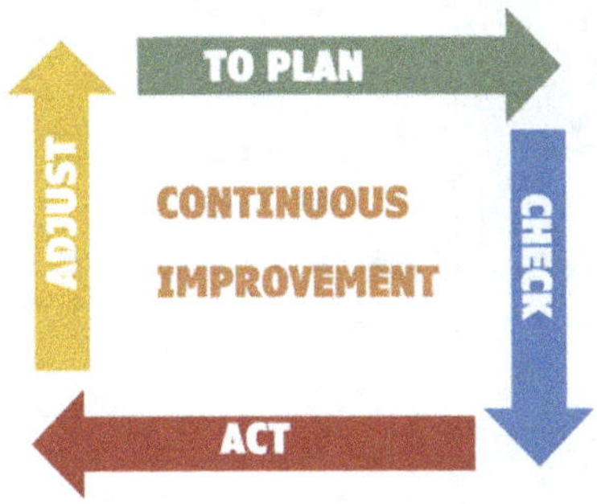

Figure 2: Application cycle of the PDCA Systematics or Deming circle. *(Source: Own interpretation of the CP National Plan [7].)*

The CP's objective is to maintain the necessary conditions to preserve heritage in use. These conditions are lost when maintenance tasks or planned restoration processes are abandoned, and deterioration may appear. It can be assured that deterioration processes are normally produced by interlocking factors that affect the enclosure or the structure of the building. Sometimes they have their origin in a lack of adaptation to improve its functionality. That is why a CP plan must define research objectives to identify risks, work methods and interventions, coordinate actions, optimize resources and provide a centralized and updated repository with the heritage available information. In any case, this process must begin with a complete survey with the most advanced digital tools [10].

In addition, a structured analysis of the available information can allow the development of diagnoses based on the data collected in different periods of the history of the building [11]. However, the simulation of these historical phases becomes a complex process due to the low level of detail of the information that can be managed and visualized through digital CAD representations [10].

Moreover, the implementation of building information modeling (BIM) methodologies in the construction sector is a significant improvement in the management of construction information. BIM pursues the generation of a centralized repository of construction information accessible to the different stakeholders and available during the building life.

The possibilities that BIM is offering for the management of building information is being extended to historical heritage through HBIM (heritage building information modeling). This discipline is offering the possibility of creating centralized repositories in which not only architectural information is managed. All the actions that have been carried out over the years on a building may be completed with information related to the management of visitors and preventive maintenance. A BIM model, reinforced by the detailed information it contains, is an excellent tool for monitoring and evaluating the performance, operation and deterioration of heritage buildings, collecting and classifying various data that can coexist in a dynamic model that can be updated over time.

The use of new information technologies is also making it possible to carry out simulations on the future buildings performance, supporting the decision-making process and analyzing future results that may affect the building life cycle [12]. These simulations can be considered in the heritage maintenance process and may be part of the CP plans (Fig. 3).

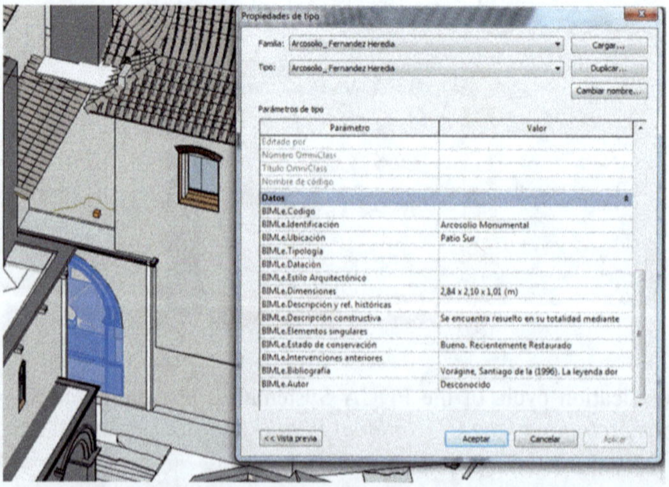

Figure 3: Partial view screen capture and H-BIM Revit model arch-burial properties box.

However, it is considered that the implementation of new technologies that belong to the HBIM environment applied to the Preventive Conservation of buildings is in an incipient state. It is still needed more scientific literature that allows supporting the development of new conservation plans based on previous experiences from other researchers.

2 OBJECTIVE AND METHODOLOGY

The objective of this research is to carry out a systematic review of the existing literature that relates the use of any type of HBIM strategies with preventive conservation plans applied to heritage. To carry out this research, a methodology has been designed that begins with the search for research examples about the above-mentioned themes. It has also been included in the search synergies among HBIM and building rehabilitation, life cycle, heritage information management and preventive conservation. During the search databases such as Web of Science, Scopus and Google Scholar were used. Keywords used during the search of articles included HBIM, LCA, conservative maintenance, building restoration, and preventive conservation.

The contributions found have been organized according to their affinity to one of these groups: HBIM and building restoration, HBIM and life cycle analysis, HBIM and heritage information management, and finally HBIM and preventive conservation. Once the different contributions have been analyzed in a systematic way, research strategies are extracted to finish with a discussion of the results and the conclusions.

3 ANALYSIS OF AVAILABLE BIBLIOGRAPHY

As it was previously mentioned, a systematic search is carried out focused on research that considers the preventive conservation of buildings and the use of new technologies in the HBIM environment. The contributions found have been divided into four groups in which the use of HBIM acts as a common element and is related to three areas present in preventive conservation such as building restoration, information management and life cycle analysis. Finally, the analysis of contributions focused on the development of HBIM with preventive conservation plans is carried out (Fig. 4).

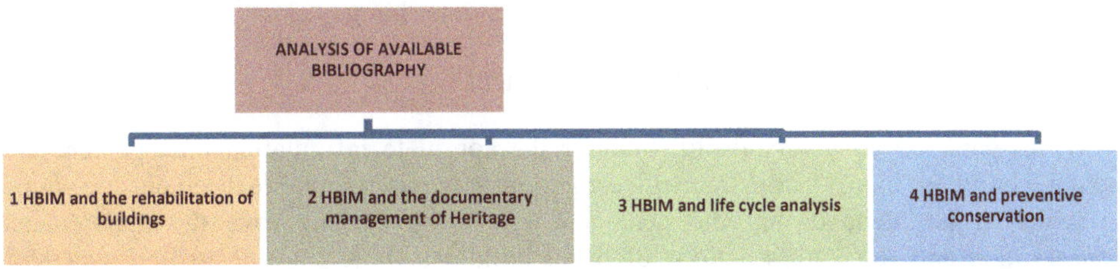

Figure 4: Outline of the search results.

3.1 HBIM and the buildings restoration

The implementation of HBIM in building restoration processes is carried out through case studies in many of the research examples analyzed. The absence of general guidelines makes researchers show their findings through applied research. In the *Heritage Building Information Modelling* book [13] the authors confirm the lack of prototypes focused on Heritage-BIM that can be used to contrast the experiences acquired and the lack of dissemination of the case studies carried out. In addition, it is worth noting the lack of research that allows distinguishing the particularities between BIM and HBIM, concluding the need for research in this area.

A special focus must be done over the interventions carried out for the restoration of buildings that try to generate synergies between different disciplines though the information management platforms of the BIM environment but adapted to the particularities of heritage buildings, such as the case of the rehabilitation of the Basilica of Collemaggio in L'Aquila (Italy) [14], [15].

It is possible to find contributions that define intervention strategies on listed buildings under restoration processes in which, based on 3D representations, it is possible to establish the interoperability of the information between the different disciplines involved on the construction process [16].

Technologies of HBIM environment have also been used to assess the historical values of buildings, such as the Cathedral of Cristo Rey in the city of Huejutla de Reyes, Hidalgo, Mexico [17].

It is also possible to find theoretical approaches in which intervention strategies and enrichment of the models are defined, even by comparing examples located in different countries [18].

These experiences not only focus on restoration actions on heritage buildings. It has been found the analysis of case studies from the mid-20th century, examples of modern architecture, such as the Glass House by Lina Bo Bardi. This experience, as occurs with older buildings, has allowed the development of a work methodology to collect, process and manage building information that could be applied in buildings with similar characteristics [19].

Among the contributions that do not focus on the analysis of a case study, those that analyze the strategies carried out during restoration interventions stand out. It is worth highlighting the integration of data collection through photogrammetry and laser scanner with BIM work methodologies. Some authors consider that this integration provides an improvement in heritage management and allows the use and analysis of spatial information by GIS [20], [21]. It is also possible to find critical analysis of the available literature that broaden the spectrum of information management in intervention processes, extending them to the evaluation and diagnosis of existing damages in buildings [22].

3.2 HBIM and the heritage information management

BIM has emerged as a very useful tool for the complete information management of a heritage building. In this sense, many investigations have been found focused on case studies in which it is intended that the generated repository can improve the maintenance, conservation and restoration actions of listed buildings. It is possible to find successful actions in buildings such as the Cathedral of Parma (Italy) [23], castles in the Piedmont region (Italy) in which the usefulness of multimedia information for Heritage management is shown [24], the Wang Stave Church in Karpcz (Poland), which demonstrates the effectiveness of document management through HBIM platforms in terms of the spatial documentation of the building [25] or the Cartuja de Jerez [26].

3D models created from point clouds surveys and their integration into HBIM modeling platforms is the first action carried out in most of the analyzed research [10] and always attending to the information management of a specific case study [27]–[29].

There are research projects in which among their conclusions it is possible to find that the 3D model is the concept that allows associating the building information management with the heritage use [30], [31]. These digital surveys are on many occasions enriched with parametric objects, introducing artistic, historical and conservation properties, allowing their exchange with other agents through open formats [32]. It is also possible to find experiences that relate the physical with the virtual by developing Digital Twin techniques for the interpretation of heritage [33].

The generated repositories can be used for other activities related to heritage management such as its use from the cultural point of view and the evaluation of the influence of visitors on heritage conservation [34]. It is very important to have models that, in addition to collecting the geometric characteristics of the building, allow different layers of information about the building state of conservation, the performed interventions and historical periods in which they have been carried out. In addition, it has been found interesting the possibility of integrating all the available information into navigable visual environments. This virtual representation makes any restoration project, preventive conservation, management or dissemination of a cultural asset become a virtual model-mock-up that allows its visit without the need to be in the building [35].

Furthermore, these 3D surveys allow obtaining crucial information to undertake maintenance and restoration works, providing information that cannot be obtained by means of 2D images [36]. It is worth mentioning the development of research that allows the management of these repositories that are characterized by their large size and offer limitations regarding the hardware to be used [37]. At the same time, it is the interest of some researchers to develop standards that allow better interoperability in many cases related to OpenBIM [38]. In addition, limitations are also identified in the management and exchange of information, developing applications that optimize access to existing documentation by the agents involved at each moment of the useful life of the building [39]. The evaluation of digital repositories that collect information on heritage buildings in a given place is also the object of research by some authors, in which strategies are also defined for the development of HBIM platforms [40].

3.3 HBIM and life cycle analysis

The life cycle assessment applied to heritage buildings has been the subject of debate due to the impact that the actions necessary to improve the buildings performance may have on protected elements [41], [42]. It is agreed the need of updating this type of buildings as any functional improvement has both social and environmental repercussions [43]. However, the need for the development of specific strategies for life cycle analysis applied to heritage building has been demonstrated in order to make the best decision in the restoration process [44]–[46]. In this sense, the evaluation of strategies to improve the maintenance of this type of buildings are developing new strategies to improve, among other areas, the sustainability of the building [47].

The tools that are integrated into the BIM platforms have allowed a better analysis of the building in the different stages of the architectural project, providing support to the specialists involved in the decision-making process [48]. The use of tools from the BIM environment for the analysis of the life cycle of buildings has also reached heritage buildings. The development of strategies based on HBIM has shown that the integration of geometric and historical data with environmental considerations supports broader issues related to risk management during the maintenance of historic buildings in the context of climate change [49].

The 3D surveys necessary for the generation of the HBIM models have been shown to be a very interesting tool in the process of a global renovation of the building, although the limitations in the libraries limit the data collection. The need for the development of non-invasive data collection has been positively demonstrated with the aim of improving the intervention on listed buildings, especially at the time of improving their energy consumption [50].

3.4 HBIM and preventive conservation

The development of strategies for the heritage conservation through PC plans is under research by the scientific community and the digitization and the generation of virtual repositories, under the HBIM environment, begin to be incorporated [51].

The repositories that are generated become the center of all the information related to the heritage building, which can be consulted in an intelligent way and can be used for the building investigation, conservation, management and administration [8], [52]. However, three main issues must be addressed: lack of standardization, insufficient interoperability, and inherent complexity of information. Balancing the geometric and non-geometric characteristics of the model, such as the level of detail precision and the amount of information linked, it is essential for the methodology to be cost-effective and therefore more attractive to end users [38]. It is possible to find experiences in which, through automation and the use of databases, responses to situations found in this type of building are automated [53].

There are experiences in which, based on case studies, protocols have been developed for the maintenance of heritage buildings based on historical documentation, evaluation of the state of conservation, planning of future actions, maintenance and cultural diffusion [54], [55]. As previously described, many of these PC plans are designed based on the integration of information and interoperability. Subsequently they are validated on a case study so that their suitability or need for improvement can be evaluated [52], [56]. On other occasions, HBIM repositories are used to help in the decision-making process by defining hierarchies in the information levels that allow organizing the importance of deterioration and favoring the establishment of priorities [57].

However, there are investigations that, aware of the limitations that a 3D repository can have for preventive conservation, design databases which are related to the repository to allow better management of the building by all types of stakeholders. These tools are tested in case studies of the importance of the Cathedral of Parma or the Ducal Palace of Mantua (Italy) [58]. It is also possible to find research projects focused on the development of strategies and guidelines for the development of PC plans based on HBIM [59].

4 DISCUSSION OF RESULTS

The analysis of the references found has allowed its organization into four large groups, although it is possible to obtain the following general conclusions (Fig. 5).

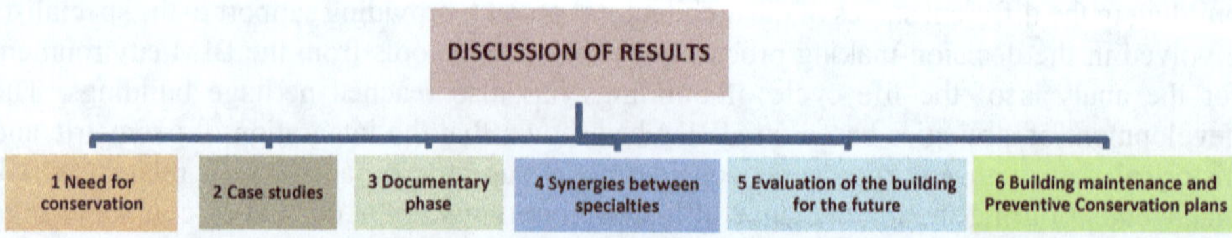

Figure 5: Outline of the discussion of results.

4.1 Need for conservation

All the obtained contributions coincide in affirming the need to preserve heritage, treating it as a social value that can only improve societies in which they are located. This situation justifies the development of present and future actions that must focus on the maintenance of this type of buildings over time, allowing access to information by different stakeholders throughout their useful life.

4.2 Case studies

There are very few examples that show strategies for the development of interventions in heritage buildings based on HBIM strategies that are the consequence or product of mature processes resulting from previous research. Most contributions are based on case studies and are a first approach to the problem, not allowing to establish a trajectory in the research process. In many cases we speak of very recent contributions in which the short trajectory has not allowed to verify the outcomes beyond the process of restoration of the building. It is worth noting the existence of experiences applied in buildings with architectural value that are not necessarily historical.

4.3 Documentary phase

Practically all the contributions have their origin in the preparation of a 3D model that collects the geometric characteristics of the building. In many cases, this is basically the work carried out. The enrichment of the model with other layers of information is proposed as options, announcing its usefulness for other tasks. It is worth noting the existence of experiences dedicated to facilitating the management of the information collected in the model through platforms and through the use of free software that can guarantee sustained accessibility over time.

4.4 Synergies between specialties

The analysis of the examples found mostly highlights the potential that HBIM has for the collaborative and multidisciplinary development of repositories in which all the information regarding the building under study is collected. In some cases, experiences are demonstrated, although in many cases they are limited to the preparation of the 3D survey and the development of restoration work.

4.5 Evaluation of the building in the future

Among the actions that are considered in the previous paragraph it can be found the life cycle analysis applied to the buildings under study. It should be noted that although there is a lot of literature that relates BIM with LCA, it is still not possible to find many experiences applied to buildings with historical values. On the other hand, it is possible to find studies that evaluate the suitability of the HBIM digital repositories generated in the decision-making process. This task is critical, especially in the early stages of the development of the project, since it can condition the viability of the actions that can affect the architectural values that define a listed building.

4.6 Building maintenance and preventive conservation plans

As it was above mentioned, most of the literature found focuses on the design of experiences and their validation on case studies. Only a few examples have been found in which the research has focused on the design of a PC plan based on an HBIM repository but which ultimately requires validation. This situation shows the lack of maturity of the discipline and the need for time so that these plans can be developed over time.

5 CONCLUSIONS

The analysis of the obtained results in the research shows the interest of a part of the scientific community to improve the intervention processes in heritage buildings and the design of procedures that guarantee their conservation over time. In this sense, the positive synergies between HBIM and the different disciplines that can act during the useful life of these buildings are confirmed. The generation of information repositories with different layers is understood as an opportunity that guarantees a multidisciplinary approach but that is encountering limitations in terms of how to manage them.

The absence of long-term research confirms the novelty of the research field and the need for new proposals that can confirm or modify those already carried out. In this sense, it is worth highlighting the need to generate proposals for Preventive Conservation plans that can integrate different disciplines and that take advantage of the advances in the digitization of construction provided by the tools of the HBIM environment.

ACKNOWLEDGEMENTS

This work is part of the first results of the research project entitled “Analysis and Development of the integration of BHIM in GIS for the creation of a tourist planning protocol of the cultural heritage of a destination (HBIMSIG-TOURISM)” granted by the Ministry of

Research of Spain under the 2020 call for R&D&I Projects oriented to the Challenges of the Society with the reference PID2020-119088RB-I00.

REFERENCES

[1] Muñoz Cosme, A., Arquitectura y Memoria. El Patrimonio Arquitectónico y La Ley de Memoria Histórica. *Patrim. Cult. España*, **1**, 2009.
[2] Chapman, E.H. & Lynch, K., The image of the city. *J. Aesthet. Art Crit.*, **21**(1), 1962. https://doi.org/10.2307/427643.
[3] Hajer, M. & Reijndorp, A., *In Search of New Public Domain*, 2001.
[4] Gehl, J., *Life Between Buildings: Using Public Space*, Danish Architectural Press: Copenhagen, 1971.
[5] de Guichen, G., Conservación Preventiva:¿en Qué Punto Nos Encontramos En 2013? *Conservación preventiva: revisión de una disciplina*, Patrimonio Cultural de España, pp 17–19, 2013.
[6] Gómez, M. & de Tapol, B., Medio Siglo de Conservación Preventiva. Entrevista a Gaël de Guichen. Entrevista Realizada Por El Comité Científico Técnico Del GEIIC (Marisa Gómez y Benoît de Tapol) En Agosto de 2009. *Ge-conservación,* pp. 35–44, 2009.
[7] Antonio Herráez, J., Durán, D. & García Martínez, E., Planes Nacionales Conservacion Preventiva, 2017.
[8] Besana, D., Cultural heritage design: Theories and methods for the project complexity management. *EGE-Expresión Gráfica en la Edif.*, 11, p. 31, 2019. https://doi.org/10.4995/ege.2019.12864.
[9] García Hernández, M. & De la Calle Vaquero, M., Geographic research of tourism in Spain. *An. Geogr. la Univ. Complut.*, **24**, pp. 257–327, 2004.
[10] Almagro, A., Medio Siglo Documentando El Patrimonio Arquitectónico Con Fotogrametría. *EGE Rev. Expresión Gráfica en la Edif.*, 2019.
[11] Martín Talaverano, R.., Cámara Muñoz, L. & Murillo Fragero, J.I., Análisis Integrado de Construcciones Históricas: Secuencia Estratigráfica y Diagnóstico Patológico. Aplicación En La Iglesia de Santa Clara (Córdoba). *Arqueol. la Arquit.*, **15**, p. 67, 2018. https://doi.org/10.3989/arq.arqt.2018.001.
[12] Galiano-Garrigós, A., González-Avilés, Á., Rizo-Maestre, C. & Andújar-Montoya, M., Energy efficiency and economic viability as decision factors in the rehabilitation of historic buildings. *Sustainability,* 2019. https://doi.org/10.3390/su11184946.
[13] Arayici, Y., Counsell, J., Mahdjoubi, L., Hawas, S. & Dewidar, K., *Heritage Building Information Modelling*, Routledge, 2017.
[14] Oreni, D., Brumana, R., Della Torre, S., Banfi, F., Barazzetti, L. & Previtali, M., Survey turned into HBIM: The restoration and the work involved concerning the Basilica Di Collemaggio after the earthquake (L'Aquila). *ISPRS Ann. Photogramm. Remote Sens. Spat. Inf. Sci.*, **2**(5), pp. 267–273, 2014. https://doi.org/10.5194/isprsannals-II-5-267-2014.
[15] Brumana, R., Della Torre, S., Previtali, M., Barazzetti, L., Cantini, L., Oreni, D. & Banfi, F., Generative HBIM modelling to embody complexity (LOD, LOG, LOA, LOI): Surveying, preservation, site intervention – The Basilica Di Collemaggio L'Aquila). *Appl. Geomatics*, **10**(4), 5pp. 45–567, 2018. https://doi.org/10.1007/s12518-018-0233-3.
[16] Nieto, J.E., Moyano, J.J., Rico Delgado, F. & Antón García, D., Management of built heritage via HBIM project: A case of study of flooring and tiling. *Virtual Archaeol. Rev.,* 7(14), p. 1, 2016. https://doi.org/10.4995/var.2016.4349.

[17] Mandujano Rodríguez, M.G., Integration of historic building information modeling and valuation approaches for managing cultural heritage sites. *27th Annu. Conf. Int. Gr. Lean Constr. IGLC 2019*, pp. 1433–1444, 2019. https://doi.org/10.24928/2019/0253.
[18] Moreira, A., Quattrini, R., Maggiolo, G. & Mammoli, R., HBIM methodology as a bridge between Italy and Argentina. *Int. Arch. Photogramm. Remote Sens. Spat. Inf. Sci.: ISPRS Arch.*, **42**(2), pp. 715–722, 2018. https://doi.org/10.5194/isprs-archives-XLII-2-715-2018.
[19] Cuperschmid, A.R.M., Fabricio, M.M. & Franco, J.C., HBIM development of a Brazilian modern architecture icon: Glass house by Lina Bo Bardi. *Heritage*, **2**(3), pp. 1927–1940, 2019. https://doi.org/10.3390/heritage2030117.
[20] Yang, X., Grussenmeyer, P., Koehl, M., Macher, H., Murtiyoso, A. & Landes, T., Review of built heritage modelling: Integration of HBIM and other information techniques. *J. Cult. Herit.*, **46**, pp. 350–360, 2020. https://doi.org/10.1016/j.culher.2020.05.008.
[21] Ferreira-Lopes, P. & Pinto-Puerto, F., GIS and graph models for social, temporal and spatial digital analysis in heritage: The case study of ancient kingdom of Seville late gothic production. *Digit. Appl. Archaeol. Cult. Herit.*, **9**, pp. 1–14, 2018. https://doi.org/10.1016/j.daach.2018.e00074.
[22] Bruno, S., De Fino, M. & Fatiguso, F., Historic building information modelling: Performance assessment for diagnosis-aided information modelling and management. *Autom. Constr.,* **86**, pp. 256–276, 2018. https://doi.org/10.1016/j.autcon.2017.11.009.
[23] Bruno, N. & Roncella, R., A restoration oriented HBIM system for cultural heritage documentation: The case study of Parma cathedral. *Int. Arch. Photogramm. Remote Sens. Spat. Inf. Sci.: ISPRS Arch.*, **42**(2), pp. 171–178, 2018. https://doi.org/10.5194/isprs-archives-XLII-2-171-2018.
[24] Osello, A., Lucibello, G. & Morgagni, F., HBIM and virtual tools: A new chance to preserve architectural heritage. *Buildings*, **8**(1), pp. 1–12, 2018. https://doi.org/10.3390/buildings8010012.
[25] Sztwiertnia, D., Ochałek, A., Tama, A. & Lewińska, P., HBIM (heritage building information model) of the Wang Stave Church in Karpacz: Case study. *Int. J. Archit. Herit.*, **15**(5), pp. 713–727, 2021. https://doi.org/10.1080/15583058.2019.1645238.
[26] Castellano-Román, M. & Pinto-Puerto, F., Dimensions and levels of knowledge in heritage building information modelling, HBIM: The model of the charterhouse of Jerez (Cádiz, Spain). *Digit. Appl. Archaeol. Cult. Herit.*, **14**, 2019. https://doi.org/10.1016/j.daach.2019.e00110.
[27] Chiabrando, F., Sammartano, G. & Spanò, A., Historical buildings models and their handling via 3D survey: From points clouds to user-oriented HBIM. *Int. Arch. Photogramm. Remote Sens. Spat. Inf. Sci.: ISPRS Arch.*, **41**, pp. 633–640, 2016. https://doi.org/10.5194/isprsarchives-XLI-B5-633-2016.
[28] López, F.J., Lerones, P.M., Llamas, J., Gómez-Garcia-Bermejo, J. & Zalama, E., Linking HBIM graphical and semantic information through the Getty AAT: Practical application to the Castle of Torrelobatón. *IOP Conf. Ser. Mater. Sci. Eng.*, **364**(1), 2018. https://doi.org/10.1088/1757-899X/364/1/012100.
[29] Chiabrando, F., Lo Turco, M. & Rinaudo, F., Modeling the decay in an HBIM starting from 3D point clouds: A followed approach for cultural heritage knowledge. *Int. Arch. Photogramm. Remote Sens. Spat. Inf. Sci.: ISPRS Arch.*, **42**(2W5), pp. 605–612, 2017. https://doi.org/10.5194/isprs-archives-XLII-2-W5-605-2017.

[30] Bulgarelli-Bolaños, J.P., Hernández-Salazar, I. & Pinto-Puerto, F., Evolución de La Producción Científica Sobre Los Conceptos HBIM y Modelado 3D En La Gestión de Obras Patrimoniales. *Rev. Tecnol. en Marcha*, pp. 89–101, 2020. https://doi.org/10.18845/tm.v33i8.5512.

[31] Brumana, R., Oreni, D., Raimondi, A., Georgopoulos, A. & Angeliki, B., From survey to HBIM for documentation, dissemination and management of built heritage. *2013 Digit. Herit. Int. Congr.*, pp. 497–504, 2013.

[32] Nieto-Julián, J.E., Lara, L. & Moyano, J., Implementation of a teamwork-HBIM for the management and sustainability of architectural heritage. *Sustain.*, **13**(4), pp. 1–26, 2021. https://doi.org/10.3390/su13042161.

[33] Dezen-Kempter, E., Mezencio, D.L., Miranda, E.D.M., de Sãi, D.P. & Dias, U., Towards a digital twin for heritage interpretation. *RE Anthr. Des. Age Humans: Proc. 25th Int. Conf. Comput. Archit. Des. Res. Asia, CAADRIA 2020*, **2**, pp. 183–191, 2020.

[34] Salvador-García, E., Valldecabres, J.L.G. & Blasco, M.J.V., Integrating HBIM models in the management of the public use of heritage buildings. *Can. J. Civ. Eng.*, **47**(2), pp. 228–235, 2020. https://doi.org/10.1139/cjce-2018-0338.

[35] Armisén, A., *HBIM Successful Case Study*, 2018.

[36] Ermenyi, T., Enegbuma, W.I, Isaacs, N. & Potangaroa, R., Building information modelling workflow for heritage maintenance. *Proc. Int. Conf. Archit. Sci. Assoc.*, pp. 220–225, 2020.

[37] Rechichi, F., Mandelli, A., Achille, C. & Fassi, F., Sharing high-resolution models and information on web: The web module of Bim3dsg system. *Int. Arch. Photogramm. Remote Sens. Spat. Inf. Sci.: ISPRS Arch.*, **41**(July), pp. 703–710, 2016. https://doi.org/10.5194/isprsarchives-XLI-B5-703-2016.

[38] Barontini, A., Alarcon, C., Sousa, H.S., Oliveira, D.V., Masciotta, M.G. & Azenha, M., Development and demonstration of an HBIM framework for the preventive conservation of cultural heritage. *Int. J. Archit. Herit.*, 2021. https://doi.org/10.1080/15583058.2021.1894502.

[39] Quattrini, R., Pierdicca, R., Morbidoni, C. & Malinverni, E.S., Conservation-oriented HBIM. The Bimexplorer web tool. *Int. Arch. Photogramm. Remote Sens. Spat. Inf. Sci.: ISPRS Arch.*, **42**(5W1), pp. 275–281, 2017. https://doi.org/10.5194/isprs-Archives-XLII-5-W1-275-2017.

[40] Fadli, F. & AlSaeed, M., Digitizing vanishing architectural heritage: The design and development of Qatar historic buildings information modeling [Q-HBIM] platform. *Sustain.*, **11**(9), 2019. https://doi.org/10.3390/su11092501.

[41] Fouseki, K. & Cassar, M., Energy efficiency in heritage buildings: Future challenges and research needs. *Hist. Environ. Policy Pract.*, **5**(2), pp. 95–100, 2014. https://doi.org/10.1179/1756750514Z.00000000058.

[42] Cabeza, L.F., de Gracia, A. & Pisello, A.L., Integration of renewable technologies in historical and heritage buildings: A review. *Energy Build.*, **177**, pp. 96–111, 2018. https://doi.org/10.1016/j.enbuild.2018.07.058.

[43] Berg, F. & Fuglseth, M., Life cycle assessment and historic buildings: Energy-efficiency refurbishment versus new construction in Norway. *J. Archit. Conserv.*, **24**(2), pp. 152–167, 2018. https://doi.org/10.1080/13556207.2018.1493664.

[44] Munarim, U. & Ghisi, E., Environmental feasibility of heritage buildings rehabilitation. *Renewable and Sustainable Energy Reviews*, 2016. https://doi.org/10.1016/j.rser.2015.12.334.

[45] Angrisano, M., Fabbrocino, F., Iodice, P. & Girard, L.F., The evaluation of historic building energy retrofit projects through the life cycle assessment. *Appl. Sci.*, **11**(15), 2021. https://doi.org/10.3390/app11157145.
[46] Atmaca, N., Atmaca, A. & Özçetin, A.İ., The impacts of restoration and reconstruction of a heritage building on life cycle energy consumption and related carbon dioxide emissions. *Energy Build.*, **253**, 111507, 2021. https://doi.org/10.1016/J.ENBUILD.2021.111507.
[47] Adegoriola, M.I., Lai, J.H.K., Chan, E.H. & Amos, D., Heritage building maintenance management (HBMM): A bibliometric-qualitative analysis of literature. *J. Build. Eng.*, **42**, 102416, 2021. https://doi.org/10.1016/j.jobe.2021.102416.
[48] Güngör, İ. & Benli, Y.N., Evaluation of building information modelling (BIM) based life cycle assessment (LCA). *ATİ, Archit. Innov. Technol.,* 2020.
[49] Prizeman, O., Pezzica, C., Taher, A. & Boughanmi, M., Networking historic environmental standards to address modern challenges for sustainable conservation in HBIM. *Appl. Sci.*, **10**(4), 2020. https://doi.org/10.3390/app10041283.
[50] Prizeman, O.E.C., HBIM and matching techniques: Considerations for late nineteenth- and early twentieth-century buildings. *J. Archit. Conserv.*, **21**(3), pp. 145–159, 2015. https://doi.org/10.1080/13556207.2016.1139852.
[51] Khodeir, L.M., Aly, D. & Tarek, S., Integrating HBIM (heritage building information modeling) tools in the application of sustainable retrofitting of heritage buildings in Egypt. *Procedia Environ. Sci.*, **34**, pp. 258–270, 2016. https://doi.org/10.1016/j.proenv.2016.04.024.
[52] Mora, R., Sánchez-Aparicio, L.J., Maté-González, M.Á., García-Álvarez, J., Sánchez-Aparicio, M. & González-Aguilera, D., An historical building information modelling approach for the preventive conservation of historical constructions: Application to the historical library of Salamanca. *Autom. Constr.,* **121**, 2021. https://doi.org/10.1016/j.autcon.2020.103449.
[53] La Russa, F.M. & Santagati, C., An AI-based DSS for preventive conservation of museum collections in historic buildings. *J. Archaeol. Sci. Reports*, **35**, 102735, 2021. https://doi.org/10.1016/j.jasrep.2020.102735.
[54] Jordan-Palomar, I., Tzortzopoulos, P., García-Valldecabres, J. & Pellicer, E., Protocol to manage heritage-building interventions using heritage building information modelling (HBIM). *Sustain.*, **10**(4), 2018. https://doi.org/10.3390/su10040908.
[55] Angulo-Fornos, R. & Castellano-Román, M., HBIM as support of preventive conservation actions in heritage architecture. Experience of the Renaissance quadrant facade of the cathedral of Seville. *Appl. Sci.*, **10**(7), 2020. https://doi.org/10.3390/app10072428.
[56] Fonnet, A., Alves, N., Sousa, N., Guevara, M. & Magalhães, L., Heritage BIM integration with mixed reality for building preventive maintenance. *EPCGI 2017: 24th Encontro Port. Comput. Graf. e Interacao*, pp. 1–7, 2017. https://doi.org/10.1109/EPCGI.2017.8124304.
[57] Acampa, G. & Grasso, M., Heritage evaluation: restoration plan through HBIM and MCDA. *IOP Conf. Ser. Mater. Sci. Eng.*, **949**(1), 2020. https://doi.org/10.1088/1757-899X/949/1/012061.
[58] Bruno, N. & Roncella, R., HBIM for conservation: A new proposal for information modeling. *Remote Sens.*, **11**(15), 2019. https://doi.org/10.3390/rs11151751.
[59] Talon, A., Cauvin, C. & Chateauneuf, A., State of the art of HBIM to develop the HBIM of the heritagecare project. Int. J. 3-D Inf. Model., 6(3), pp. 33–43, 2018. https://doi.org/10.4018/ij3dim.2017070103.

REEXAMINING THE OLD DEPOT MUSEUM IN SELMA, ALABAMA, USA

JUNSHAN LIU[1*] & DANIELLE WILLKENS[2†]
[1]McWhorter School of Building Science, Auburn University, USA
[2]School of Architecture, Georgia Institute of Technology, USA

ABSTRACT

Housed within the distinctive architectural fabric of the former L&N Railroad Depot and listed on the Water Avenue Historic District National Register of Historic Places, the Old Depot Museum depicts life in Alabama and the histories that made Selma the "Queen City of the Black Belt." Formally known as the Selma/Dallas County Museum of History and Archives, collections range from the times of pre-historic American Indians to the Civil War, continuing through the Depression and the historic path of the Civil Rights Movement, illustrating how a site that was once home to a Confederate foundry became a critical repository for the preservation and interpretation of key documents, artifacts, and photographs related to "Bloody Sunday", as one of the most significant incidents of American Civil Rights Movement, and its aftermath. Through enlightening and enriching displays, Selma's Old Depot Museum provides unparalleled educational opportunities for the general public and area schools, with particular respect to the path from slavery to voting rights for African Americans. This paper presents a research project that brought the architectural fabric and archives of the Old Depot Museum in Selma to life, both on-site and remotely through Heritage Building Information Modelling (HBIM) and other technology, including LiDAR, 360° photography, photogrammetry, and UAVs. The findings of this project offer a workflow for digitally documenting and showcasing other cultural heritage sites, underscoring the importance of redundancy in digital captures and platform flexibility for information integration and dissemination.

Keywords: 360° photography, building information modelling, cultural heritage, heritage building information modelling, LiDAR, photogrammetry, terrestrial laser scanning, UAV, virtual tour.

1 INTRODUCTION

Housed within the distinctive architectural fabric of the former L&N Railroad Depot and listed on the Water Avenue Historic District National Register of Historic Places, the Old Depot Museum (Selma, Alabama, USA) depicts life in Alabama and the histories that made Selma the "Queen City of the Black Belt." The red brick, stone trimmed building in the Romanesque Revival style was built around 1,890 (Fig. 1). Its unique survival and the broad aesthetic appeal make it one of twelve railroad depots in the southeast of the US designated by *Southern Living* as one of architectural and historical importance. The museum's collections range from the times of prehistoric American Indians to the Civil War, continuing through the Depression and the historic path of the Civil Rights Movement, illustrating how a site that was once home to a Confederate foundry became a critical repository for the preservation and interpretation of critical documents, artifacts, and photographs related to "Bloody Sunday" and its aftermath. The Old Depot Museum provides a greater context for understanding both the struggle for Civil Rights and Selma's current demographics and socioeconomics. However, with such a strong educational mission, the museum and its collections are known to few outside of Selma [1].

* *ORCID: https://orcid.org/0000-0003-4476-5673*
† *ORCID: https://orcid.org/0000-0003-3271-6890*

WIT Transactions on The Built Environment, Vol 205, © 2021 WIT Press
www.witpress.com, ISSN 1743-3509 (on-line)
doi:10.2495/BIM210141

Figure 1: (a) A historic photo (c.1905) from the L&N Railroad archives; and (b) A current view showing the front of the Old Depot Museum, Selma, Alabama, USA. *(Source: https://www.archdsw.com/odm.html.)*

In order to proceed towards a full preservation and restoration project for this significant building, and to make the rich collections of the museum more visible to the general public, this paper outlines a research project that brought the architectural fabric and historic archives of the Old Depot Museum to life for historians and visitors, both on-site and remotely through Heritage Building Information Modelling (HBIM) and other building digital documentation technology, including LiDAR, 360° photography, photogrammetry, and UAVs.

2 LITERATURE REVIEW

According to UNESCO, cultural heritage includes tangible culture (such as buildings, monuments, landscapes, etc.), intangible culture (such as folklore, traditions, language, and knowledge), and natural heritage. All of them reflect universal values that must be a legacy [2]. Architectural documentation is considered one of the first processes that must take place for the preservation of heritage buildings, including both public and private buildings. The emergence of the digital revolution and how it leverages modern technologies provide new ways to document, interpret, and restore historic structures. Due to the historical value of these buildings and their status within both local and national heritage contexts, it was deemed important for these locations to be archived in a way that expands the breadth of dissemination and facilitates ease of access for scholars and researchers [3]. Documentation records the past and current conditions of resources at a level of detail for a variety of uses. Based upon these records, resources may be reconstructed, restored, and studied by professionals and academics [4]. Through a review of the published literature, the researchers have identified two types of digital technologies: one used to digitally capture and record the historic buildings and heritage sites, another used to digitally reconstruct them.

2.1 Digitally capturing and recording of heritage buildings

Over the last few decades, a technological revolution brought with it new tools to improve and accelerate the topographic techniques of spatial data acquisition to generate accurate construction information models [5]. Today's most influential techniques are photogrammetry, terrestrial laser scanning (TLS), and 360° photography [5], [6].

2.1.1 Photogrammetry

Photogrammetry is a precise, non-invasive and contactless 3D measurement technique where high-quality image captures facilitate the collection of semantic and spatial data of a building or object to be accelerated [5]. Photogrammetry is a technique based on triangulation, where the lines of vision for the cameras, which are located in several places, are joined at a common point in the object. The results obtained with the photographic survey are orthographic images or models of triangulated and textured surfaces. However, due to the intensive requirement of manually processing of data, photographic reconstruction can be disadvantageous when the texture of the object is poor, or its shape is very complex [5].

2.1.2 Laser scanning

Laser scanning technologies are adopted for their ability to expedite the spatial data collection of existing buildings or complex surfaces, as well as for the accuracy and precision of the acquired data [5]. Laser scanners are subdivided into two specific areas: aerial and terrestrial. Each one has an associated range and precision that are suitable for the use to which it will be destined. Terrestrial Laser Scanning (TLS) technologies work via a laser beam that travels towards the area being scanned and back, measuring angles and distances with accuracies from millimetres to centimetres (Fig. 2(a)). In this context, TLS can obtain a precise and detailed geometric reproduction of three-dimensional objects in a short time, in the form of millions of points (a cloud of points), with geometric coordinates (X, Y, Z) (Fig. 2(b)), in a digital environment, with both metric and radiometric information [5].

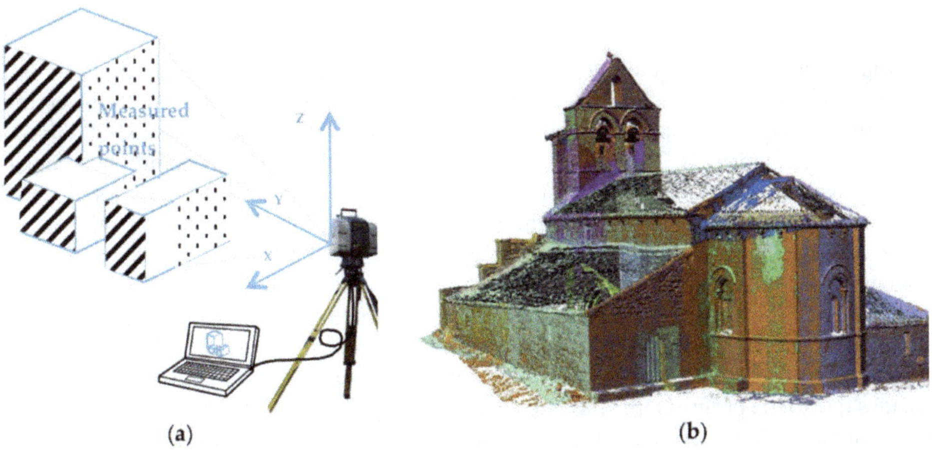

Figure 2: (a) Laser scanning process to create a point cloud; and (b) Georeferenced point cloud, (X, Y, Z) [5].

According to López et al., TLS plays a very important role in a wide range of purposes that can be directly extrapolated to cultural heritage, including the follow-up of an intervention, defect detection, or the reproduction of existing models [5]. However, the high cost of these technologies, the need for specialized management, and the pre-processing of the information acquired can be considered as challenges and disadvantages. Also, it is not generally possible to capture all the geometric information of some objects or external surfaces from a single scanning position, since there may be elements that block the scanner's view. For this reason, it is necessary to place the scanner in different positions with respect to the object. The data extracted from the different scans creates partial-point clouds, and then each one of these point clouds is merged through a process called registration to generate a global cloud that will describe the site [4]. TLS can be complicated by difficult-to-reach parts of a buildings, such as the rooftop; therefore, another technology, UAV photogrammetry, can be deployed to capture images that can be used to create point clouds to create a comprehensive documentation of the building [7], [8].

2.1.3 360-degree (or 360°) photography

The term 360-degree or 360° technology refers to the concept of working with 360° images, which are often called photo spheres (or photospheres) [9]. Many manufacturers offer 360° cameras, yet the basic idea always remains the same: when such a camera is triggered, it captures one photo through each of its two (or more) lenses at the same time. Then, the camera 'stitches' the photos together to one 360° photo. This way everything around the photographer is thoroughly captured. Thus, 360° technology fills the gaps for typical, single aperture analogue or digital photographic captures, providing more immersive documentation for a building or a site. The 360° photos can be uploaded online as a "virtual tour" of a built structure which allows public users to tour the building's interior and exterior. A main advantage of virtual tour is that viewers can easily orient themselves within the virtual site, understanding both where they are and what they are going to see next [9].

2.2 Digitally reconstructing heritage buildings through heritage building information modelling (HBIM)

Building Information Modelling (BIM) is a widespread technology in the new-build construction sector; however, it is not commonly used when it comes to heritage preservation [10]. Because BIM can represent the actual conservation state of the analysed building(s) in a virtual environment [6], researchers have argued that BIM can become an invaluable decision-making and management tool for historical assets. According to Adami et al., HBIM represents a very promising tool for the management of Cultural Heritage, both for daily operations and for the planned preservation of the asset itself and it requires a specific effort to adapt tried and tested tools and methods for new construction to existing Cultural Heritage buildings [10]. There are many software tools to perform the modelling and complete three-dimensional documentation of the intervened structures. However, the Architecture, Engineering and Construction (AEC) sector adopted the BIM standard over the last few decades due to the progress that has been made in its qualities and capabilities. The complex modelling of cultural heritage through commercial BIM software leads to the consideration of the concept of Historic/Heritage BIM (H-BIM or HBIM), which pursues the modelling of architectural elements, according to artistic, historical, and constructive typologies. In addition, HBIM is an emerging technology that enables researchers to understand, document, advertise, and virtually reconstruct the built heritage [11].

However, the digital reconstruction procedures associated with historic buildings are not straightforward: the objects associated with historic models consist of components whose heterogeneous, complex, and irregular characteristics and morphologies are not represented in the BIM software libraries. Therefore, it is essential to introduce the historical and technical approaches into the BIM environment, as well as the TLS or photogrammetry point clouds and 360° photography, to model the different virtual parametric components and achieve a "BIM as-is model" of the architectural heritage [5], [12]–[14]. Researchers have proposed a method to work in a 3D environment throughout the modeling process and to develop semantics during the construction phase, which enables geometric accuracy and reliability of visualization outputs [15].

3 METHODOLOGY AND FINDINGS

The development of comprehensive digital documentation for the Old Depot Museum consists of both the built fabric of the building and the archival digitization of documents to make them more accessible to researchers and general public across the globe. Due to the large-scale collections (e.g., canons) and integrated (i.e., wall affixed) displays traditional, analogue surveying is precarious. The research team used both traditional Historic American Buildings Survey (HABS) methods [16] and technological tools for digital documentation and restoration. These methods allowed the team to try different techniques for a geomatics approach, integrating TLS points of view through a LiDAR system as well as UAVs to monitor the site in different seasonal and lighting conditions. For example, repetitive 3D LiDAR scans and photogrammetry captures generate point cloud data which facilitate iterative studies of the building's weathering, fostering a foundation for a much-needed historic structures report that will lead to essential, large-scale preservation and restoration projects. The researchers also determined an appropriate Level of Knowledge (LOK) for this project, in order to ensure the HBIM model to provide suitable alphanumerical and graphical outputs for strategic decision-making on the major guidelines of heritage management: research, protection, conservation and dissemination [17]. A list of the digital documentation technological tools that have been utilized to date by the research team includes:

- TLS LiDAR: to capture 3D point cloud and panoramic photographs of built structures and site mapping of the area.
- UAV Aerial Photogrammetry: to complement the TLS LiDAR scans, especially to capture the condition and character of roofline.
- HBIM: to digitally reconstruct an 'as-built' model of the historic structure and store historical archives.
- 360° Photogrammetry: to capture a virtual tour of the building and some of the archival assets of the museum. Data is used to help generate the HBIM model and published to the public audience.

3.1 Data capture through TLS (point cloud)

The data capture by TLS was carried out with a FARO Focus3D X130 HDR (FARO X130) laser scanner (Fig. 3(a)). FARO X130 is considered one of the higher-end laser scanners available on the market for ACE industries. It has a range from 0.6 m to 130 m, with a measurement speed up to 976,000 points per second and an accuracy of ±2 mm [18]. It also has capability of High Dynamic Range (HDR) photo recording. The TLS point cloud capture task was divided in two stages to accurately capture the data onsite and then correctly register the scans [12].

- On-site scanning using the FARO X 130 laser scanner
- Processing the scans to create the point cloud

3.1.1 On-site scanning

The research team spent time developing a "scanning map" prior to the field work (Fig. 3(b)), in order to ensure a good coverage of the building's exterior and interior and to have sufficient overlaps among the scans, which was critical for post-processing of the scans and developing a global point cloud. A total of eighty-four (84) scans were captured on three different trips to the Old Depot Museum. All of these scans were performed using a medium resolution (between 12 to 20 million points captured by each scan) with a complete field of view (360°

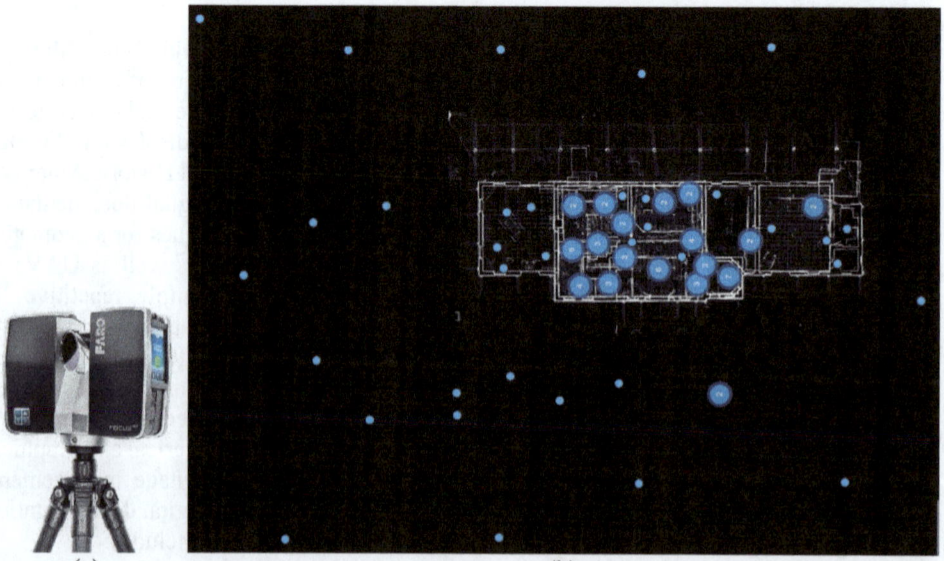

(a) (b)

Figure 3: (a) FARO Focus3D X130 HDR laser scanner; and (b) An overview map showing locations of the TLS scans on the ground level.

in the horizontal plane and 270° in the vertical plane). Panoramic photos were also taken by the scanner for colorizing the point clouds. On average, each scan took about 15 minutes including scan station setup, laser scanning, and capturing panoramic photos.

3.1.2 Processing scans and creating point clouds

Two software programs were used to process the scans: FARO SCENE and Autodesk ReCap Pro. The research team used both software programs previously on other projects. With the large number of scans captured for the Old Depot Museum, the team decided to use FARO SCENE to conduct initial data processing and register all of the 84 scans together into one single point cloud [19], and then transfer the data into ReCap Pro for further cleaning, manipulation, and preparation of the data for BIM modelling in the next stage [20].

Each scan was imported into FARO SCENE, the points were colorized using the panoramic photos during the processing. Due to the large number, the 84 scans were divided based on their locations into the following five 'clusters' to help manage the scans and expedite the registration: basement, level-1, level-2, attic and exterior. Table 1 shows the details of the scan 'clusters' and the registration results. The mean point error in the scan registration reports indicated the mean distance separation between two scans or scan clusters. The successful registration result of the majority of the scans was a result of the relatively high density of scans captured (as evidenced by high value in scan overlaps) and a good scanning plan; however, the registration of scans in the basement had relatively large errors due to the low scan overlap caused by the cumbersome materials stored in the space. Overall, the size of the processed point cloud in FARO SCENE format was over 140 GB.

After scan registration in FARO SCENCE, the point cloud data was exported and transferred to Autodesk ReCap Pro where it was cleaned by removing unusable or errant points from both the interior and exterior of the building. Following an optimization process, a 'clean' point cloud for the whole building consisting of over 688 million points was created. The file size of this point cloud is 26 GB. The research team also used ReCap Pro to create still images of the point cloud for visualization purposes. Fig. 4(a)–(d) illustrate the cleaned point cloud in ReCap Pro. To further decrease the point cloud file size, it was exported to a 'unified' project in ReCap Pro with 5 mm spacing. The final point cloud had a file size of 5.2 GB and was ready to be used for BIM modelling.

Table 1: Details of the field scans and registration results.

Scan clusters	Building square footage covered	# of scans captured	Scan registration result	
			Mean point error	Min scan overlap
Basement	552 ft²/51 m²	6	0.31"/7.9 mm	9.8%
Level-1	4,163 ft²/387 m²	19	0.04"/0.9 mm	24.9%
Level-2	1,823 ft²/169 m²	23	0.04"/0.9 mm	16.1%
Attic	1,823 ft²/169 m²	13	0.05"/1.3 mm	16.3%
Exterior	-	23	0.06"/1.6 mm	9.0%
Total:	8,361 ft²/776 m²	84		

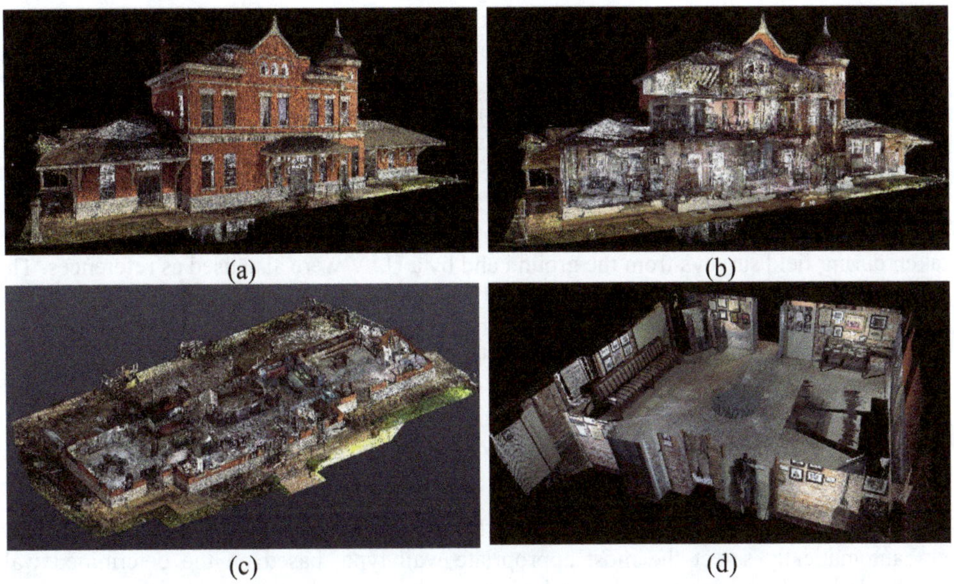

Figure 4: Photos of the registered point of the Old Depot Museum in FARO SCENE. (a) Front view; (b) Clear front view; (c) Level-1 sectional view; and (d) Plan cut of the Civil Rights room.

3.2 Point cloud to heritage BIM (HBIM)

There are a number of BIM platforms but few interface with existing structures, especially those of historic importance, because it is difficult to express unique construction details, settlement and irregular structural issues, and changes over time. For these reasons, the virtual and detailed reconstruction of historic structures has revealed some limitations of BIM platforms, such as the unavailability of historical parametric object libraries and the lack of tools for managing complex, irregular, and uncertain shapes that are obtained from point clouds [5], [12].

Autodesk Revit was selected as the BIM platform for this project by the researchers due to their familiarity with this software – one of the team members is an Autodesk Revit Certified Professional and has many years of experience of using and teaching Revit. The approach adopted by the research team for developing a HBIM model of the Old Depot Museum was to fully replicate the critical current structural data of this historical building, including the building's geometry and materiality. However, due to the limitations of Revit software, historical marks such as patina (i.e., wear and tear on the building), structural and aesthetic markings from settlement and renovations, and other damage to the structure might not to be fully addressed in the HBIM model. Also, it was the researchers' understanding that the information that was unable to be captured through field surveys or not visually exposed would not be included in the model (e.g., floor joists, subfloors, foundations, concrete reinforcements, etc.). As a HBIM model, it is required that every object introduced is not only a geometric representation of real shapes or technical components as in the CAD system, but it is a multi-data collector on its own, with specific significance to the user of the model. These elements provide an abstract computer representation of a physical world: a wall, as an object, represents an actual wall in the physical world as well as a door as a digital object represents a real door, perhaps connected to the hosting wall [12].

To create a real HBIM model of the museum, the research team associated the additional information of historic fabric such as photos and narratives to specific elements in the BIM model. This was accomplished through the creation of a set of new BIM Project Parameters and then assigning information (i.e., materials and specific finish of walls, hardware, photos, notes, etc.) to these Project Parameters of the specific elements.

Besides the point cloud captured by TLS technology that was used as the main source of information for the BIM modelling, one historic plan of the building (Fig. 5), and photos taken during field surveys from the ground and by a UAV were also used as references. The workflow for developing the HBIM model of the Old Depot Museum is illustrated by Fig. 6. Fig. 7(a)–(h) demonstrate detailed steps of creating a small section of the HBIM model. An add-on application, FARO As-Built for Revit (As-Built), was used by the research team during the early stages of the BIM modelling to improve productivity and accuracy. As-Built was produced by FARO Technologies, Inc. for modelling 3D building models from 3D point cloud data [20]. The two As-Built's point cloud modelling functions that the research team utilized on this project included:

- Create walls quickly and precisely directly in the point cloud. The As-Built software can automatically select the most appropriate wall type, based on the determined wall thickness. If no appropriate wall type can be found, a new adapted wall type is automatically created by As-Built.
- Fitting beams and columns. As-Built can place standard structural elements in the point cloud. Common beam and column profiles can be automatically recognized.

Due to the unique historic fabric and form of the structure, the researchers spent a large amount of time on creating individual model elements using Revit's 'Model In-Place Component' command to precisely digitally reconstruct the building in the HBIM model. The estimate for this project is that 70% of the modelling time was spent on those customized elements such as wall surface details and decorative elements such as trims, railings, and mouldings that account for less than 20% of the items in the model. One of the greatest advantages of Revit as a tool for the design of new buildings is its large library of standard

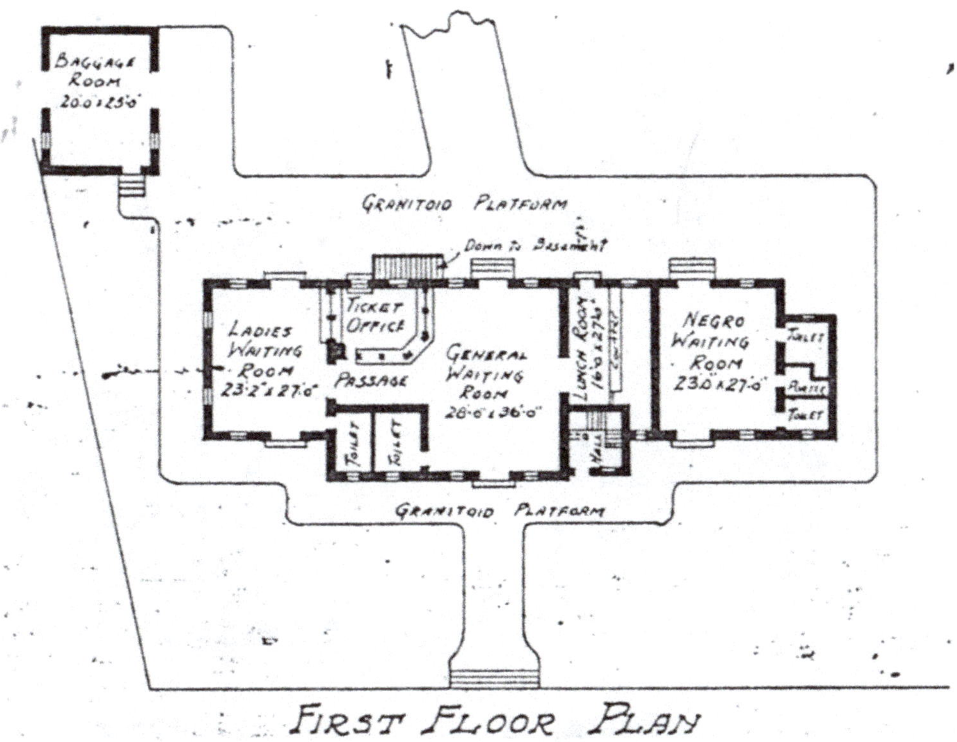

Figure 5: An original plan of the Old Depot Museum from the Louisville & Nashville archives, showing the segregated waiting rooms. It was used as a supplemental reference of HBIM modelling.

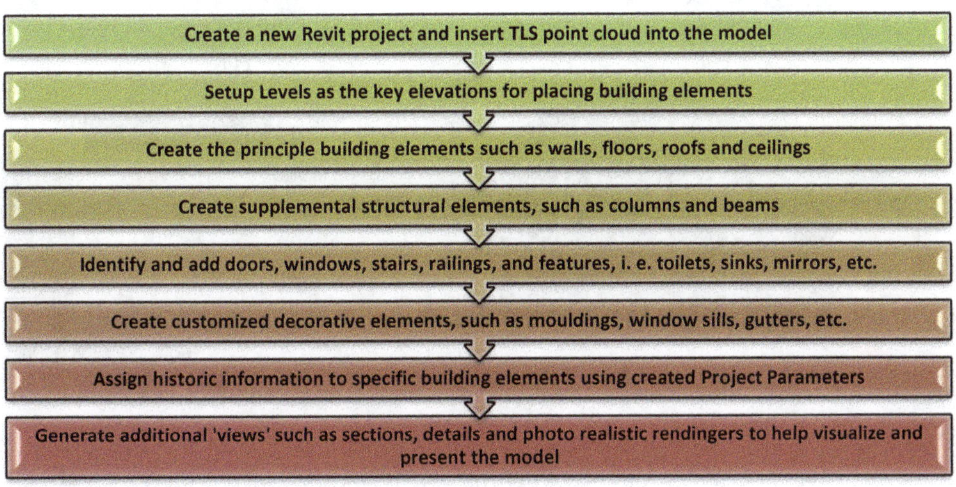

Figure 6: Workflow of HBIM model development.

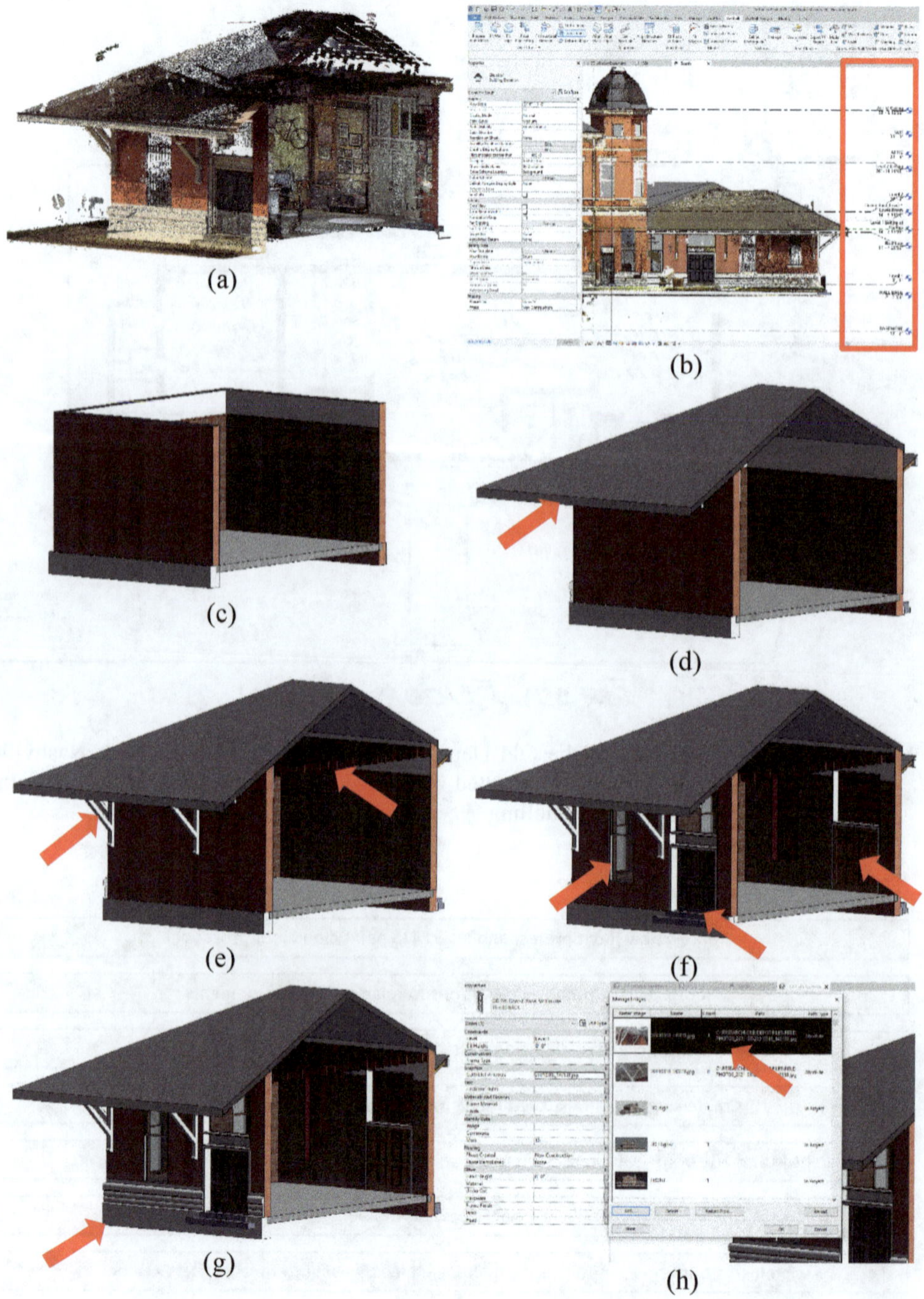

Figure 7: Detailed steps of the development of a small portion of the HBIM model. (a) Insert point cloud into a new Revit model; (b) Set up levels (key elevations); (c) Create walls and floors; (d) Create roofs; (e) Create structural elements; (f) Add doors, windows and stairs; (g) Create wall sweeps and moldings; and (h) Assign information and attributes to project parameters.

model elements (Revit family files). Conversely, this is Revit's biggest disadvantage for heritage building modelling since the element library needs to be highly customized for a specific HBIM model. A full HBIM model of the Old Depot Museum can store data of the building's geometry, materiality, and some of its historic fabric. This model can also be used as a basis for possible future development of complementary documentation that can be linked to necessary conservation and maintenance tasks and preservation planning for the museum [12]. Fig. 8(a)–(c) display a comparison of the Old Depot Museum as seen in a photo, BIM model rendering, and the actual model in Revit. Fig. 9(a)–(c) demonstrate details of the model of the ground floor.

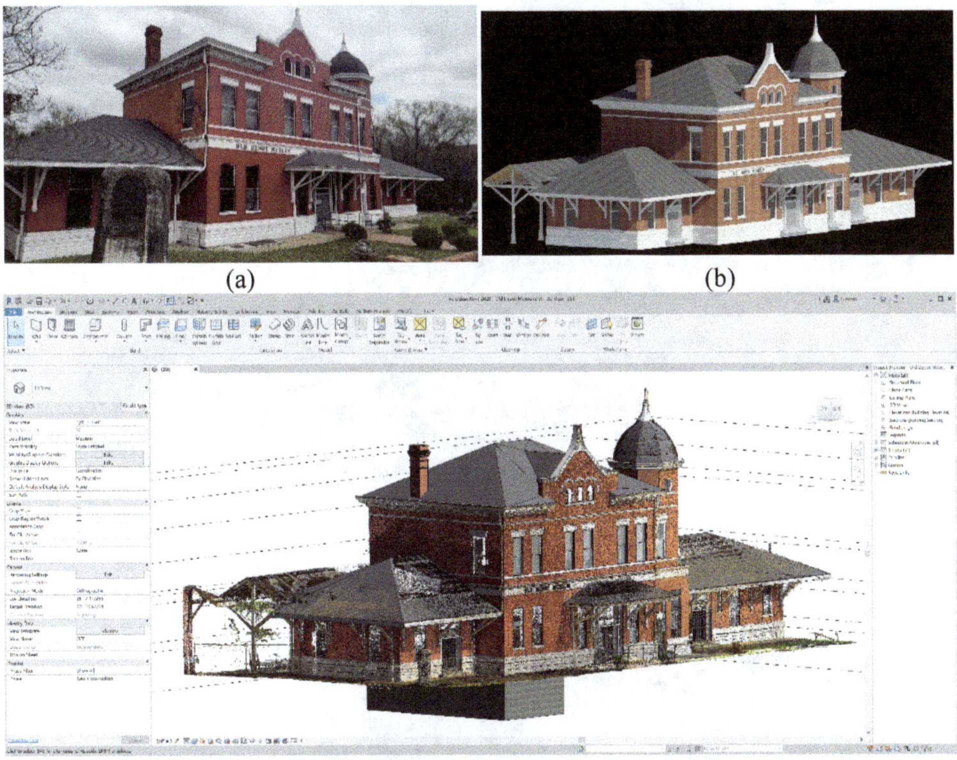

(a) (b)

(c)

Figure 8: Comparison of the building's front façade, the HBIM model rendering, and its view in Revit. (a) Photo of the front view; (b) Rendering of HBIM model; and (c) A view of the HBIM model in Revit with the inserted TLS point cloud.

3.3 360° photography (virtual tour)

A Matterport Pro2 3D Camera was used by the research team to capture a 360° virtual tour of the Old Depot Museum. Matterport is a three-dimensional camera system that creates realistic, fully immersive virtual experiences, or a 'virtual tour' [21]. Using a virtual tour, viewers can explore the site from a variety of viewpoints and get a feel for the space without having to travel. This technology is an excellent choice for cultural spaces, such as a museum,

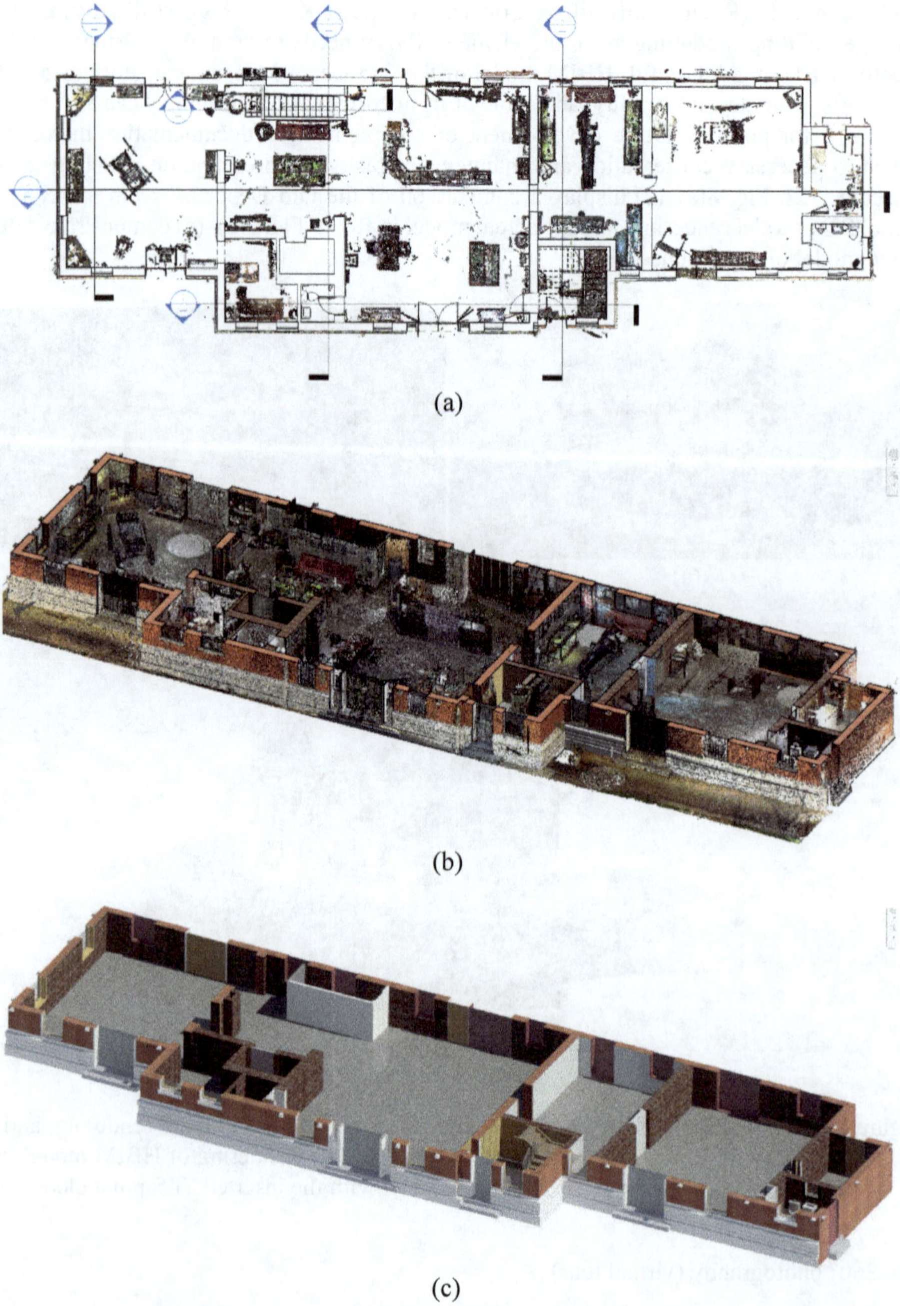

Figure 9: Details of the first floor of the HBIM model in Revit. (a) Ground-level floor plan view of the HBIM model in Revit with TLS point cloud displayed; (b) Ground-level 3D sectional view of the HBIM model in Revit with TLS point cloud displayed; and (c) Ground-level 3D sectional view of the HBIM model in Revit with TLS point cloud display turned off.

to offer the public an online exhibition and this has been particularly beneficial during the travel and physical restrictions associated with the Covid-19 pandemic. A few of the benefits of using Matterport's 360° photography technology as listed on their website [21] include:

- High Quality: Matterport can create realistic virtual environments that make the viewer feel like they're really in the location. The renderings are also interactive. The viewers can view the space from virtually any angle and get a 360° view while exploring from a first-person viewpoint.
- Precision: the camera system allows the users to scan up to 10,000 square feet per project. The results of the scan are dimensionally accurate, within 1%, and can produce spatially accurate schematic floor plans.
- Versatility: Matterport is versatile and allows users to add additional information to the photography and share it.

However, there are several disadvantages of Matterport's 360° photography technology, including: high service subscription cost, limited compatibility with many 3D cameras available on the market, relatively lower image quality for outdoor environments, significant time required for capturing and editing the virtual tours, and a lack of capability of working offline or extracting the data to other platforms. Despite the disadvantages, the research team was able to develop a 360° virtual tour of the museum with a good coverage of both of its inside and outside, as shown in Fig. 10. This virtual tour has been published on the newly updated Old Depot Museum's website (https://selmamuseum.org/virtual-tour). This virtual tour has also assisted the research team in the development of the HBIM model by providing high-quality visual references of the building.

The research team evaluated the possibility of using the 'raw' point cloud data and the Revit model to create a virtual tour. However, many issues surfaced from this approach, including: the public's interest in photorealistic imagery vs. the ephemeral views of the point cloud data, difficulty of hosting the large model, lack of interactive features for the viewers to navigate independently and explore certain items in detail. It is also worth noting the reluctance of private or governmental owners (the Old Depot Museum is owned by the City of Selma) to share highly accurate and detailed data (point clouds or HBIM models) of historic sites with the public due to security concerns. Photographic tours, however, are typically viewed with less hesitancy.

4 CONCLUSION

As an adaptive reuse project that transformed the L&N Railroad Depot, listed on the Water Avenue Historic District National Register of Historic Places, into the Old Depot Museum, this site in in Selma, Alabama is a significant historical and architectural structure. A research team consisted of an architectural historian, architectural designer, construction scholar, and digital documentation experts using Terrestrial Laser Scanning (TLS), UAVs, HBIM, photogrammetry, and 360° photography technologies on a research project that digitally documented and reconstructed a digital twin the Old Depot Museum for preservation planning and the building's architectural fabric and historical archives were brought to life through a separate virtual tour.

TLS LiDAR scanning precisely documented the geometry of the Old Depot Museum. The researchers' field measurements and verifications augmented the 3D scanning data; the resulting drawings and structural studies were also supplemented by aerial images captured by an UAV, a documentation process not yet fully explored within the Secretary of the Interior's Standards for heritage documentation. This method provided detailed views of

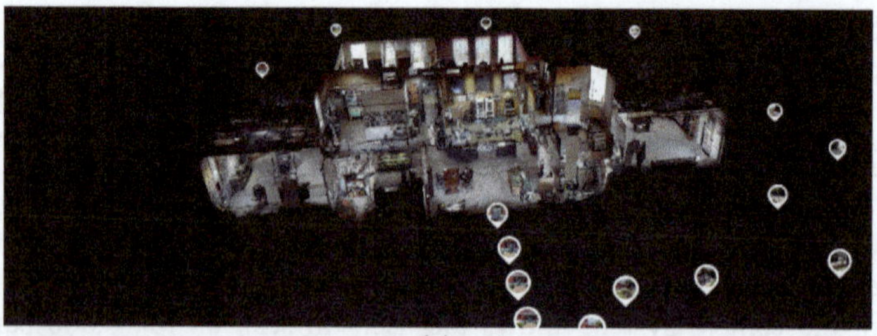

(a)

(b)

(c)

(d)

Figure 10: A virtual tour of the Old Depot Museum presented by Matterport 360° photography technology. (a) Dollhouse view of the virtual tour providing an overall representation of the building; (b) Ground-level plan view in the Matterport virtual tour; (c) A full-screen view of the virtual tour that allows the viewers to explore from a first-person point of view; and (d) A view of the building's primary façade as seen in the Matterport virtual tour.

hard-to-access areas without invasive or costly procedures. Between the point clouds generated from the 3D TLS LiDAR scans and UAV aerial photogrammetry, the digital documentation techniques yielded a model that was accurate to less than one eighth of an inch [1]. Then, the processed point cloud data was inserted into a Building Information Modelling (BIM) software platform, Autodesk Revit, for development of a 3D BIM model. With the precision of the 3D point cloud data and the assistance of an artificial intelligence (AI) aided modelling add-on, FARO As-Built for Revit, the researchers were able to create an accurate BIM 'as-built' model to restore the geometry and material data of the Old Depot Museum.

A substantial issue for BIM applications at historic structures is the irregularity of historical built fabric. Existing BIM tools focus on designing and buildings from scratch, such as Revit, and they are much less effective when working with existing, historic assets [9]. Because Revit's libraries and tools focus on the design and construction of new buildings with simple, regular, and standardized objects [4], the researchers had to spend a majority of their time on creating a relatively small amount of customized building elements, to fully replicate the genuine condition of the building. Furthermore, additional information was assigned to specific building elements to make the model a real Heritage BIM (HBIM) model. One of the disadvantages of LiDAR point cloud data and HBIM models is that neither of them can be easily presented to the public. Due to the large files and graphics card intensive processing, only professionals have access to the necessary hardware and software. To supplement the digital documentation and make the data more accessible to public audience, Matterport technology was incorporated as a tool to capture a 360° photography virtual tour. The virtual tour also served as a visual reference for developing the HBIM model.

The findings of this project can be implemented to help digitally document, restore, study, and showcase other heritage structures. In future projects, the researchers plan to develop a more streamlined process to fully integrate the strengths of hardware and software, and increase the compatibility of the data captured and created to minimize time lost due to software transfers and redundant form building. They will also experiment with additional artificial intelligence (AI) aided applications similar to the FARO As-Built for Revit software to improve the efficiency of HBIM modelling. Finally, when the museum reopens following the Covid-19 pandemic, they will use 3D printing to augment on-site information displays for the public.

REFERENCES

[1] Willkens, D.S., Haley, H.M. & Liu, J., Race, space, and digital interpretation at Selma's Old Depot Museum. *Arris: The Journal of the Southeast Chapter of Architectural Historians*, **31**, pp. 108–119, 2020.

[2] Vecco, M., A definition of cultural heritage: From the tangible to the intangible. *Journal of Cultural Heritage*, **11**(3), pp. 321–324, 2010. DOI: 10.1016/j.culher.2010.01.006.

[3] Alsabban, R., Utilizing virtual reality technology in the preservation of architectural heritage: An empirical study of the local architecture of Hijazi identity in the Mecca region of Saudi Arabia. *WIT Transactions on the Built Environment*, vol. 191, WIT Press: Southampton and Boston, p. 113, 2020. DOI: 10.2495/STR190101.

[4] Gogo, M.E., Digital documentation for historic resources, a thesis submitted to the Graduate Faculty of the University of Georgia, 2011.

[5] López, F.J., Lerones, P.M., Llamas, J., Gómez-García-Bermejo, J. & Zalama, E., A review of heritage building information modelling (H-BIM). *Multimodal Technologies and Interaction*, **2**(2), p. 21, 2018. DOI: 10.3390/mti2020021.

[6] Pocobelli, D.P., Boehm, J., Bryan, P., Still, J. & Grau-Bové, J., BIM for heritage science: A review. *Heritage Science*, **6**(1), p. 30, 2018. DOI: 10.1186/s40494-018-0191-4.
[7] Abbas, M., et al., Three-dimensional data quality assessment: Unmanned aerial vehicle photogrammetry and mobile laser scanner. *Engineering Journal*, **25**, pp. 143–154, 2021. DOI: 10.4186/ej.2021.143.
[8] Liu, J., Jennesse, M. & Holley, P., Utilizing light unmanned aerial vehicles for the inspection of curtain walls: A case study. pp. 2651–2659, 2016. DOI: 10.1061/9780784479827.264.
[9] *Construction Junkie*, How 360° images take construction documentation to the next level. https://www.constructionjunkie.com/blog/8 Aug. 2017/how-360-images-take-construction-documentation-to-the-next-level. Accessed on: 26 Apr. 2021.
[10] Adami, A., Bruno, N., Rosignoli, O. & Scala, B., HBIM for planned conservation: A new approach to information management, p. 41, 2018.
[11] *Strelka Magazine*, BIM for heritage: How to develop a historic building information model. https://strelkamag.com/en/article/bim-to-the-rescue-of-heritage-sites. Accessed on: 26 Apr. 2021.
[12] Rodríguez-Moreno, C., Reinoso-Gordo, J.F., Rivas-López, E., Gómez-Blanco, A. Ariza-López, F.J. & Ariza-López, I., From point cloud to BIM: An integrated workflow for documentation, research and modelling of architectural heritage. *Survey Review*, **50**(360), pp. 212–231, 2018. DOI: 10.1080/00396265.2016.1259719.
[13] Koehl, M., Schneider, A., Fritsch, E., Fritsch, F., Rachedi, A. & Guillemin, S., Documentation of historical building via virtual tour: The complex building of baths in Strasbourg. *Remote Sensing and Spatial Information Sciences*, **XL-5/W2**, pp. 385–390, 2013. DOI: 10.5194/isprsarchives-XL-5-W2-385-2013.
[14] Napolitano, R., Scherer, G. & Glisic, B., Virtual tours and informational modelling for conservation of cultural heritage sites. *Journal of Cultural Heritage*, **29**, 2017. DOI: 10.1016/j.culher.2017.08.007.
[15] Quattrini, R., Malinverni, E.S., Clini, P., Nespeca, R. & Orlietti, E., From TLS to HBIM. High quality semantically-aware 3D modeling of complex architecture. *Remote Sensing and Spatial Information Sciences*, **XL-5-W4**, pp. 367–374, 2015. DOI: 10.5194/isprsarchives-XL-5-W4-367-2015.
[16] Historic American Buildings Survey/Historic American Engineering Record/Historic American Landscapes Survey – Digitizing the Collections, 1930. www.loc.gov/pictures/collection/hh/digitizing.html. Accessed on: 27 Apr. 2021.
[17] Castellano-Román, M. & Pinto-Puerto, F., HBIM oriented towards the master plan of the charterhouse of Jerez (Cádiz, Spain). *Remote Sensing and Spatial Information Sciences*, **XLII-2/W15**, pp. 285–290, 2019. DOI: 10.5194/isprs-archives-XLII-2-W15-285-2019.
[18] FARO Laser Scanner Focus3D X 130 HDR Tech Sheet.pdf powered by box. https://faro.app.box.com/s/lz4et2dd6zxk2dwtijmxgvu7yi3m9tve/file/441635448354. Accessed on: 27 Apr. 2021.
[19] SCENE Software. https://www.faro.com/en/Products/Software/SCENE-Software. Accessed on: 27 Apr. 2021.
[20] ReCap Software Get Prices and Buy ReCap Pro 2022 Autodesk. https://www.autodesk.com/products/recap/overview. Accessed on: 27 Apr. 2021.
[21] Complete Guide to Matterport What is Matterport and More. https://www.designblendz.com/blog/guide-to-matterport. Accessed on: 28 Apr. 2021.

BIM AND CLOUD PLATFORMS FOR FACILITY MANAGEMENT OF ROMAN TEMPLE "HADRIANEUM": CHAMBER OF COMMERCE IN ROME – DIGITALIZATION AS A SOLUTION FOR HISTORICAL HERITAGE MANAGEMENT

DAMIANO DI CIACCIO, EDOARDO MARODER, SARA AMBROSIO & FRANCESCO PACCANICCIO
Rilo Digital Planning S.r.l., Italy

ABSTRACT

Facility and maintenance management can be made much more efficient and cost-effective by using a digital twin of the building. In this article we present the GIM protocol which constitutes an innovative approach for maintenance management through the creation of a digital twin interconnected to the existing building. It uses a BIM model that is constantly updated on the characteristics and status of the building. To show the interest of our proposal we present a case study concerning the digitization process of the Hadrianeum in Rome, one of the most representative buildings in the city and headquarter of the Rome Chamber of Commerce.

Keywords: BIM, facility and maintenance management, cloud computing, building digital twin.

1 INTRODUCTION

The diffusion of the BIM methodology is profoundly transforming the different phases of the building process. This transformation involves professional skills, equipment and organizational structure.

On the other hand, the digitalization process that has involved many other sectors, in the construction sector is still very backward and there is a widespread lack in terms of archiving and consultation of documents, monitoring of structures and systems, of collections. of data concerning the history of the building and the management of essential services.

This paper intends to present an innovative management approach, characterized by ease of use, the intuitiveness of the graphic interface and the immediacy of the results, expressed in schematic form through visualization dashboards. The paper takes up the concepts already exposed in [1] providing further details.

We will initially introduce a general framework of the problems involved, then the proposed solution and the results that can be achieved will be illustrated with a case study concerning the digitalization process of the Hadrianeum in Rome, one of the most representative buildings in the city and headquarter of the Rome Chamber of Commerce. In the conclusions we will outline some possible developments.

2 THE DIGITAL TRANSITION

Senaatti's experience in the digital management of Finnish public assets [2] has demonstrated the usefulness of using parameterized digital projects. A more limited experience is the English one which, albeit on a smaller scale, has also led to the definition of a standard dedicated to the management of assets and their maintenance, the COBie [3]. These experiences have clearly shown that through digitization it is possible to dynamically monitor all the technical and economic characteristics of the project.

The underlying problem for such techniques is interoperability. IFC (Industry Foundation Classes) is an open data format, created to facilitate interoperability between the various

WIT Transactions on The Built Environment, Vol 205, © 2021 WIT Press
www.witpress.com, ISSN 1743-3509 (on-line)
doi:10.2495/BIM210151

operators. This standard aims to allow the sharing of an information model without loss or distortion of data or information. The definition of IFC is aimed at the coding of all building information, throughout the life cycle, from the feasibility analysis to its construction and maintenance, passing through the various design and planning phases

However, digital technologies are emerging and will certainly have a great impact in the construction and management of a building. The advent of 5G and the increasingly widespread use of sensors will allow for constant monitoring of the building in terms of component movement, temperature, ventilation, air quality, etc. Such a large and continuous collection of building data will allow the application of automatic collection and surveillance systems based on artificial intelligence. The advantage of these advanced procedures is in the evident saving, in terms of time and data consistency, compared to the use of current BIM-databases which, although dynamic, require a substantial intervention of the operator in the insertion of the measures or in the extrapolation of the data.

On the other hand, the evolution towards an interconnected digital system, in a context such as the European one, where most of the building stock is existing or historical, and the companies are mainly small and not very innovative, requires quite a long time

3 OUR PROPOSAL

Our proposal consists of an Integrated Maintenance Management (GIM). This provides a protocol that includes an application methodology focused on the digitalization of architectural-engineering elements, aimed at their management. In a first phase, this procedure is based on the verification of the real state of the construction in question and consequent reproduction through digital BIM modeling. The final result is the creation of a digital twin of the property, containing all the technical, construction and documentary information relating to its current state, which populate a database in a dynamic and continuously updated BIM environment.

Once the computer database has been created, the next step is to use and enhance the digital information available, through the use of customized methodologies, which allow the traditional ordinary and extraordinary maintenance process to be managed electronically. The tools defined in this way lead to a continuous collaboration and integration of the computer data on the building with the dynamic procedures of the management area, through the use of external databases connected in a bidirectional network with the current state of the building.

The first fundamental step for the creation of the protocol is given by the evaluation of what are the real expectations and needs of the client compared with the starting information and technical material available. Starting from these considerations, a preliminary overview of the actual development of the protocol is obtained in terms of times, needs, costs and benefits. By balancing these four fundamental parameters to summarize the correct relationship from time to time, the scope of work and the procedures necessary to meet the highlighted needs are clearly defined.

Once any gaps found in terms of data and/or documentation have been filled with the tools available, we proceed to integrate what is necessary for the development of the protocol according to the established purposes, through various types of activities (non-invasive surveys analysis, laser scanner, thermographic surveys, due diligence, etc.).

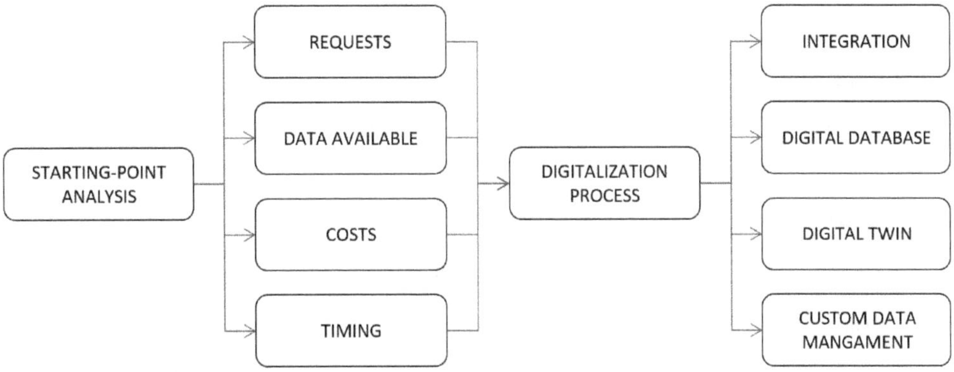

Figure 1: Functional diagram: Analytical phase.

4 CREATION AND MANAGEMENT OF THE DIGITAL MODEL IN THE CLOUD

The next step is the digitization of all the information assets. This process represents the fundamental aspect on which the entire protocol is based both from a conceptual and operational point of view: building and hierarchizing in an appropriate way our "data container" is undoubtedly the necessary condition to implement advanced methodologies aimed at managing dynamic and interactive data.

Parallel to the creation of an IT archive of the property, a digital twin is produced, by means of computerized modelling B.I.M. able to combine technical and documentary contents with the geometric and performance data of the individual construction elements that are part of the building. In this way, the consultation of any information relating to the property in question is standardized and immediate; in the same way, it is possible to catalogue and use all the data implemented in a totally customized way, for example through the strengthening of a general structure of work division, created according to the needs of the case

The result to which the previous phases have led can be summarized as the creation of a database that is consistent, complete and updated with the current state of the property in question. The natural development of what has been prepared is the configuration of methods and tools for using the data itself; this allows, on an operational level, to program customized solutions in relation to the customer's needs, which can be combined, if necessary, with existing standards and procedures.

These data, therefore, are integrated and put in a one-to-one relationship within a space in the cloud that acts as a data store, designed to communicate with suitably configured IT tools and processes. The digital structure thus created allows you to connect, within the platform, the information related to ordinary and extra-ordinary maintenance processes to the elements making up the BIM., based on the level of detail required. This leads to enriching the static information modelling with dynamic data. In fact, within these procedures, management workflows are implemented that take into account the interaction of the various stakeholders (maintenance technician, building manager, expense manager,), who have access to permissions and customized methods of use depending on the role that they have.

The platform, thus populated by computer links between the current state of the property (static data) and the related management procedures (dynamic data), allows the analysis of

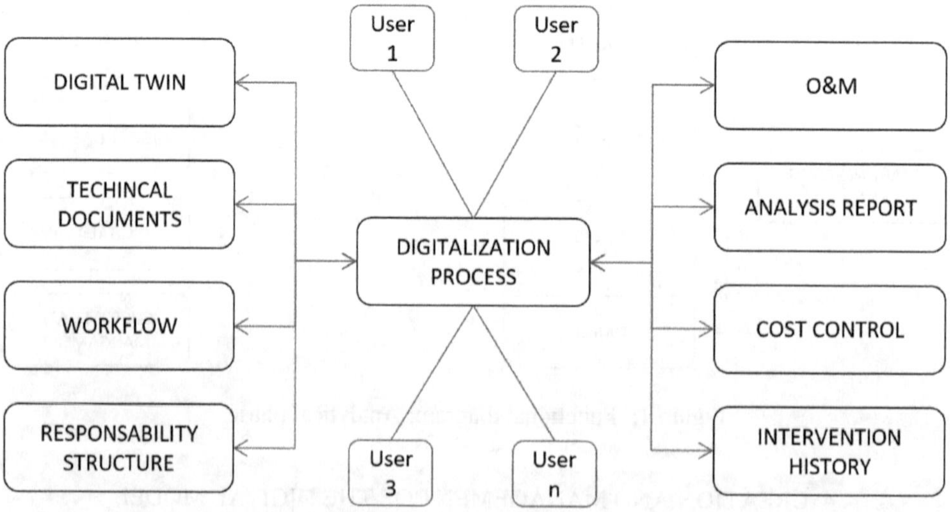

Figure 2: Functional diagram: Cloud platform.

the information collected through the creation of customized dashboards. This tool therefore aims to increase, in terms of transparency and process control, consultation and action by stakeholders. This therefore configures a common workspace capable of encouraging interaction and the definition of responsibilities, thus preventing inconsistencies and therefore costs and complications related to the ex-post resolution of problems.

The development of the IT structure described has led to the highlighting of four main areas, strongly correlated with each other, where direct or indirect advantages were found in the use of the digital tools associated with them:

- Reduction in the number of overall emergency interventions: the digital programming of ordinary maintenance procedures with automatic remind/alert systems has led to a stricter compliance with those which are the interventions foreseen on each component. It was highlighted how to punctually fulfill the deadlines set by the maintenance plan, an activity that is often complex to organize and carry out, is able by its nature to prevent some unexpected failure situations.
- Increase in the effectiveness of the operations: the analysis of the technical-construction nature of the property in question together with the subsequent continuous updates and additions with the platform containing the digital model, led to the configuration of a new unique and complete consultation tool. This prevents waste of time and economic resources, in light of possible quantitative and/or qualitative errors that may occur during the ordinary and extra-ordinary maintenance phases (for example where the need to replace a component occurs).
- Speed of processes: accessibility to the large amount of IT data in the cloud, necessary for carrying out maintenance procedures, has been made enormously faster and more direct, avoiding situations of inconvenience due to the unavailability of documentation or non-uniqueness of the data. Similarly, the communication systems developed have led to a direct information exchange structure, consistent with the procedures set out above and within them.

- Control and transparency: the use of tools shared by the various stakeholders, within which it is possible to find all the information of the management procedures, has made it possible to carry out the tasks assigned to each of the parties in a more defined and clear way. The digital definition of responsibilities on a temporal and quantitative scale has also led to a significant increase in the transparency of the procedures in relation to the contractual obligations between the parties. The automatic generation of historical data relating to ordinary and extra-ordinary interventions have allowed a dynamic and constantly updated control both on the costs of the processes and on any anomalies, which can thus be detected and corrected in the shortest possible time.

From an economic point of view, in the face of a greater initial investment and longer initial times for the definition and development of the IT structure, a reduction in the costs of maintenance activities is obtained, as well as an increase in the quality of the service itself, compared to a "traditional" process (not integrated and/or digitized).

5 APPLICATION TO THE CHAMBER OF COMMERCE BUILDING

The case study concerns the digitization process of the Hadrianeum in Rome, one of the most representative buildings in the city and headquarter of the Rome Chamber of Commerce. The activities were divided into three main phases: digital survey and technological census, BIM digitalization, integration of FM processes within a cloud platform capable of connecting dynamic data to the various elements of the digital model. The result is a unique wired container customized to the needs of the various operators who daily enrich the digital database by carrying out their duties. Business intelligence tools have been integrated into the dynamism of the system to provide the property with a structured control over the costs and activities of FM, as a basis for making "well-informed-decisions".

Figure 3: Exterior view: Hadrianeum.

The enormous construction and system complexity of the building was managed thanks to the use of innovative technologies, such as mobile laser scanning, which made it possible to carry out activities without interfering with the normal management of the Institution.

The process thus managed to trace 108 autonomous plants, 12 centralized systems, linked by 4.2 km of pipes. The transition from traditional practices to digital tools connected to the network and accessible remotely has also allowed a reduction in the carbon footprint of FM's activities, leading to a 100% reduction in the paper used (about 22,000 sheets per year) and a waste reduction of material thanks to the simple and rapid availability of the technical information of each component.

Figure 4: Point cloud model.

6 CONCLUSION

The case study, conducted on a historic building with "tertiary" use, made it possible to obtain the expected results. The objectives of the work, such as the recognition of the fault, the organized management of maintenance, the clear identification of all systems and components in the building, have been fully achieved. It also emerged that this protocol is more useful when the building has plant complexity. Further experiments with different buildings are underway, in this way it will be possible to evaluate the potential and reliability of the outputs in different contexts.

REFERENCES

[1] Di Ciaccio, D., Rossini, F.L. & Maroder, E., From sheets to models: Innovative approaches for the facility and maintenance management. *3D Modeling & BIM Data Modeling & Management for AECO Industry*, pp. 287–299, 2020.

[2] Bortoluzzi, B., Efremov, I., Medina, C., Sobieraj, D. & McArthur, J.J., Automating the creation of building information models for existing buildings. *Automation in Construction,* **105**, Elsevier: The Netherlands, 2019.

[3] Rossini, F.L., Fioravanti, A., Novembri, G. & Insola, C., Holobuild: Process optimization by the introduction of mixed reality in construction site. *Back to 4.0: Rethinking the Digital Construction Industry*, eds. Ciribini et al., Maggioli Editore, pp. 279–288, 2016.

RETROSPECTIVE STUDY ON THE ITERATIVE METHODS OF ARCHITECTURAL VISUALIZATION FROM THE NEOLITHIC TO THE DIGITAL BUILDING INFORMATION MODEL

MATTHIAS STANGE
HafenCity University Hamburg, Germany

ABSTRACT

The current application of building information modelling (BIM) is supposed to be a promising design methodology and plays an essential role in visualizing the architectural process from the early design phase through planning and construction up to building usage and recycling. The application of BIM and its associated processes are seen as key to increasing productivity in the architecture, engineering and construction (AEC) industries. With the help of BIM, all building information is available in real time to all stakeholders involved throughout the entire lifecycle. However, there is a large research gap on the developmental steps of architectural visualization from the beginnings of civilization in the Neolithic through the outstanding discoveries of the Renaissance to the digital building information model we use today. In particular, in response to the questions: What fundamental developments in architectural visualisation have made the currently used building information model possible, and what are the prospects? With the help of an extensive literature review, this study examines the developmental stages of architectural practice and architectural visualisation, as well as their communicative achievements and culture-creating or culture-changing effects. The analysis focuses on the triggers of significant developments, their importance for subsequent methods and the transmission of knowledge across epochs. In particular, with regard to the main barriers to the implementation and use of BIM, which are still present despite the fact that BIM has been on the market since the 1980s.
Keywords: architecture engineering construction (AEC) process, visualization, development stages, knowledge transfer, building information modelling.

1 INTRODUCTION

Architectural visualisation has been practised for as long as there have been buildings. Visualisation generally refers to the process of translating abstract data and relationships that are difficult to formulate linguistically or logically into a graphical, visually comprehensible form. During the design phase, architectural visualisation initially serves the aesthetic verification and visualisation of the developed idea. During further design development, visualization helps to represent the geometric, structural and semantic information of the future building for construction and use, including all spatial dimensions, materials, key figures and technical characteristics. Builders have always been interested in how their building will look and operate before it is built. Only the prior visualization of the structures allows timely influence on the further development of the buildings in terms of cost, time and quality. In this respect, architectural visualization has always represented an early (and only) basis for decision-making in the construction process. Towards the end of the 19th century, the traditional, analogue methods of representation were replaced by computer-aided, digital methods of representation. However, these fundamental changes in the production of space and new forms of representation have not emerged incoherently. Looking at the history in terms of the media and methods of architectural visualisation, it becomes clear that this development was an iterative process that successively had an impact on the production of space and the theory of space. This study focuses on the basic research of architectural

WIT Transactions on The Built Environment, Vol 205, © 2021 WIT Press
www.witpress.com, ISSN 1743-3509 (on-line)
doi:10.2495/BIM210161

practice from the beginnings of civilization in the Neolithic, through the outstanding discoveries of the Renaissance, to the digital three-dimensional building information model we use today. The triggers of further developments, their significance for subsequent techniques, and the epoch-spanning transmission of knowledge are of particular interest, as well as their communicative achievements and their culture-creating and culture-changing impact. The analyses focus geographically on the two regions: Ancient Orient and Egypt in pre-Antique times as well as Europe in antiquity (Greece and Roman Empire), in the Middle Ages mainly France and Germany, in the modern times and modernity mainly Italy and partly north of the Alps. The cultural areas of China, Southeast Asia, Africa, India, the Islamic world, the advanced civilizations of South America, and the European-influenced America and Australia of the early modern period were not included. The division by epochs and cultural areas was chosen because building knowledge has historically consistently provided specific solutions to tasks that were fundamental to the natural and social needs of a particular society or culture. However, the knowledge of the media and methods of architectural visualization underlying the great building achievements of the past has hardly been made the central subject of cross-epochal investigations.

2 THE ANTIQUITY

Antiquity is a term used in the historical sciences and, for the Mediterranean and Near Eastern civilizations, covers the period from the beginning of the Neolithic (around 5600 BC) to the beginning of the Middle Ages (around 600 AD). Spatially, antiquity refers to the Mediterranean region and the Near East and also includes classical antiquity both temporally (depending on the delimitation, from 1700 or 800 BC to about 600 AD) and spatially (Ancient Greece and the Roman Empire). At the beginning of antiquity there is the formation of the ancient oriental civilizations of the Near East: Mesopotamia, Persia, Levant, Asia Minor and Egypt. Antiquity ends with the downfall of the Roman Empire during the migration of peoples around 375 AD until the invasion of the Lombard's in Italy around 568 AD.

2.1 Ancient Oriental civilizations

Some of the most significant changes in human history took place during the Neolithic period. While in the Old Stone Age (Palaeolithic) people made their living by hunting animals and gathering plants, around 4000 BC agriculture and animal husbandry increasingly prevailed as the basis of the economy. The change in subsistence strategy, from food appropriation to food production, is considered an essential feature of Neolithic cultures and established for the first time the need for a sedentary way of life in permanent settlements (instead of caves as before) and thus to produce buildings. According to Kurapkat [1] the Australian-British archaeologist and archaeological theorist Vere Gordon Childe (1892–1957) coined the term “Neolithic Revolution” to describe these pivotal changes. In the Neolithic period, the decisions about the size and orientation of the building and its layout, which were made mentally before the actual start of construction, already constituted the building plan. The use of the natural measures of length, foot and cubit, made it possible to transfer an imagined plan defined by proportions directly to the building ground, for example by laying out the ground plan with stones or by staking out alignment lines. Later, these imagined plans, defined by proportions, were applied to small wax tablets, or carved into clay tablets to be made available for construction. Renn and Valleriani [2] report that the invention of cuneiform writing in Mesopotamia, texts initially also served as a tool for construction planning, since most ancient Near Eastern buildings were based on a canonical construction method that both builders and workers could imagine without construction drawings.

Fauerbach [3] reports that plans on papyrus, with graphic elements and explanatory inscriptions as well as isolated measurements, have survived from ancient Egypt for the first time. Likewise, detailed models of clay, for example of door jambs and column capitals, have survived.

2.2 Ancient Greece

The outstanding achievements of ancient Greece in the fields of poetry and philosophy, as well as the knowledge gained in the fields of mathematics, geometry and optics, laid the foundation for the cultural development of the Occident. In architecture, the Doric temple was developed to maturity and planning was applied beyond the individual building for the first time, such as urban planning as an overall architectural work. The architectural drawing occupies a special position in ancient Greece. On the one hand, as a medium in which the actual design work took place in order to develop determining proportions in the floor plan and elevation through geometric construction. On the other hand, as a medium to visualize the design idea to the client and the executing craftsmen. The most important source for the analysis of ancient design procedures are the buildings themselves. Of particular importance in this context are the building drawings, which have survived in the form of incised drawings on a few building elements or walls. With regard to their significance for Greek architects, a distinction must first be made between the function of these drawings. On the one hand, drawings can be the medium in which the actual design work took place, in order to develop determining proportions in the ground plan and elevation through geometric construction (e.g., compass strokes, triangular constructions or golden section). On the other hand, drawings can be the medium to visualize the design idea to the client and the executing craftsmen. In addition to the drawings, the inscriptions of the buildings also provide important clues to the working methods of the Greek architects. Osthues [4] interprets these inscriptions as a work outline representing the profile from which the stonemasons could take the individual measurements with the chisel. For the Hellenistic period, it is documented that inscriptions not only designated the outline of the work for the executing stonemason, but also a construction drawing for the purpose of form-finding by the architect.

2.3 Roman Empire

In Roman architecture, the Greek model had a striking effect, especially in architectural details and decoration. The Romans adopted the Greek column orders and temple forms. Independent achievements lay in the development of new types of buildings, such as forums, basilicas, thermae, amphitheatres, podium temples and triumphal arches. But also, the typical Roman developments of road construction, viaducts and aqueducts. Roman architecture made great strides in floor plan and room design. The principles of symmetry and axiality prevailed and were equally effective in representative buildings, monumental plazas and sacred buildings. The Romans also made an important contribution to urban architecture. The development of axiality improved spatial organization, which was unmistakably expressed in the system of the *Via Appia* (long-distance roads) and aqueducts for the transport of water. Beginning in the Augustan period (30 BC to AD 14), Greek Hellenistic architecture in much of the Roman Empire was increasingly absorbed into established Roman architecture.

To illustrate the way in which a Roman architect presented his ideas to a client, reference is made to the Roman architect Vitruvius (c. 80 BC to 15 BC). Vitruvius requires the architect to be able to represent his designs graphically and lists the three types of architectural drawing representation: *Orthographia*, *Ichnographia*, and *Scaenographia* (elevation, floor plan, and perspective view). The Greek terms used by Vitruvius indicate that the design of buildings from drawings was first developed by Greek architects. It is probably a development from the time of Hellenism, because for the classical period almost all researchers now assume that the designs were worked out using other methods. From Vitruvius's point of view, the drawing representation of the design idea is the core competence of the architect. He describes this clearly in a passage from his sixth book:

> "[...] that if a building is successful in terms of craftsmanship, it is to be attributed to the craftsman in charge of the construction. That if the building is splendidly equipped, it is to be attributed to the financial strength of the builder. And if the building, however, turns out gracefully, it is the glory of the architect."

Vitruvius supports this statement in the same place:

> "[…] that only the architect has the ability to form an accurate and concrete idea of the building even before the building takes shape in reality, so that he can reliably assess design decisions in advance. Therefore, the architect must present his client with an idea of the proposed building design, and sketches and drawings serve this purpose."

The design of buildings on the basis of proportions can be traced not only in the designs of Vitruvius, but also in later measurements of Roman buildings. Mark Wilson Jones (2000), in his comprehensive study of the design practice of Roman architects, concludes that certain proportions would even have established themselves as standard designs. Consequently, the main dimensions of many buildings were established in simple multiples of five, ten or twelve feet. It is also certain that many detailed forms of building elements were constructed geometrically by drawing. However, detailed forms were also constructed with the help of so-called mixed forms, through computational and geometric approach.

2.4 Transmission of the ancient building knowledge

Wesenberg [5] reports that in ancient times, there was a vast literature on Greek architecture, however, but all these writings are lost. In the introduction to the seventh of his ten books on architecture, Vitruvius notes that there are writings about individual Greek buildings whose authors he mentions by name. Among others, he names Theodoros as author of a writing about the Hera temple of Samos, Pytheos as author of a writing about the Athena temple of Priene, and Chersiphron and Metagenes as authors of a writing about the Artemision of Ephesus. Among the Romans, architectural writings, with the exception of Vitruvius' *De architectura libri decem*, were hardly known. However, we know through Vitruvius that there were individual texts on architectural topics before his work, but not an overall presentation. It is also known that the Roman polymath Marcus Terentius Varro (116 BC to 27 BC) – without being an architect himself – also wrote a book on architecture for his *Disciplinarum libri IX* (*Nine Books on the Subjects*), which is, however, lost. Only one architectural writing, also lost, is known from later times, that of the Greek writer Apollodor of Athens (c. 145 BC to 120 BC) about the construction of the so-called Trajan's Bridge.

3 THE MIDDLE AGES

In European history, the Middle Ages refers to the period between the end of antiquity and the beginning of the modern era. The European Middle Ages extend from the end of the Migration Period in 568 to the beginning of the Early Modern Period around the middle of the 14th century. Spatially, the following section mainly refers to the European Middle Ages in France and Germany. From a humanistic point of view, the Middle Ages were an *aetas obscura* (dark age) in which the social, technological and cultural level of development of Greco-Roman antiquity was completely lost. The understanding of reality in the Middle Ages was not focused on the sensually perceptible world, but predominantly on unreal and two-dimensional world views. For example, the understanding of the earth as a disk over which God rules in heaven, which was spread by the church. In the Middle Ages, church building was the most essential building task. The initiative for the construction of a building came exclusively from a theologically educated builder, who was also the source of ideas. Architecture was considered a secret science in the Middle Ages and was banned. The technical construction was led by master craftsmen who had collected building knowledge passed down to them during their travels and used it in the new buildings.

3.1 The medieval drawing

The shape of medieval buildings initially existed only in the imagination of the clerical builders, and the buildings were designed *opus in mente conceptum* (in the mind). The ground plan was then drawn and marked out by the master craftsmen directly on the building site, similar to the procedure in the Neolithic period. The lack of construction plans requires a precise verbal list of the required services by the master craftsmen. According to Binding [6] small sketches on wax tablets could capture the image of the planned building in the mind. Complete floor plans have survived as coloured pen drawings in manuscripts only since the middle of the 9th century. The St. Gallen monastery plan also belongs in this context (Fig. 1).

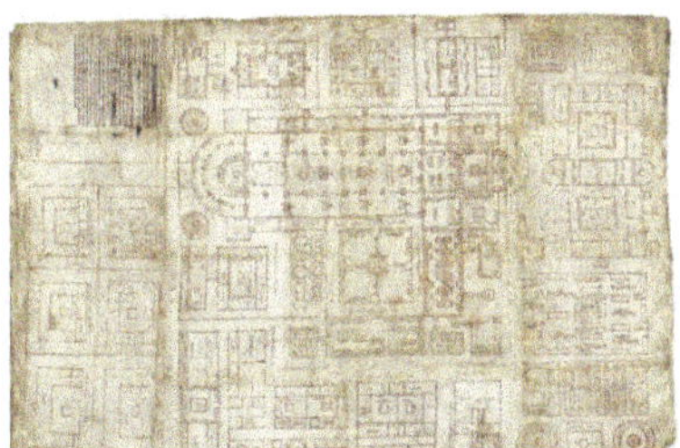

Figure 1: St. Gallen monastery plan.

View drawings have also been known in handwritten form as illustrations since the turn of the millennium, for example in the Psychomachia of the Christian poet Prudentius in the Vatican (around 1000), in the Hillinus Codex of Cologne Cathedral (around 1025) or the water supply plan of the Canterbury monastery (around 1160). Around 1260, the oldest drawing was made for a partial elevation of the west facade of the Strasbourg cathedral, influenced by the facades of the Paris cathedral. Drawings of other building sites also date from the last quarter of the 13th century. This is also the case with the oldest preserved drawing of the west facade of Cologne Cathedral from 1248, presumably drawn by Master

Arnold. The plan drawn on parchment is one of seven surviving medieval architectural drawings of Cologne Cathedral. The drawing of the west facade of Cologne Cathedral stands out unmistakably from the drawings of the late 13th century due to its size of 4.05 metres and its precision. The knowledge of the Greco-Roman art of measurement was never lost in the Middle Ages. Geometry provided the basic knowledge for measuring the building and in this way made it possible to realise the form. The first building plans to scale were made in the Middle Ages. Information about the tasks of a High Medieval craftsman can also be found in the only surviving book on the building trade from the Middle Ages, whose drawings were made by Villard de Honnecourt (c. 1200–1250) on his travels, partly from reality and partly from pictures. His sketches provide valuable information on the planning techniques of the time. Especially for Reims Cathedral and for the development of planning procedures in the High Gothic period is of utmost importance. He depicted the cathedral of Reim on six sheets with interior and exterior views as well as detailed drawings (Fig. 2). The plans do not serve as real building plans but represent diagrams describing the form concepts. The views of Reims Cathedral are presented in a curious folding view that does not require spatial foreshortening. However, the laws of perspective, which were known to the Romans, were not to be rediscovered until the Renaissance. The frequently expressed view that he was more of a theorist is rather unlikely in view of the multitude of details in his pattern book.

Figure 2: Cathedral Reims, Honnecourt.

3.2 Transmission of the medieval building knowledge

Hardly anything is known about building knowledge and its transmission for the early and high Middle Ages. The master craftsmen could neither read nor write, they were so-called *illitteratus* (did not know Latin). They were therefore unable to access important textbooks such as *De architectura libri decem* by Vitruvius or *Geometria* by Boethius. In analogy to God's creatorship, the medieval builders and master craftsmen shaped every item as a form according to the same law by which God created the world: *Omnia mensura et numero et pondere disposuisti* (all things are ordered by measure, number and weight). Of particular interest were the numbers, especially the geometric interpretation of the numbers. Geometry thus provides the basic knowledge for measuring buildings and in this way enables the spatial realisation of the form. Innovative changes in the planning and construction process of buildings appeared in the first half of the 13th century, such as building plans or templates and, last but not least, the possibility of prefabricating the templates over the winter months in a so-called "Bauhütte". The pattern book of Villard de Honnecourt illustrates this change. Nevertheless, the design process, the knowledge of materials and construction techniques as well as the organisation of logistical construction processes were still based on oral instructions and experience during the building process at that time. A normative architectural theory in the sense of antiquity or later the Renaissance, as exemplified by Alberti, Filarete or Serlio, did not exist in the Middle Ages.

4 THE MODERN ERA

In European history, the modern era refers to the period between the late Middle Ages, around the middle of the 14th, and the transition from the 18th to the 19th century. In European history, the modern era refers to the period between the late Middle Ages around the middle of the 14th century and the transition from the 18th to the 19th century. The early modern period is considered to be the beginning of a new era in which a new image of human spread throughout Europe. The focus was on the self-determination of humans and their abilities. The developments of the modern era can first be identified in Italy, where it began as early as the 14th century and reached its cultural heyday in Florence in the following 15th century. From there it spread throughout Europe until the beginning of the 16th century. The predominant artistic and architectural styles are Renaissance and Baroque.

With the end of the 14th century, the ancient building typologies, proportions, and structural systems became the model for the formal language of the buildings of the early modern period. With the paradigm of antiquity, the figure of the architect became recognized again, the cultural man trained theoretically and, in many subjects as Vitruvius describes him. Leon Battista Alberti is more pragmatic on this point and sees the qualification of the architect not only in the perfect ability in many subjects, but also emphasizes the importance of drawing and model making as well as knowledge of painting and mathematics. In 1563, under the leadership of Giorgio Vasari and Vincenzo Borghini, the *Accademia et Compagnia dell'Arte del Disegno* (Academy of Design) became founded. The Italian term *Disegno* (to draw) is a fundamental concept in the Renaissance art theory, interpreting drawings as both an artistic idea and an intellectual concept. Schlimme et al. [7] report that during the Renaissance, painters and sculptors made up a substantial proportion of Italian architects, such as the sculptors Filippo Brunelleschi, Antonio di Pietro Averlino (called Filarete) and Gian Lorenzo Bernini, or the painters Donato Bramante and Giorgio Vasari. Architecture encompassed increasingly fields of activity that could clearly belong to the arts of *Disegno*, primarily the actual architectural design. But there were also activities from which academics wanted to clearly distinguish themselves, such as construction. The visible result of the architect's work was more and more the creative design idea illustrated in drawings, which was then realized in a second separate step.

4.1 The discovery of perspective

Even in ancient times, people had a concept of perspective. The ancient concept of perception through vision is based on the assumption of the Greek mathematician Euclid (c. 365 to 300 BC): vision occurs through visual rays that connect the eye to the object being viewed by the shortest route. In the Middle Ages, the beginnings of perspective representation fell into oblivion. During the Renaissance, perspective was rediscovered and promoted by patrons in Florence and other Italian cities. The Florentine sculptor Filippo Brunelleschi (1377–1447) is considered the discoverer of perspective. After a series of experiments at the beginning of the 15th century, Brunelleschi developed the central perspective projection. Around 1420, Brunelleschi demonstrated central perspective in an experiment in the cathedral square of the Baptistery of Santa Maria del Fiore in Florence by creating a lifelike image of the building using a mirror apparatus and a panel painting. Brunelleschi's practical experiments were theoretically substantiated by Leon Battista Alberti (1404–1474) in his painting treatise *Delle pittura* (The Painting) from 1435 and written down for the first time on the basis of his scientific investigations into vanishing points and horizons. Alberti abbreviates Brunelleschi's perspective method for the practical use of painters. According to Alberti's principles, the optical image is created by rays of vision in the form of a cross-section of a visual pyramid, whereby the size and shape of the objects as they appear in the optical image are determined by the relative position of the rays of vision. In 15th century Italian art, the innovations in painting and architecture seem to support each other in a remarkable way. The fact that architecture and painting entered a particularly close relationship in the early Renaissance is exemplified by Alberti's painting treatise *Delle pittura*. In it, Alberti speaks conspicuously often of architecture. To explain the perspective construction of the pictorial space, Alberti refers, for example, to a regularly structured floor (the so-called *pavimento*), which he constructs with the help of the vanishing point and distance method (Fig. 3).

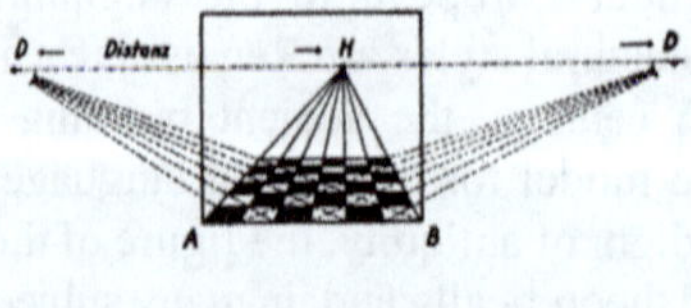

Figure 3: Albertis's "pavimento".

What is indicated in Alberti's treatises can also be observed in Italian painting of the Quattrocento. Architectural forms and structures are fundamental to a new form of representation that emerged particularly during the establishment of perspective and made a decisive contribution to Renaissance painting. As in the case of the Italian painter and architect Raffaello Sanzio da Urbino (1483–1520) in his work "School of Athens" (Fig. 4). The painting is a fresco that Raphael made for Pope Julius II in the *Stanza della Segnature* of the Vatican. This painting, as well as numerous others from Renaissance painting, is an example of central perspective. In addition to the lines running to the central vanishing point, it is easy to see how the foreshortening of similar objects intensifies the effect. Especially when the columns behind relate to vanishing lines.

Figure 4: School of Athens, Raffaelo.

In the 15th century, Alberti demanded the use of orthogonal projection. In contrast to the painter's picture, which attempts to represent spatiality through perspective, the architect's ground plans and views represent spatiality in scaled reduction and true angles. Alberti thus detaches perspective from Vitruvius' *Ichnographia*, *Orthographia* and *Scaenographia*, criticising the conflation between orthogonal projection and perspective. By the beginning of the 16th century, orthogonal projection was then a firmly established planning tool, as for example with the Italian architect and fortress builder Antonio da Sangallo the Younger (1484–1546). The role of Brunelleschi has only been emphasised in retrospect, who contributed significantly to the perfection of representational techniques with his experiments in perspective in front of the Florentine Baptistery. The decisive novelty of linear perspective was not to depict existing objects, but to represent invented objects and spaces. The mastery of perspective, but above all the perfection of orthogonal projection, promoted the development and diversity of use of architectural drawing.

4.2 The illusionist appearance in the Baroque

As the 17th century progressed, an increased artistic ambition in architectural representation became more and more apparent. Coloured washed surfaces in the presentation drawings clarified materials, construction phases or cut surfaces. An appealing design of the drawings was intended to win over the clients for the design. Gerber [8] reports that the term "appetite sketch" is documented from the Baroque period. Baroque is generally associated with the idea of illusionistic appearance in contrast to the representational reality that dominated the Renaissance. According to Wölfflin (1915), the Renaissance still adheres to real substance and conveys a "tactile image". The Baroque, on the other hand, creates a pure "visual image" that is rooted only in the eye and addresses only the eye to create an optical effect. The illusionistic appearance of the Baroque is always associated with Gian Lorenzo Bernini (1598–1680), one of the most important Italian sculptors and architects of the Baroque. Two examples are demonstrable evidence that Bernini saw the effect of his architecture as a component in its own right. One example concerns the decoration of St Peter's Church in Rome. In his sketch, Bernini records how one can see the Cathedra Petri (the Pope's chair) in the choir through the canopy above the high altar. The other example concerns his design of St Peter's Square in Rome, where he was confronted with the problem that the facade of St Peter's appeared too flat and broad (Fig. 5).

Figure 5: St. Peter´s Square Rome, Bernini.

For this reason, he placed colonnades around the square and tried to visually elevate the facade by contrasting the lower members. With the help of this optical illusion, Bernini succeeded in giving the facade of St Peter's a more sublime visual effect. Seen from a distance, the facade of St Peter's rises majestically above the colonnaded square and is crowned by the dome of St Peter's Basilica. Colonnades were already known in ancient Greece and Rome and served to enclose public squares. For example, the Agora of Athens or the colonnaded streets of Roman city foundations.

4.3 Transmission of modern era building knowledge

In the early modern era, building knowledge was recorded in tracts, books, manuscripts or engraved publications. The architectural treatises and annotated translations of Vitruvius' writings made it possible to deal with the questions of building in a way that had not been known before. The architectural treatises of the Renaissance, such as those by Leon Battista Alberti (1485) or Vincenzo Scamozzi (1615), but also most of the annotated editions of Vitruvius, presented the basic principles of architecture and were intended to convey the abstract idea of architecture and the cultural as well as practical value of architecture. Quality concepts and design principles of antiquity is the central theme of these treatises. Other tracts of the Renaissance, such as that of Sebastiano Serlio (1537), were to be regarded as sets of rules and reports of experience from planning and building practice. In the 17th century, the tendency to publish buildings in architectural tracts increased. This was also the case with Guarino Guarini (1624–1683), one of the most important architects of the late Baroque or Sicilian Baroque, in his tract *Architettura Civile* (Civil Architecture), published posthumously in 1686. His writings *Placita Philosophica* (1665), *Euclides Adauctus* (1671) and *Architettura Civile* (1686) play a pioneering role in descriptive geometry. In the 17th and early 18th centuries, modern Roman architectural culture in particular was published, for example the publications of Valerianus Regnartius (1650) or the *Nuovo Teatro delle Fabbriche* edited by Giovanni Giacomo de Rossi (1665) and engraved by Giovanni Battista Falda (1699).

5 THE MODERN AGE

In the *Querelle des Anciens et des Modernes* (1687), an intellectual-historical debate in France at the turn of the 17th and 18th centuries, modernity was still the antithesis of antiquity. The question was to what extent antiquity can still be the model for contemporary art. In the history of architecture, modernism refers to an epoch that cannot be easily defined. In the classification schemes of architectural history, a distinction is made between Early Modernism, Classical Modernism and Postmodernism, whereby a multitude of styles developed within these boundaries, sometimes simultaneously. The beginning of modernism saw the development of Art Nouveau (c. 1880–1914), which was characterised by decorative ornamentation and the abandonment of symmetries. Classical Modernism (around 1920 until today) initially begins with minimalist and functional tendencies. Afterwards, this style became more and more prevalent. In architecture, this refers in particular to those purpose-driven works that later became famous as Bauhaus style or Bauhaus architecture. Postmodernism generally refers to the expression of Western society, art and culture after modernity.

5.1 The expanded concept of architecture in the early 20th century

Architecture and space on the one hand, and movement and time on the other, were not discussed in a single context until well into the 19th century. Yet the movement of people in architectural space was a central theme of architectural designs in this period. According to Noell [9] the progress of industrialisation and the associated technical possibilities arose the question of a changed relationship between space and time in the 19th century. In 1904, the Dutch physicist Hendrik Antoon Lorentz (1853–1928) replaced the previous understanding of time with that of a time dependent on the state of motion for the first time with the so-called "Lorentz transformation". The German physicist Albert Einstein (1879–1955) published his work "Electrodynamics of Moving Bodies" in 1905 and several books on his special and general theory of relativity in the following years. Around 1907, the German mathematician and physicist Hermann Minkowski (1864–1909) realised that the work of Lorentz and Einstein could be understood in a (Non-)Euclidean space. Minkowski [10] argued that space and time are connected in a four-dimensional space–time continuum. Almost all painters, sculptors and architects of the early 20th century who sought an artistic redefinition of space and time referred to these three scientists in deriving or justifying their positions.

The Dutch painter and architect Theo van Doesburg (1883–1931) attempted to develop a new definition of spatiotemporal architecture around 1919. In 1923, together with the Dutch architect Cornelius van Eesteren (1897–1988), he recognised that axonometry, as opposed to perspective, was no longer fixed to the viewer's point of view and a vanishing point. Axonometry enabled a separation from the individual viewer and thus from a fixed position in space. Axonometry thus deviates from the spatial representation of perspective, which is considered finite, since it allows for an infinite and seemingly universally valid representation of space. Moreover, it opens up the possibility of all-round observation, which includes a rotation in space. Van Doesburg argued that the new architecture (or rather the representational method of axonometry) made front and back, and possibly top and bottom, equal. Axonometry thus laid the foundation for today's digital architectural visualisations.

5.2 The computer-based production of space in the late 20th century

After the invention of the world's first fully automatic, programmable computer by Konrad Zuse in 1941, the first geometric formalisations in the 1950s and the first drawing approaches in the 1960s, the so-called computer-aided design (CAD) entered the planning practice of architects and civil engineers in 1985. Moelle [11] refers to the introduction of CAD as the first digital revolution in construction and then later the shift from two-dimensional CAD to 3D modelling as the second digital revolution in construction. In the 20th century, the computer became the dominant tool in contemporary architectural production. The computer is still used today for the production of architectural drawings. In addition to the possibility of a three-dimensional visualisation of the design idea, through which extended spatial ideas can be gained, the benefit of the computer is seen above all in the acceleration of the entire planning and building process.

Increasingly, architects discovered the possibility of using modern computer technology to deviate from basic geometric forms and bring about a change in conventional architectural forms. In the 1990s, the computer became a suitable instrument for the precise planning and construction of such free forms. Until then, computer-aided design systems had aimed to produce a coherent order in response to complex conditions. Peter Eisenman, however, introduced the computer into architectural production in order to reject holistic structures of order, to decouple form from function. For him, the computer was an effective instrument for "overcoming the metaphysics of architecture" – as he calls it. Frank O. Gehry, on the other hand, is an outstanding representative of those architects who carry out and monitor spatial complexity with the computer, but do not create it with it. His buildings of the late 1990s were often misunderstood as so-called "computer architecture" due to their sleek appearance and computer-based production.

While Gehry, who designed in analogue and built digitally, turned to the computer in an application-oriented way and only used it when it came to the detailed planning implementation of his designs, Peter Eisenman already used the computer as a conceptual design medium in the form-finding process. Characteristic of his design process was that the architect triggered the process of form formation and set in motion an automatism behind which he (the architect) seemingly disappeared. With the use of axonometry as a method of representation, Eisenman followed the pioneers of modernism, such as van Doesburg. Eisenman used the visual properties of axonometry to express his idea of "a-perspectival" and "disorganised" space without a preferred direction. Höfler [12] reports that one of the main advantages of axonometry is that, unlike perspective, it provides a geometrically more exact view of the object, as it conveys measurable relationships of lines, surfaces, and bodies.

5.3 The digital building information model in the 21st century

The digital building information model in the 21st century is based on the fundamental idea of product modelling to abstract and describe reality. An integrated product model contains all relevant product characteristics that arise in the individual life cycle phases of the product. The theoretical roots of product modelling can be seen in the conceptual modelling of database applications on the one hand and in object orientation on the other. An integrated product model contains all relevant product characteristics that arise in the individual life cycle phases of the product. Junge [13] describes product models as a structured digital representation of information of manmade objects, with the main purpose of a computer-aided processing of this information throughout the life cycle of the represented object. The basis for this is a uniform, general, cross-lifecycle and redundancy-free basic structure. Wender [14] reports that product models in construction are based on an object-oriented modelled conceptual scheme that describes the data required across the life cycle in its entirety in a structured way. With the succinct words "Begin with the End in Mind", Charles E. Mies (member of the Autodesk AEC Business Development Team) made it clear that the feasibility and ideal use of the building should be the top priority right at the beginning of every construction project. Building information modelling (BIM) seems to be the key to virtually mapping multidimensional problems across disciplines and solving them across the life cycle. Mies [15] argues that the more information contained in this digital image of the building, the greater the added value for the entire project.

6 CONCLUSION

From the first civilizations over 6000 years ago to today's digital building information model, the visual representation of a proposed structure has been formed by ground-breaking social and technological developments. The outstanding discoveries of ancient Greece in the fields of mathematics, geometry and optics laid the foundation for the development of perspective representation in the early modern period. These changed the perception and representation of architecture and had a significant influence on the magnificent pictorial art of the "Quattrocento". As a result of the advancing industrialisation in the modern age and the associated technical possibilities, the question of a changed understanding of space and time arose at the beginning of the 20th century. Inspired by this, Theo van Doesburg and Cornelius van Eesteren developed a new "spatiotemporal" architecture around 1919 and recognised that axonometry, in contrast to perspective, enables detachment from the individual viewer and thus from a fixed position in space. Axonometry, in contrast to finite perspective, enables an infinite and seemingly universal representation of space as well as its all-round observation. Axonometry thus laid the technological foundation for the later computer-assisted architectural representation in an infinite three-dimensional space, the so-called Computer Aided Design (CAD). For the first time in architectural history, architectural objects could be modelled, visualised and, if necessary, modified in an active three-dimensional space. The possibility of object-based modelling through three-dimensional CAD forms the basis for today's building information modelling (BIM). With the help of BIM, specific attributes (properties) and methods (functions) can be assigned to each of the architectural objects in the model. This information is collected centrally in a database and is thus accessible to everyone involved in real time. BIM, whose origins lie in axonometry, makes it possible for the first time in the history of architectural representation to make the necessary information about a building available to all stakeholders in real time throughout the entire life cycle.

REFERENCES

[1] Kurapkat, D., Bauwissen im Neolithikum Vorderasiens. *Wissensgeschichte der Architektur. Band I: Vom Neolithikum bis zum Alten Orient,* eds J. Renn, W. Osthues & H. Schlimme, Edition Open Access: Berlin, pp. 57–127, 2014.

[2] Renn, J. & Valleriani, M., Elemente einer Wissensgeschichte der Architektur. *Wissensgeschichte der Architektur. Band I: Vom Neolithikum bis zum alten Orient*, eds J. Renn, W. Osthues & H. Schlimme, Edition Open Access: Berlin, p. 26, 2014.

[3] Fauerbach, U., Bauwissen im alten Ägypten. *Wissensgeschichte der Architektur. Band II: Vom alten Ägypten bis zum antiken Rom,* eds J. Renn & W. Osthues, Edition Open Access: Berlin, pp. 7–124, 2014.

[4] Osthues, W., Bauwissen im antiken Griechenland. *Wissensgeschichte der Architektur. Band II: Vom alten aÄgypten bis zum antiken Rom*, eds J. Renn, W. Osthues & H. Schlimme, Edition Open Access: Berlin, pp. 127–261, 2014.

[5] Wesenberg, B., Zu den Schriften der griechischen Architekten. Bauplanung und Bautheorie der Antike. Wasmuth: Berlin, 1984.

[6] Binding, G., Architectus, Magister operis, Wercmeistere: Baumeister oder Bauverwalter im Mittelalter. *Mittellateinisches Jahrbuch,* **34**, pp. 7–28, 1999.

[7] Schlimme, H., Holste, D. & Niebaum, J., Bauwissen im Italien der Frühen Neuzeit. *Wissensgeschichte der Architektur. Band III: Vom Mittelalter bis zur Frühen Neuzeit*, eds J. Renn, W. Osthues & H. Schlimme, Edition Open Access: Berlin, pp. 97–334, 2014.

[8] Gerber, A., Metageschichte der Architektur. Ein Lehrbuch für angehende Architekten und Architekturtheoretiker. Transcript: Bielefeld, 2014.

[9] Noell, M., Bewegung in Zeit und Raum. Zum erweiterten Architekturbegriff im frühen 20. Jahrhundert in *Raum-Dynamik: Beiträge zu einer Praxis des Raumes*, eds F. Hofmann, S. Lazaris & J. Sennewald, Transcript: Bielefeld, pp. 301–314, 2004.

[10] Minkowski, H., Raum und Zeit. *Das Relativitätsprinzip. Eine sammlung von Abhandlungen*, eds H.A. Lorentz & A.M. Einstein, Teubner: Leipzig/Berlin, pp. 56–73, 1913.

[11] Moelle, H., Rechnergestützte Planungsprozesse der Entwurfsphasen des Architekten auf Basis semantischer Modelle. Dissertation, Technische Universität München, 2006.

[12] Höfler, C., Form und Zeit: Computerbasiertes Entwerfen in der Architektur. Dissertation, Humboldt-Universität zu Berlin, 2009.

[13] Junge, R., Interoperabilität mit IFC: Entwicklung, Grundlagen, Schema. Vortrag zum BIM Workshop, Bauhaus-Universität Weimar, 6 Nov. 2008.

[14] Wender, K., Das virtuelle Bauwerk als Informationsumgebung für die Planung im Bestand. Dissertation, Bauhaus-Universität Weimar, 2009.

[15] Mies, C.E., Begin with the end in mind: A guide to process transformation. *BIM for ICS*, eds C. Achamer & I. Kovacic, Technische Universität Wien, pp. 22–29, 2013.

STUDY ON FACTORS AFFECTING BIM IMPLEMENTATION IN EUROPEAN RENOVATION PROJECTS

FATEMEH DANESHVARTARIGH[1,2,*] & SERGIO ROSSI[1]
[1]Metropolia University of Applied Sciences, Finland
[2]HTW Berlin University of Applied Sciences, Germany

ABSTRACT
New technologies and applications have radically altered construction techniques in recent years. In order to anticipate how the building will act, perform, and appear, these technologies encompass a wide range of visualization, simulation, and analytic tools. These new technologies and applications have a considerable impact on completing construction projects in today's AEC industries. The rate of changes in BIM-related topics is different worldwide, and it depends on many factors, e.g., the national policies of each country. Therefore, there is a need for comprehensive research focused on a specific area with common characteristics. One of the necessary measures to increase the use of this new approach is to examine the challenges and obstacles facing it. In this research, based on the Delphi method, at first, the background and related literature are reviewed. Then, using the knowledge obtained from the literature, a primary questionnaire is generated and filled by experts who are selected using snowball sampling. It covered the experts' attitudes towards implementing BIM in renovation projects and their view of the benefits and obstacles in this regard. By analyzing the primary questionnaire, the second group of experts is selected among the participants to be interviewed. The results are analyzed using Theme analysis. Six themes, including Management support, Staff resistance, Client willingness, Cost of software and implementation, the difficulty of implementation, and other reasons, are obtained. Then a final questionnaire is generated from the themes and filled by the same group of experts. The result is analyzed by the Fuzzy Delphi method, showing the exact ranking of the obtained themes. The final results show that Management support, Staff resistance, and Client willingness are the most critical barrier to BIM usage in renovation projects.
Keywords: *building information modeling, BIM, BIM implementation, BIM barriers, BIM in renovation.*

1 INTRODUCTION

Nowadays, the construction industry, like other industries, has undergone many growths and transformations, which has caused it to be divided into various categories. From residential to commercial or industrial areas, all have experienced fundamental changes. Construction trends in recent years show that intelligence, productivity, and profitability are the main factors that are improved. Understanding the developments in the construction industry and recognizing solutions to overcome business barriers will help construction companies to scale their company effectively for the upcoming decades.

New technologies and applications have radically altered construction techniques in recent years. In order to anticipate how the building will act, perform, and appear, these technologies encompass a wide range of visualization, simulation, and analytic tools. These new technologies and applications have a considerable impact on completing construction projects in today's architectural, engineering, and construction industries.

ORCID: https://orcid.org/0000-0003-2963-427X

WIT Transactions on The Built Environment, Vol 205, © 2021 WIT Press
www.witpress.com, ISSN 1743-3509 (on-line)
doi:10.2495/BIM210171

Business owners, contractors, consultants, architects, engineers, and subcontractors use communication platforms to standardize business procedures to manage and exchange information.

Building information modeling is a digital depiction of a facility's physical and functional characteristics and a shared database for information about decisions made throughout its life cycle [1]. Simulating and analyzing the structure with the aid of modern modeling technologies allows project stakeholders to see how a building behaves, how it will function, or how it will look far more realistic than ever before.

However, it is critical to determine the shareholders' duties, the range of information to be exchanged in the project, and the supporting software utilized in the construction execution procedures. To summarize, the technology and use of BIM Systems, aid in the completion of building projects in less time, at a lower cost, and in a more sustainable manner.

2 LITERATURE REVIEW

According to Gökgür's research findings in 2015 [2], the potential benefits of BIM for new construction are widely recognized by experts. However, there are still hurdles such as low client demand, the complexity of renovation projects, and a shortage of qualified labor to adopt BIM for renovation projects. The study's findings emphasize the potential for increased and improved BIM utilization in future renovation projects.

An extensive literature review in 2017 carried out by Joblot et al. found out that few scholarly references are devoted to this issue, highlighting that the renovation sector is not a key target of BIM publications (less than 2% specifically addressed renovation). It stated that the current technology and tools, which are sometimes quite expensive, have not shown their profitability for this business type.

It confirms the conclusions of Ilter and Ergen [3] that obtained about five results among 500 articles by querying other databases with these same research topics. It shows that nearly 60% of the publications focused on energy optimization, and there is a very fair coverage of the entire scope defined. Finally, it suggested that it is necessary to map the different observable renovation processes and highlights some gaps that need future works [4].

Based on the review of previous works, most of the efforts have focused either on case studies where BIM is implemented in renovation projects or the development of BIM tools. Previously, the research is conducted by focusing on the specific individual benefits of using BIM in a renovation project; however, the overall and actual percentage of BIM use in renovation companies has not been studied yet. Based on the current situation of renovation projects and focusing on a specific market, analyzing the experts' ideas about the applicability of BIM in renovation projects seems essential. And due to the rapid advancement in BIM technologies, there is a need for continued research to clear the path for future improvements in this area.

3 METHODOLOGY

The study's goal is to conduct an exploratory investigation. The research philosophy is pragmatism because it focuses on current practical issues. It has a deductive approach, and it follows the grounded theory strategy. The research enjoys Multiple-methods where qualitative and quantitative data collection methods are used. Snowball sampling is used to acquire data. The research is a cross-sectional study in terms of time horizons. The overall research method is based on the Delphi method, and the data analysis is performed by the Fuzzy Delphi method.

4 ANALYSIS OF THE RESULT

The main barriers and factors affecting incorporating BIM in renovation projects are identified by using the Delphi method. This method includes a basic questionnaire, an interview, and a final questionnaire.

4.1 The first questionnaire

In this section, the authors found the main barriers and factors affecting incorporating BIM in renovation projects. Using the Delphi method, a basic questionnaire is generated to collect general information about the research. The first questionnaire is designed and conducted through the online software tool SurveyMonkey™. The scope of research is limited to northern and western Europe. The questions are in accordance with the checklist of questions for designing a survey method by Cresswell [5]. It is filled by seventy-four experts whose education and jobs are related to BIM, and results are collected out of the questionnaire. A primary conclusion obtained from the first questionnaire shows the general opinions of the participants about the barriers. The validity and reliability of the questionnaire are checked by using Cronbach's alpha formula shown in eqn (1). The calculated value is 0.87 and according to Table 1, the internal consistency of the questionnaires is good.

$$\alpha = \frac{k}{k-1}(1 - \frac{\sum_{i=1}^{k} \sigma_{y_i}^2)}{\sigma^2 x}). \quad (1)$$

Table 1: Acceptable Cronbach's alpha test of reliability range (in conformity with [6]).

Cronbach's alpha coefficients	Internal consistency
$\alpha \geq 0.9$	Excellent
$0.9 > \alpha \geq 0.8$	Good
$0.8 > \alpha \geq 0.7$	Acceptable
$0.7 > \alpha \geq 0.6$	Questionable
$0.6 > \alpha \geq 0.5$	Poor
$0.5 > \alpha$	Unacceptable

4.2 The interview

After analyzing the first questionnaire and as a qualitative part of the research, according to the derived data, interviews are performed with 18 experts selected according to their relevant education/experience among the questionnaire's initial respondents.

The interview findings were classified and open coded using theme analysis suggested by Corbin [7] for Grounded Theory research. Open coding is an analytical procedure through which concepts are identified and their properties and dimensions are discovered. At this stage, Grounded Theory forms the primary categories of information about the phenomenon under study by segmenting the information. The researcher then classifies the collected data based on their categories. The classified data forms the themes. The researcher considers meaningful patterns and topics related to the research. This analysis involves a continuous flow between data sets and chosen patterns, and finally, related patterns are extracted and coded. The researcher, using theme analyses, extracted 34 basic open codes out of interviews along with their frequencies, which are mentioned in Table 2.

Table 2: Basic concepts and open codes of interviews.

Open code	Basic concepts	Frequency at interview
A1	Lack of knowledge of managers	9
A2	Management support	8
A3	High cost of software	5
A4	Insufficient understanding of the benefits	6
A5	Use of traditional method	5
A6	Complexity	7
A7	Not being demanded by the clients	9
A8	Management attitude	7
A9	ICT Implementation problem	6
A10	Most of staffs are old and not willing to use systems	3
A11	Lack of expertise	4
A12	High costs	8
A13	Fear of failure	5
A14	Projects are too small to use BIM	3
A15	High cost of implementation	7
A16	Lack of relevant software	5
A17	Decision makers don't provide full support	3
A18	Legal barriers exit	2
A19	Unavailability of proper training on BIM	6
A20	Lack of long-term thinking	8
A21	Resistance to change	8
A22	End-user does not realize the benefits	8
A23	Lack of guidelines and standards	3
A24	Hard to implement in renovation projects	5
A25	Only being hired for the first phases	4
A26	Employee's avoidance of changing their way of working	8
A27	Cost of new software	7
A28	Cost of updating the system	7
A29	Cost of training the team	7
A30	Lack of experience in BIM projects	9
A31	Lack of collaboration	8
A32	BIM licensing	5
A33	Maintenance costs	4
A34	Lack of BIM risk insurances	3

Next step, the researcher identifies potential themes to be aggregated. In other words, it must be decided which first-level code has a proper semantic relationship with others to be placed in a category or theme. Table 3 shows part of the extracted main and sub themes.

Table 3: Part of extracted main and sub themes.

Code	Themes	Basic concepts	Open code	Freq at interview	Total frequency	Rank
B1	Management support	Lack of knowledge of managers	A1	9	43	1
		Management support	A2	10		
		Management attitude	A8	8		
		Decision makers don't provide full support	A17	8		
		Lack of long-term thinking	A20	8		
B2	High cost	High cost of software	A3	3	27	4
		High Costs	A12	2		
		High cost of implementation	A15	3		
		Cost of new software	A27	3		
		Cost of updating the system	A28	4		
		Cost of training the team	A29	3		
		BIM licensing	A32	3		
		Lack of relevant software	A16	3		
		Maintenance costs	A33	3		

4.3 Final questionnaire

To finalize the research finding, the final themes, which are the results of the interviews, are used to create the final questionnaire having a Five-Point Likert Scale based on Delphi's approach. The ranges of the answers are from strongly agree to strongly disagree. Output of the final questionnaire is summarized in Table 4.

Finally, the extracted scores are converted to the TFN using the Fuzzy Delphi method. This process enabled the authors to get the exact impact of each barrier and better understand the qualitative data of linguistic ideas of experts. The conversion of five-point Likert Scale to Triangular Fuzzy Numbers [8] are summarized in Table 5.

The associated membership function is calculated by eqn (2) [9], in which $l \leq m \leq u$ and where l, m and u are real numbers. Therefore, TFN number w_{ij} is TFN of expert i for question j.

$$\mu_{\bar{N}}(x) = \begin{cases} 0, x < 1 \\ \frac{(x-1)}{(m-1)}, 1 \leq x \leq m, \\ \frac{(u-x)}{(u-m)}, m \leq x \leq u, \\ 0, x > u, \end{cases}. \tag{2}$$

Finally, the generated TFN is stored in a table. A sample of generated TFN for four experts is presented in Table 6. As it is shown in the table, there are three L, M, and U numbers for each answer of every expert.

Table 4: Output of the final questionnaire.

Experts	Management support is effective on BIM usage	High cost of software and implementation is a barrier to use BIM	How client willingness effects on BIM incorporation	Staff resistance has great impact on BIM usage	Difficulty of implementation is a barrier to use BIM	Other reasons	Total
Expert1	4	3	3	5	3	1	19
Expert2	5	5	4	3	1	1	19
Expert3	3	4	4	4	2	2	19
Expert4	5	2	3	4	4	2	20
Expert5	4	3	4	5	2	3	21
Expert6	5	3	4	3	3	2	20
Expert7	2	4	5	4	2	2	19
Expert8	4	2	2	5	4	3	20
Expert9	3	5	4	5	3	2	22
Expert10	2	4	3	4	3	4	20
Expert11	5	2	4	4	2	3	20
Expert12	5	3	4	3	4	3	22
Expert13	5	4	5	4	3	4	25
Expert14	4	3	5	4	3	4	23
Expert15	3	4	4	1	2	3	17
Expert16	4	3	2	4	4	2	19
Expert17	4	5	2	3	4	2	20
Expert18	4	2	3	5	3	2	19

Table 5: Triangular fuzzy numbers for five-point scale.

Likert scale					
Linguistic terms	Strongly disagree	Disagree	Neither agree nor disagree	Agree	Strongly agree
Linguistic value	1	2	3	4	5
TFN	0.0, 0.0, 0.2	0.0, 0.2, 0.4	0.20, 0.4, 0.60	0.4, 0.6, 0.8	0.60, 0.8, 1.00

After fuzzification and experts' opinions aggregation, to get understandable numbers, the resulted values should be defuzzified. The defuzzifier receives the fuzzy input and converts it to the crisp output. There are several methods for defuzzification. The authors selected the one mentioned in eqn (3), which is one of the suggested methods in qualitative research [10].

Table 6: Score of 4 experts converted to TFN (as an example).

Expert 1			Expert 2			Expert 3			Expert 4		
L	M	U	L	M	U	L	M	U	L	M	U
0.40	0.60	0.80	0.60	0.80	1.00	0.20	0.40	0.60	0.60	0.80	1.00
0.20	0.40	0.60	0.60	0.80	1.00	0.40	0.60	0.80	0.00	0.20	0.40
0.20	0.40	0.60	0.40	0.60	0.80	0.40	0.60	0.80	0.20	0.40	0.60
0.60	0.80	1.00	0.20	0.40	0.60	0.40	0.60	0.80	0.40	0.60	0.80
0.20	0.40	0.60	0.00	0.00	0.20	0.00	0.20	0.40	0.40	0.60	0.80
0.00	0.00	0.20	0.00	0.00	0.20	0.00	0.20	0.40	0.00	0.20	0.40

Table 7: Result of defuzzification (crisp scores).

#	Questions	L	M	U	Fuzzy average	Rank
1	Management support is effective on BIM usage	7	21	14	1.76	1
2	High cost of software and implementation is a barrier to use BIM	5	17	12	1.40	4
3	How client willingness effects on BIM incorporation	6	19	13	1.55	3
4	Staff resistance has great impact on BIM usage	7	21	14	1.74	2
5	Difficulty of implementation is a barrier to use BIM	3	14	10	1.09	5
6	Other reasons	2	11	9	0.85	6

$$x_{ij}^{defuzzy} = \frac{(l_{ij}+2m_{ij}+u_{ij})}{4}. \tag{3}$$

The results of defuzzification using the defuzzifier is mentioned in Table 7.

As shown in the table, management support has a 1.76 fuzzy average score and stands as the first rank. Staff resistance with 1.74 scores is the second barrier, client willingness has a 1.55 score and stands as the third, and high cost of software and implementation with 1.4 Fuzzy average scores is the fourth barrier from the point of view of the experts. The fifth one in the table belongs to the Difficulty of implementation that has a 1.09 fuzzy average. Other reason is the last one with 0.85 fuzzy average scores.

5 CONCLUSION

In this paper using the Delphi method, a basic questionnaire filled by 74 BIM experts, collected some general information related to the research. Then, the interviews with 18 selected experts are analyzed by theme analysis method, 34 open codes are extracted, and six themes are created. In order to get the exact weight of the selected themes and according to the Delphi method, the final questionnaire created from the extracted themes, is sent to the same 18 experts. As a result, the detected barriers are management support, staff resistance, client willingness, high cost of software and implementation, difficulty of implementation, and other reason respectively. Finally, the extracted scores are converted to the triangular fuzzy numbers using the Fuzzy Delphi method. This enabled the authors to get the exact impact of each barrier and better understand the qualitative data of linguistic ideas of experts. Then the obtained TFNs are defuzzificated and the Fuzzy average and rank of each theme are calculated. The interesting point is that the output of two calculations of classic and Fuzzy Delphi are identical.

REFERENCES

[1] Jensen, P.A. & Maslesa, E., Value based building renovation – A tool for decision-making and evaluation. *Building and Environment*, **92**, pp. 1–9, 2015.

[2] Gökgür, A., Current and future use of BIM in renovation projects, 2015.

[3] Ilter, D. & Ergen, E., BIM for building refurbishment and maintenance: Current status and research directions. *Structural Survey*, **33**(3), pp. 228–256, 2015.

[4] Joblot, L., Paviot, T., Deneux, D. & Lamouri, S., Literature review of building information modeling (BIM) intended for the purpose of renovation projects. *IFAC-PapersOnLine*, **50**(1), pp. 10518–10525, 2017.

[5] Creswell, J.W., *Research Design: Qualitative, Quantitative, and Mixed Methods Approaches*, 4th ed., SAGE Publications: Thousand Oaks, 2014.

[6] Cronbach, L.J., Coefficient alpha and the internal structure of tests. *Psychometrika*, **16**(3), pp. 297–334, 1951.

[7] Corbin, J., *Basics of Qualitative Research: Techniques and Procedures for Developing Grounded Theory*, 4th ed., SAGE Publications: New York, 2014.

[8] Cheng, C., Group opinion aggregation based on a grading process: A method for constructing triangular fuzzy numbers. *Computers & Mathematics with Applications*, **48**(10–11), pp. 1619–1632, 2004.

[9] Shahyamal, A.K. & Pal, M., Triangular fuzzy matrices. *Iranian Journal of Fuzzy Systems*, **4**(1), pp. 75–87, 2007.

[10] Talon, A. & Curt, C., Selection of appropriate defuzzification methods: Application to the assessment of dam performance. *Expert Systems with Applications*, **70**, pp. 160–174, 2017.

SECTION 5
DIGITAL TWIN AND ARTIFICIAL INTELLIGENCE FOR BUILDINGS AND INFRASTRUCTURES (SPECIAL SESSION ORGANISED BY F. CUMO)

DIGITAL TWIN FOR CRITICAL INFRASTRUCTURES: THE VENTOTENE ISLAND PORT CASE STUDY, ITALY

FABRIZIO CUMO
CITERA Research Center University of Rome La Sapienza, Italy

ABSTRACT

The digital twin has been used for decades in industries and in this last twenty years in the construction industry as a model that creates a connection between the physical systems and the digital systems. Thanks to IoT and AI, a static model evolves as a constantly updated system that is in connection with the real environment, thanks to data provided by sensors placed in it. In this paper the possibility of applying this system to complex infrastructures, such as port areas, is analysed. It has been chosen to apply it to a compact port area in order to be able to use this system, to investigate the different possibilities that it offers and to then be able to use it for larger port areas. The case study is located in the Ventotene Island, a small island located in the Tyrrhenian Sea, in front of the coast of central Italy, which is characterized by a considerable number of visitors during the summer season and a significantly reduced attendance during the rest of the year. It has been created an as-built digital model developed in BIM environment. It contains the architectural, structural and MEP models, both of port area and the building which is used for different services. Studies and simulations were carried out using the model, integrated with monitoring system to increase the security of the area. Finally, the model was connected to sensors installed in the real environment, obtaining the digital twin. Using this system, it is possible to achieve the control of facility management interventions, towards predictive maintenance, but also the control of the security and energy management, in order to verify the efficiency and the profitability of the operations and of the whole infrastructure.

Keywords: BIM environment, facility management, predictive maintenance, security management, energy management, digital twin.

1 INTRODUCTION

Contemporary society is witnessing the fourth industrial revolution where everything is connected: people and machines, machines with machines.

The 21st century is characterized by an increasing interaction between the physical/real world and the virtual world. This interaction is made possible by increasing digitization and the pervasiveness of computational devices (Internet of Things – IoT) with high learning capabilities (machine learning).

Digital twins make it possible to test and understand how systems and products will behave in a wide variety of environments, using virtual space and simulation. This is done by combining different technologies into a single database that will contain all the design data for the system or product, simulation software, real-time data from the production environment and much more. The benefits are many, starting with the ability to easily access data from many different sources, to aggregate it and visualize it through a single centralized, synchronized, and shared dashboard, with the ability to add contextual information.

Digital twins can use machine learning and AI systems to process data and produce new knowledge. As evolved digital clones of the physical world, they lead to new opportunities for collaboration between product experts and data scientists whose job is to analyze data collected from a variety of sources in order to understand trends and to generate interpretative models that bring value to organizations. By standardizing data and metadata (information about data extracted from corporate knowledge, such as the position of a machine on the line and its function), other sources of information can be easily structured and linked to the digital twin, such as programming codes, documentation, articles, videos, or scientific projects. Ideally, a digital twin contains all the information of the physical object through a three-dimensional representation of its aspects at the mechanical level, at the geometrical level and at the electronic level, i.e., embedded software, micro software, product data, data associated with increasingly pervasive sensors and actuators. With this information, it is possible to develop an experimental activity, as in the case of product design, which saves on the creation of an expensive physical prototype, or a predictive activity, as in the case of the creation of a process, making it possible to predict anomalous behavior, risks, and errors in advance.

2 DIGITAL TWIN, CONCEPT AND ARCHITECTURE

The idea of accurately modelling the physical world came from National Aeronautics and Space Administration's Apollo program, to perform simulations of spacecraft from Earth [1].

DT can be defined as "a virtual mirror of the physical counterpart integrating multiple physics and scales, and it employs both dynamic sensor data and historical data from the product's life cycle". This means creating a complex virtual model that is the exact counterpart of a physical thing [2].

The concept of the digital twin was first used in 2001 by Michael Grieves, now Chief Scientist for Advanced Manufacturing at the Florida Institute of Technology, who during a Product Lifecycle Management (PLM) course at the University of Michigan described the digital twin as the virtual and digital equivalent of a physical product.

In his approach to PLM, Grieves emphasised a Mirrored Spaces Model referring to a highly dynamic representation: the real and virtual dimensions remain linked throughout the entire life cycle of the system, passing through all phases of creation, production (manufacturing), operation (support/support) and disposal. A necessary condition for the realisation of a digital twin is the existence of:

- physical products in real space;
- virtual products in virtual space;
- data and information flow linking systems that connect physical space to virtual space and virtual sub-spaces.

Digital twins make it possible to test and understand how systems and products will behave in a wide variety of environments, using virtual space and simulation. This is done by combining different technologies into a single database that will contain all the design data for the system or product, simulation software, real-time data from the production environment and much more. The benefits are many, starting with the ability to easily access data from many different sources, aggregate and visualise it through a single centralised, synchronised, and shared dashboard, with the ability to add contextual information [1]. Models, data, connections, and services are the core parts of the DT, even though there are various understandings of it. Said parts are introduced, respectively.

Models: The models, with the aid of the data collected by sensors, have four main functions. The first one is to produce an exact replica of the physical counterpart, including its properties, behaviour, and rules. The second one is to work in an autonomous way in the digital part, and to generate stimulated behaviours, also known as "ideal", that can be followed while operating on the physical counterpart. The third one is to predict problems that could occur to the physical object, in order to formulate preventive strategies. Lastly, the fourth one is to validate the performance or a system or product before its actual completion.

Data: It's the information provided by the data that allows the DT to perform continuously. Said data can be acquired from both digital and physical spaces, such as stimulated data from digital models or product lifecycle data from physical entities, and it drives the operations of the DT. For example, digital model construction, decisions in the related information systems, and operation of the entities, are all driven by data obtained from different sources. The DT works and provides further analysis and optimisation with the aid of the data, especially if it is real-time data, as it is continuously generated.

Connections: Connections allow the various entities of the DT to interact with each other, and they can be classified in connections within physical space, connections within virtual space, and connections across physical and virtual space. With the aid of said connections it's possible to interconnect physical objects for data exchange, and also to link models and information systems to obtain comprehensive information flows. The connections also allow the physical and digital components to be synchronised and consistent, in order to achieve a higher level of collaboration and optimisation.

Services: Since users may not have professional knowledge of related fields, standard services containing the DT's functions (such as prediction, evaluation, validation and optimisation) have been implemented for usage. The services are presented in their standard formats in order for them to be understandable and user-friendly. Users only need to insert parameters such as cost, time or quality to the services, to obtain the required results. The usage of the DT therefore becomes easier and more intuitive.

3 METHODOLOGY OF APPLICATION OF DT FOR INFRASTRUCTURE

The digital twin is a process applicable to a wide variety of human activities, both in the management of the environment in which we live and in the narrower but equally important area of industry and civil and industrial construction.

Table 1 shows the main areas of application for DT:

Table 1: Main application areas for digital twin.

Areas	Topics
Electric power generation	Smart grid planning, operation and maintenance
Infrastructures – roads/ports/railway	Reducing fuel consumption Fault prediction and maintenance
City management	Real time monitoring Planning and decision-making optimization
Construction	Progress monitoring Work schedule and budget adjustment Resource allocation and waste tracking Safety monitoring for workers Quality assessment for construction Improving usage rate for equipment
Security and Emergency	Mitigating existing risks or hazards Security network and infrastructure protection Catastrophic disaster prevention

By restricting the scope to the management and control of critical infrastructures, in the field of road and freight transport, such as roads, bridges, harbor quays, railway lines, etc.

The first important operation is the breakdown of the whole structure into elements, difficult because there are no manuals indicating how many pieces an infrastructure is made up of. This is why it is necessary to find the correct synthesis, so that the number of elements is neither too high nor too low. Once this is done, we start by associate the desired information deduced by projects, monitoring, surveys etc., and we connect it to the element and to the elementary component of the infrastructure, building an initial database that already completely digitalizes the inspection process. In fact, the inspecting technician will know exactly where the maintenance/reparation/inspection activities made by the colleague who intervened before him was located and how it was classified.

The next step is to create the Point Cloud, i.e. the digital image of a structure with a scan of the infrastructure and then the so-called image texturing, i.e. "spreading" the photographs on a georeferenced base of the infrastructure. The most interesting phase is when an element, a group of points or x y z coordinates, is related to the object and component of the bridge itself. In this way, by touching a point on the bridge, it is possible to identify the defect, which is automatically brought into the database precisely because it was previously associated with the asset element. The last step in this process is the use of artificial intelligence techniques to track the evolution of the single equipment/part of the structure by sending drones to the same point to take a picture or use satellite images in order to follow the evolution of the whole structure after a predefined period.

4 CASE STUDY: DIGITAL TWIN OF THE PORT AREAS OF VENTOTENE

The methodology described has been applied and verified in the port area of small island Ventotene, off the western coast of Italy, between Rome and Naples.

The main model uses are energy management [5], [6], area safety and security controls, facility management optimization and traffic pollution reduction in the dock [7].

The port area (Fig. 1(b)) is made up of a breakwater pier with a quay wall and of a square which is the main tourist and commercial docking point for ferries and hydrofoils. There is also a building, the Maritime Station which contains various services including the offices of the Port Authority, the mooring companies and the Municipality, as well as ticket offices, toilets and an infirmary.

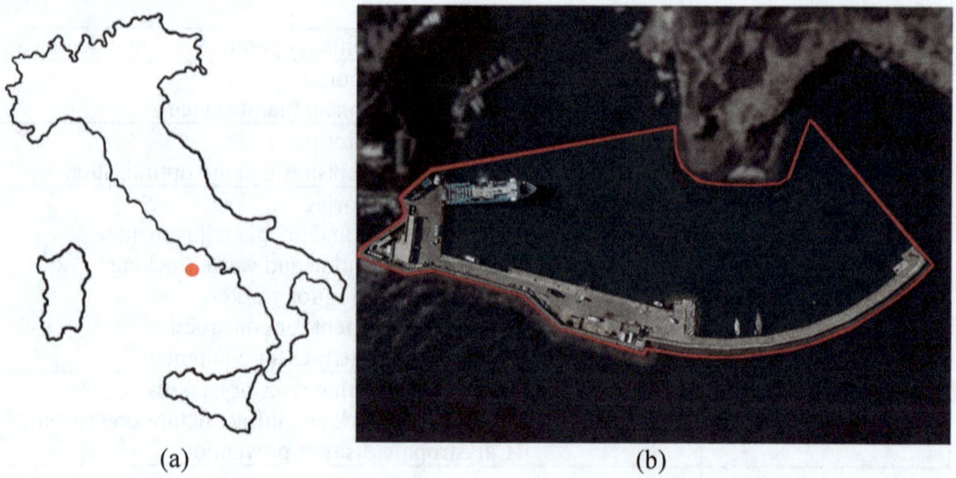

(a) (b)

Figure 1: (a) Ventotene Island in Italy; and (b) The port area.

In order to model the Masterplan, in addition to the documentation provided by the port Authority, it was necessary to carry out surveys directly on the project area. Thanks to the surveys it has been possible to find numerous inconsistencies and missing port equipment that have been updated and integrated into the model. This created a valid starting point for the following phases of the project.

Loadable families have also been created and customized according to the characteristics of the objects present in the port area. In particular have been realized special equipment dedicated to the port such as anti-reflective formwork, breakwaters, different types of steel and iron plates and bollards for docking boats.

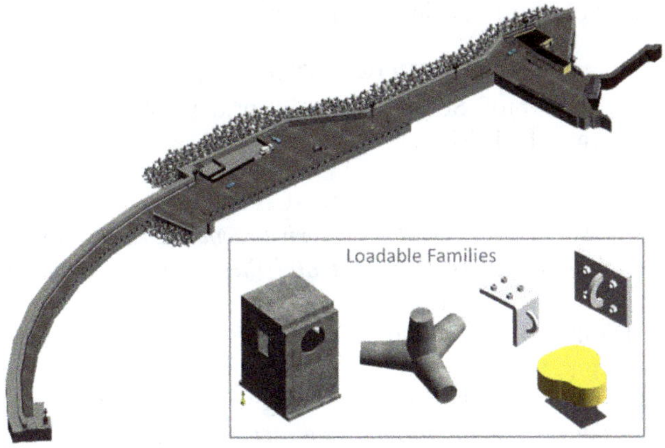

Figure 2: BIM model of the area and loadable families.

The Digital twin will be completed at the end of 2021 but the structural and technical plant [8] model has already been digitalized in BIM. This model was then imported into a GIS environment using Naviswork software [9] and studied to assess the impact on vehicular traffic (critical in the section adjacent to the Roman port) of the construction of a parking area to be used at the moment of ferry embarkation/disembarkation.

The sensor system will soon be implemented, essentially based on video cameras for image recognition, so as to be able to manage the safety and security of the area with the help of artificial intelligence systems to support the Port Authority.

5 CONCLUSIONS

The digitization process in the AEC sector is very important to keep the built environment efficient and safe, especially critical infrastructure such as ports.

Digital technologies can be used for different goals: increasing safety, optimizing the management of logistics, of mobility, of energy, of maintenance interventions.

Upgrading and adapting port areas according to the digitalization process allows for an improvement in the reliability of services and significant savings in management time and energetic costs [10].

The implementation of digital technologies requires specific skills from employees and managers of port areas. Promoting training in the use of these new tools is very important to ensure that the objectives of greater efficiency, quality and sustainability of port areas are achieved [11].

In fact, the enabling technologies underpinning DT are not substantially new and have gradually developed as a heterogeneous set of modelling systems: geometric and physical in its individual and various components (static, kinematic, dynamic, fluidic, electrical, magnetic, energy, thermal, etc.). The integrated set of all these models, plus the various modes of simulation of the individual phenomena modelled, plus the various optimization systems, all open to communication with an increasingly IOT external reality, constitutes the architecture, the substance and the disruptive potential of the Digital Twin.

The obvious advantages of using DT in critical situations are only a question of simulation response time and optimization compared to the time available to decide what to do (and not predicted in advance). But let's remember that the exponential evolution of technologies, which underpin the whole world of simulation, plays resoundingly in favor and at zero cost.

ACKNOWLEDGEMENT

The author wishes to thank prof. Sofia Agostinelli for the strong and continuous support in all Digital Twin activities of CITERA.

REFERENCES

[1] Tao, F., Liu, A. & Hu, T., *Digital Twin Driven Smart Design*, Google Books edition.

[2] Batty, M., Digital twins. *Environment and Planning B: Urban Analytics and City Science*, **45**, pp. 817–820, 2018.

[3] Centre for Digital Built Britain, 2020. https://www.cdbb.cam.ac.uk/what-we-do. Accessed: 11 Apr. 2021.

[4] Mohammadi, N. & Taylor, J.E., Smart city digital twins. *Proceedings of the 2018 IEEE Symposium Series on Computational Intelligence, SSCI 2018*, pp. 1–5, 2018. DOI: 10.1109/SSCI.2017.8285439.

[5] Acciaro, M., Ghiara, H. & Cusano, M.I., Energy management in seaports: A new role for port authorities. *Energy Policy*, **71**, pp. 4–12, 2014. DOI: 10.1016/j.enpol.2014.04.013.

[6] Iris, Ç. & Lam, J.S.L., A review of energy efficiency in ports: Operational strategies, technologies and energy management systems. *Renewable & Sustainable Energy Reviews*, **112**, pp. 170–182, 2019. DOI: 10.1016/j.rser.2019.04.069.

[7] Acciaro M. et al., Environmental sustainability in seaports: A framework for successful innovation. *Maritime Policy & Management*, **41**(5), pp. 480–500, 2014. DOI: 10.1080/03088839.2014.932926.

[8] Hautala, K., Järvenpää, M.E. & Pulkkinen, P., Digitalization transforms the construction sector throughout asset's life-cycle from design to operation and maintenance. *Stahlbau*, **86**(4), pp. 340–345, 2017. DOI: 10.1002/stab.201710474.

[9] Coronado Mondragon, A.E., Coronado, E.S. & Coronado Mondragon, C.E., Defining a convergence network platform framework for smart grid and intelligent transport systems. *Energy*, **89**(2015), pp. 402–409, 2015. DOI: 10.1016/j.energy.2015.05.117.

[10] Lam, J.S.L., Ko, M.J., Sim, J.R. & Tee, Y., Feasibility of implementing energy management system in ports. *IEEE International Conference on Industrial Engineering and Engineering Management, 2017*, pp. 1621–1625, 2018. DOI: 10.1109/IEEM.2017.8290167.

[11] Woo, J.K., Moon, D.S.H. & Lam, J.S.L., The impact of environmental policy on ports and the associated economic opportunities. *Transportation Research Part A: Policy and Practice*, **110**, pp. 234–242, 2018. DOI: 10.1016/j.tra.2017.09.001.

ACTIONABLE FRAMEWORK FOR CITY DIGITAL TWIN-ENABLED PREDICTIVE MAINTENANCE AND SECURITY MANAGEMENT SYSTEMS

SOFIA AGOSTINELLI
Department of Astronautics, Electrical and Energy Engineering, Sapienza University of Rome, Italy

ABSTRACT

The growing potential offered by the use of ICT (Information Communication Technology)-based approaches for process management in the built environment is increasingly used to configure effective digital twins of products and processes. Focusing on Smart Cities, the realization of integrated three-dimensional BIM (Building Information Modeling) and GIS (Geographic Information Systems) provides microscopic and macroscopic geometric databases containing static, dynamic, geometric and semantic data representing the information centre for an efficient management of the lifecycle in vertical and horizontal systems. Therefore, processing data and information about the lifecycle of city assets through a Digital Twin Model ensures considerable support to the process management of built environments. Through the integration of these models with Artificial Intelligence (AI) systems it becomes possible to increase the optimization and progressive functional automation of the activities interconnected to the assets' lifecycle. The proposed research presents a full-digital actionable framework based on pervasive and ubiquitous systems for security and facility management based on City Digital Twins. An integrated digital system for security management is introduced through the use of self-learning systems for unmanned security based on image recognition and AI. Finally, even maintenance operations can benefit from the configuration of predictive maintenance systems aimed at reducing costs and failures through a digital information management system combined with analytics based on data flows from sensors and historical data.

Keywords: digital twin, BIM, artificial intelligence, machine learning, security management, predictive maintenance, facility management, smart cities.

1 INTRODUCTION

Modern challenges proposed by the increasingly digitization of industrial processes, give the opportunities of developing new digital scenarios and management tools, which combined with the potential of Artificial Intelligence (AI) and Machine Learning (ML) systems, allows to optimize and automate the construction process through Cyber-Physical systems (CPS).

CPS can be defined as systems of systems, composed of heterogeneous, collaborating and ubiquitous components both physical and virtual as a Digital Twin (DT) of reality [1].

Defining a *Digital Management Framework* is the key objective for an effective full-digital building process and ecosystem based on three-dimensional information models, replicating physical objects such as buildings, managing and monitoring their interactions with reality.

The growing potential offered by ICT (Information Communication Technology)-based approaches opens renewed scenarios for process management in the built environment through BIM (Building Information Modeling) and GIS (Geographic Information Systems) integrated approaches. In fact, it becomes possible to microscopic and macroscopic geometric databases containing static, dynamic, geometric and semantic data representing the core information for an optimized management phase in the life cycle of vertical and horizontal systems. Through the integration of BIM models with AI systems it is possible to increase the optimization and progressive functional automation of the activities interconnected to the life cycle of assets, configuring Digital Twin (DT) systems.

WIT Transactions on The Built Environment, Vol 205, © 2021 WIT Press
www.witpress.com, ISSN 1743-3509 (on-line)
doi:10.2495/BIM210191

DT is a concept deriving from the high-value manufacturing industries, and it is going to be gradually used also in the construction sector. DT of buildings reproduce their geometric and semantic information as 3D models configured and structured according to databases criteria, where data are contained within objects composed of specific parameters and attributes describing the components themselves, thus enclosing useful information for the definition and simulation of processes.

The DT system is able to improve and enriching its knowledge and data receiving input and signals from sensors, constantly monitoring buildings, developing self-learning capabilities and predictivity through the integration with AI systems.

Therefore, the concept of DT is extremely transversal and applicable to various and diversified areas.

The so-called Key Enabling Technologies allow the configuration of Digital Twins combining sensors for detecting and transmitting data, computing power for data processing, and the hardware and software for analyzing and simulating specific scenarios. This framework allows to configure systems for managing urban metabolism, as well as configuring predictive maintenance of building systems defining intelligent human-scaled urban environments.

From literature analysis different applications of City Digital Twins emerged: the KIRA-digi pilot project is aimed at producing the DT of the Kalasatama area near to Helsinki [2], as a smart city development configuring a platform for designing, testing, applying and servicing the entire lifecycle of the built environment; Virtual Singapore is a dynamic three-dimensional city information model and collaborative data platform of Singapore [3], aimed at public, private and research uses enabling users to join simulation, planning and decision-making tools, solving emerging challenges for the city; even the DT of the port of Rotterdam [4], using AI analyzing data, is able to do accurate real-time predictions about moor and departures, reducing waiting times and costs.

As described in Fig. 1, the present paper identifies specific experimental domains for City Digital Twin such as security management, i.e., the physical security of places through advanced computer vision-based surveillance cameras, and facility management, i.e., the management of the built environment through predictive maintenance.

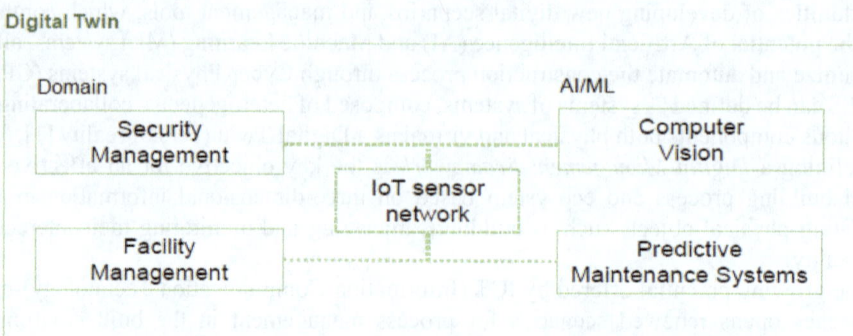

Figure 1: DT-based domain and technology integration.

2 CITY DIGITAL TWIN ENABLING TECHNOLOGIES

The smartification of digital systems towards defining Collaborative CPS as a combination of virtual and physical assets, has determined the need for specific digital design approaches.

A framework has been developed achieving generic scenarios bridging them on promising real cases. Considering the integration of physical and virtual worlds as a significant support for people daily experiences, from business to healthcare or elderly care. An integration of data flows coming from different sources as well as predictive models based on historical data and current performances of the physical system is necessary to achieve effective DTs purposes.

Therefore, the development of a Digital Twin Model (DTM) is supported from the configuration of a multidisciplinary integrated information model (BIM), containing useful process information and communicating data coming from different sources. Moreover, AI provides DTs with learning capabilities, implementing autonomous actions based on the performed elaborations.

The configuration of autonomous systems is aimed at analysing and processing data, implementing corrective actions, reducing errors, waste and systems malfunctions.

The described construction process of the DT at macroscopic level meets the synergy between Big Data, i.e. a huge amount of data about the city and AI systems, capable of processing and analysing them to implement strategies. Therefore, the very first DT configuration phase is related to the production of structured data and information, coming from their collection in databases and information models, up to data processing and analysis.

Thus, configuring the DT of a residential district can be a key tool for the storage, visualization, analysis and creation of data useful for the management of urban life. Moreover, given the absolute centrality of data systems in the configuration of digital processes, the integrated use of GIS and BIM technologies allows information management both at the building and territorial scale.

Therefore, although they share the essential concept of describing the real world through the combination of visual representations and information, BIM and GIS are conceived and developed as belonging to different domains. In fact, they present divergences in the level of detail, as BIM can be used to create, manage and share lifecycle data of vertical structures, e.g. buildings, while GIS can store, manage and analyze data describing the horizontally distributed urban environment.

2.1 Internet of things for smart cities

The Internet of Things (IoT) allows connections between physical world and virtual information at any time, in any place and between anything [5], defining the configuration of smart systems through heterogeneous and ubiquitous objects [6] originating the development of interactive object automation to exchange information [7].

However, one of the main components of IoT systems are the Wireless Sensor Networks (WSN) [8] which interconnect sensors in order to obtain data stored in a server to analyze and automate tasks. Actuators allow the execution of actions on motors, machines, and so on. Moreover, WSN and actuator systems can be connected in systems known as Wireless Sensor and Actuator Network (WSAN). Also, Smart Objects are considered as core IoT components, as they can perform actions having sensors capabilities and being able to process information as well as performing actions using an internet connection.

In the configuration of smart urban systems and Smart Cities, different types of sensors can be distributed to gather information about city management processes. The integration of ubiquitous communication networks, WSN, WSAN, and intelligent systems, allow Urban Intelligence, improving citizens daily life [9]. Some European Smart Cities like Luxemburg defined the Smart City concept basing on six indicators and a European project called

SmartSantander is another example [10], proposing an IoT architecture for Smart Cities, showing and offering different urban services.

2.2 Computer vision for enhanced security management

Modern security management systems for buildings, residential complexes and urban areas are increasingly based on technological "unmanned" systems, capable of acquiring, interpreting and finally suggesting methods to ensure the safety of the inhabitants.

Being able to have a digital model allows security managers to virtually test the effectiveness of the positioning of control IoT sensors, as well as the surveillance systems, and the timeliness of the implementation of consequent actions to secure the monitored scene. Current security systems are based on video cameras, drones and perimeter and volumetric motion IoT sensors, but the use of image recognition algorithms allows to detect potentially dangerous situations, without the human presence in operating rooms.

AI applications based on self-learning capabilities can benefit from a DT system that allows to enrich the database with the results of simulations aimed at obtaining a virtual replica of the controlled area.

Computer vision is an emerging approach to advanced digital security management using facial recognition applications. The use of computer vision as a public safety tool is still under exploration, focusing on the possible integration of advanced surveillance cameras computers for a possible inexpensive solution for digital security management. Computer vision offers the main benefit of substituting human monitoring with automated analysis, where the human assumes a supervisory assessment and decision role.

Computer vision digitizes a visual image into a grid of pixels where a numerical value is assigned to each pixel, representing its colour. This process generates a two-dimensional grid of numbers for each video frame.

The assigned numbers are fundamental for enabling the computer vision analysis, interpretation and labelling capabilities, defining a picture-to-number process making computer vision possible.

In essence, computer vision processes determine a numerical amount of pixel values providing information about an image or frame, such as identifying people and places, face recognition and image stabilization for automated real-time video stream analysis or even post-hoc searches of archived video files.

Live real-time video analysis involves the possibility to promptly identify and correctly respond to ongoing events, as the capability and effectiveness of a surveillance system is in reducing crime through real-time intervention, also reducing the amount of time currently spent reviewing and searching video frames.

An online dictionary learning approach to detect abnormalities is being pursued to cope with these challenges, cutting long videos into small clips and comparing detected elements with those in a dictionary for the identification of abnormalities.

Computer vision capability is also to automatically analyze video frames and screen out irrelevant information [11]. An identified solution approach is to use semantic indexing in computer vision, deep learning and foreground object detection to determine a small set of real-time suspicious video clips from multiple camera feeds or from a video archive [12].

The proposed DT system is based on the configuration of an integrated digital system for unmanned security management using self-learning systems and AI based on image recognition, which increase security levels in urban areas, minimizing the risks of aggression, theft, intrusion and forcing, gatherings etc. [13].

Therefore, the objective is the configuration of a completely unmanned security system, through the integration of control systems with self-learning and AI based on image recognition.

This allows to increase the overall security levels by reducing the risk of attacks, thefts, intrusions and break-ins through specific algorithms capable of recognizing images through a network of video surveillance cameras integrated with AI systems.

Through the DTM it is possible to evaluate data coming from the risk database based on a BIM Model which contains a specific analysis of the areas subjected to computer vision, according to the risks associated to the current security conditions. The DTM also provides analysis and simulation for different design solutions suggesting camera types according to the lighting system necessary to ensure high levels of security.

Then the result of the DTM-based risk analysis is the definition of a level of danger for each area evaluated through a security system that can identify the nature of the area as well as its current level of security, assigning a tag for the digital monitoring of the entire area.

2.3 Predictive maintenance systems

Building Management systems based on predictive maintenance [14], [15] are going to radically change the way buildings and complex areas are going to be maintained. The DTM will help facility managers to easily access data and promptly carry out operations and maintenance works even remotely, having data available to ensure constant monitoring [16]. Consequently, the DTM can be used to evaluate the current condition of the asset [17]; to predict future behaviours, both in conditions of normal obsolescence or referring to extreme events, through the improvement of performance control [18], [19] and the possibility of constantly optimizing operations.

In fact, the added value of predictive maintenance [20] consists in the ability to determine the remaining useful life of a plant, as well as of a component or a material, to establish what will be the most appropriate time to perform maintenance or replacement of individual elements [21].

The DTM will allow facility managers to make predictions about what will happen to a building or infrastructure in the future [22], [23] creating new possibilities for optimized activity planning and scheduling, allowing maintenance managers to get real-time detailed reports about each individual system, accurately monitoring performances over time.

Thus, AI is the enabler of prediction according to DTM-based simulations [24] and analysis, depending on the amount of available data as time series [25].

In fact, before the advent of AI and DTMs, facility managers used to spend hours in evaluating costs and benefits of any intervention before implementing it, without having support from real data to evaluate and simulate the actual success chances and the consequences of their choices [26]. Otherwise, DTM-based predictive maintenance allows facility managers to test every single element before activating it, and to analyze in advance what will be the impact of each decision [27], promptly responding to any failures [28]. Thanks to the amount of data continuously collected through sensors, more detailed analysis can be performed highlighting new management possibilities [29].

2.4 Digital twin-based advanced building management systems

As an example, the operations and maintenance of a residential district can benefit from the configuration of predictive maintenance systems aimed at reducing operating costs and malfunctions [30], through the definition of a digital maintenance system based on IoT [31] and historical data, continuously updated through the DTM. Fig. 2 shows the DT framework architecture.

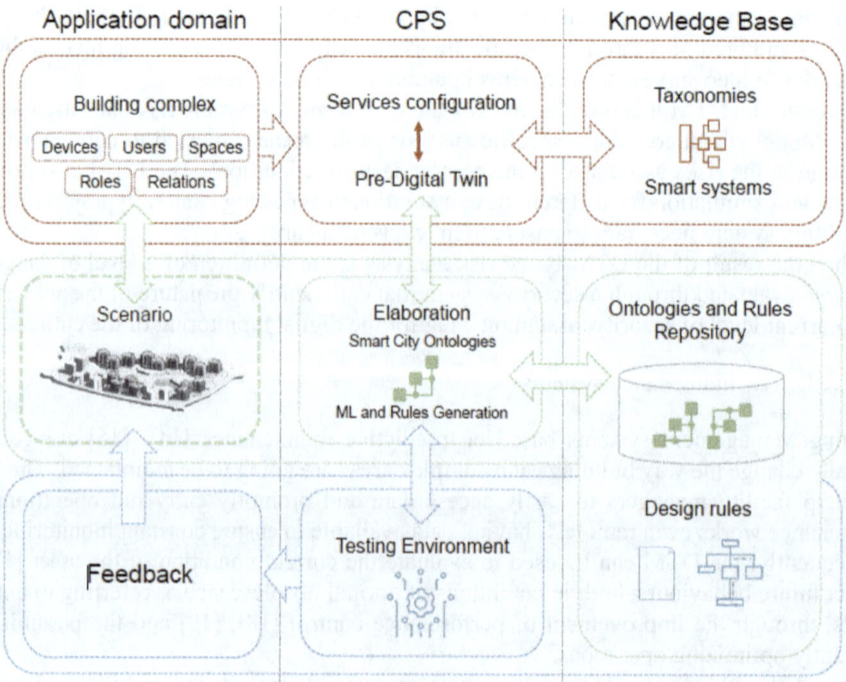

Figure 2: DT framework architecture.

Moreover, the techniques of Machine and Deep Learning combined with mechanisms of automatic reasoning allow to strategically exploit the large amount of information available, through features for (1) the acquisition of heterogeneous data streams; (2) prediction of failures; (3) profiling of users' behaviors and operating states of machinery; (4) optimized planning of raw material supplies and processing shifts; (5) real-time monitoring of systems and infrastructures; (6) multidimensional analysis of information.

The proposed solution is based on automatic reasoning, which is complementary to the inductive models, monitoring the scheduled maintenance activities using rule-based approaches for the definition of malfunctioning scenarios, aiming at a progressive decrease of breakdown interventions.

The BIM Model has a central role in the architecture of the information management system, as a physical and 3D spatial information container aimed at optimizing operations and maintenance interventions, integrating ML systems using rule-based approaches such as association rule mining [32].

Therefore, in the present framework it seems useful to find a hierarchy in the classification of spaces using clustering [33] machine learning techniques, which automatically identify groups/classes of similar spaces for their digital representation through a scalable and modular approach.

The implementation of the proposed system is based on the integration of three components:

- data from BIM three-dimensional information containers;
- data flows automated through visual programming systems (e.g. Autodesk Dynamo) for a bidirectional connection of the WSN, controlled by AI systems, to the BIM Model, configuring a reliable DTM;
- ML systems for data processing using AI based on Phyton, due to efficiency, reliability and compatibility reasons with flows programmed through BIM-based visual programming systems; the AI system could be implemented through mixed techniques of hybrid reasoning and regression analysis [34].

Then, in the context described above, closed circuit television (CCTV) systems represent an additional source of data useful even for optimizing monitoring as well as for the identification of breakdowns, which can be associated and integrated with the alerts generated by Ticketing and Building Management System (BMS) as described in Fig. 3.

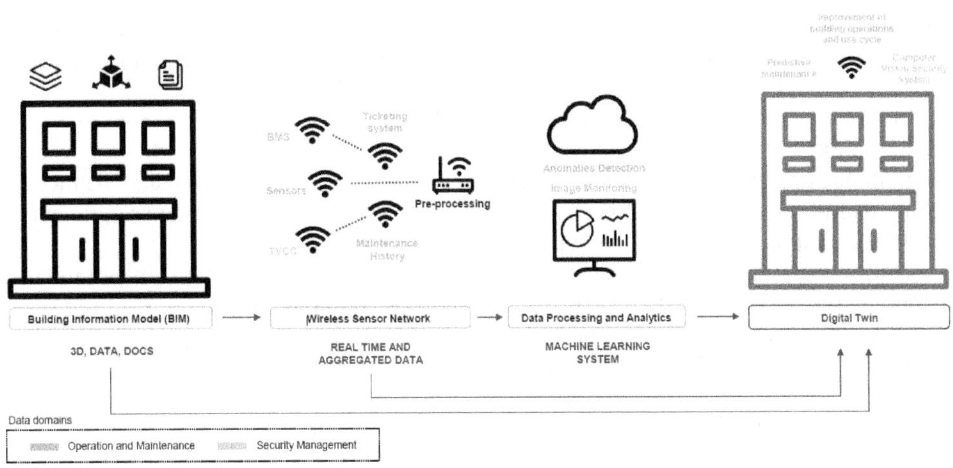

Figure 3: Functional system architecture.

In fact, cameras capture day and night images for security reasons even in closed or unreachable environments. Consequently, they can be considered as a source of useful data for machine learning and computer vision systems to process and compare the chromatic range variation [35] as the performance of the lighting fixtures decreases, or the state of conservation of elements/components that are difficult to reach (roofs, facade elements, etc.).

As described, the main objective is the reconfiguration of the operations and maintenance model traditionally limited to plant operation and fault interventions, by implementing digital and intelligent systems for predictive maintenance based on sensors and historical data [36].

As defined, the virtual monitoring through digital systems can reduce maintenance interventions allowing a considerable saving on the overall costs, eliminating breakdown-maintenance, and reducing inefficiencies for the users.

The first step is the definition of a database of the annual maintenance operations overall duration for each type of intervention, analyzing the maintenance sheets and integrating data within the DTM.

Therefore, the overall sum of the calculated interventions gives a cost that can be considered as a fixed annual fee for the single user, without variations coming from emergency interventions. From a preliminary analysis on a residential district composed of 16 buildings, it emerges how the use of drones integrated with image recognition systems for building monitoring through the same cameras used for security systems, could lead to a decrease in working hours with a consequent annual saving of about €25,000.

3 CONCLUSIONS

The proposed framework shows how the integration of AI in DTM-based management systems can develop cognitive building systems with predictive capabilities, implementing autonomous decisions based on the analysis performed, representing a methodological approach replicable for different areas and uses.

Moreover, the potential of DT systems combined with AI can build an efficient virtual ecosystem of control and management in urban residential areas, as well as for complex infrastructures or machines. In particular, the configuration of a full-digital and integrated system for security management defines an ecosystem of unmanned security potentially integrated with technologies for the optimization of maintenance operations, with the objective of reducing costs and malfunctions, configuring digital systems for predictive maintenance and aiming at the progressive elimination of failures.

Computer vision can address issues associated with human errors in current public space security systems. In fact, an automated security camera system only detects what is programmed to view, reducing errors and ineffectiveness coming from humans' distractions during real-time monitoring. Moreover, AI will not be focused on detecting specific events resulting in important events going unnoticed.

Thus, integrating AI into facility management processes means using software that can autonomously make decisions and solve problems through deduction and reasoning processes by collecting and analyzing large amounts of relevant data, supporting decisions and providing insights [37]. ML elaborates information and provide the ability for intelligent systems to learn automatically and improve themself basing on experience, without being explicitly programmed to do so. Therefore, a major advantage in this regard is the ability to identify and eliminate routine and repetitive tasks [38] that can be easily automated. Therefore, facility management using service automation systems benefits from automating tasks such as recurring scheduled maintenance and mandatory inspections of critical equipment. AI and ML then take automation to the advanced level of determining which tasks to automate to achieve increased efficiency, making the system smarter gradually over time as the amount of analyzed data increases.

Further developments on this research will be focused on implementing the proposed framework on an experimental residential compound, determining results in cost savings and automated processes.

REFERENCES

[1] Nazarenko, A. & Camarinha-Matos, L.M., Basis for an approach to design collaborative cyber-physical systems. *Technological Innovation for Industry and Service Systems*, Costa de Caparica, Portugal, 8–10 May 2019, vol. 553, Springer: Cham, Switzerland, pp. 193–205, 2019.

[2] Ruohomäki, T., Airaksinen, E., Huuska, P., Kesäniemi, O. & Martikka, M. & Suomisto, J., Smart City Platform Enabling Digital Twin. *International Conference on Intelligent Systems (IS)*, pp. 155–161, 2018. DOI: 10.1109/IS.2018.8710517.

[3] Singapore experiments with its digital twin to improve city life. https://www.smartcitylab.com/blog/digital-transformation/singapore-experiments-with-its-digital-twin-to-improve-city-life/. Accessed on: Jun. 2021.

[4] Rotterdam's Digital Twin Redefines Our Physical, Digital, and Social Worlds. https://smart-cities-marketplace.ec.europa.eu/news-and-events/news/2019/rotterdams-digital-twin-redefines-our-physical-digital-social-worlds. Accessed on: Jun. 2021.

[5] International Telecommunication Union, Overview of the Internet of Things, 2012, 14. https://www.itu.int/rec/T-REC-Y.2060-201206-I.

[6] Jara, A.J., Sun, Y., Song, H., Bie, R., Genooud, D. & Bocchi, Y., Internet of Things for cultural heritage of smart cities and smart regions. *2015 IEEE 29th International Conference on Advanced Information Networking and Applications*, IEEE: Gwangiu, pp. 668–675. DOI: 10.1109/WAINA.2015.169, 2015.

[7] Benalla, M., Achchab, B. & Hrimech, H., A distributed intelligent system for emergency convoy. *International Journal of Interactive Multimedia and Artificial Intelligence*, **4**, p. 42, 2016. DOI: 10.9781/ijimai.2016.418.

[8] Delamo, M., Felici-Castell, S., Pérez-Solano, J.J. & Foster A., Designing an open source maintenance-free environmental monitoring application for wireless sensor networks. *Journal of Systems and Software*, **103**, pp. 238–247, 2015. DOI: 10.1016/j.jss.2015.02.013.

[9] González García, C., García-Bustelo, C.P., Espada, J.P. & Cueva-Fernandez, G., Midgar: Generation of heterogeneous objects interconnecting applications. A domain specific language proposal for internet of things scenarios. *Computer Networks*, **64**, pp. 143–158, 2014. DOI: 10.1016/j.comnet.2014.02.010.

[10] Sanchez, L., et al., SmartSantander: IoT experimentation over a smart city testbed. *Computer Networks*, **61**, pp. 217–238, 2014. DOI: 10.1016/j.bjp.2013.12.020.

[11] McCarthy, O. & O'Mahony, M., End user response to an event detection and route reconstruction security system prototype for use in airports and public transport hubs. *Transportation Research Board 95th Annual Meeting*, Washington, DC. http://amonline.trb.org/trb60693-2016-1.2807374/t001-1.2823436/254-1.2823593/16-5450-1.2980693/16-5450 1.2993283?qr=1. Accessed on: 1 Jul. 2020.

[12] Mitchell, S., Villa, N., Stewart-Weeks, M. & Lange, A., The Internet of everything for cities: Connecting people, process, data, and things to improve the "livability" of cities and communities. pp. 1–21, 2013. http://www.cisco.com/web/strategy/docs/gov/everything-for-cities.pdf.

[13] Ayub Khan, A., Ali Laghari, A. & Ahmed Awan, S., Machine learning in computer vision: A review. *ICST Transactions on Scalable Information Systems*, 2021.

[14] Mihai, S. et al., A Digital Twin Framework for Predictive Maintenance in Industry, 4.0, 2021.

[15] Melesse, T.Y., Di Pasquale, V., Riemma, S., Digital twin models in industrial operations: A systematic literature review. *Procedia Manufacturing*, 2020.

[16] Barricelli, B.R., Casiraghi, E. & Fogli, D., A survey on digital twin: Definitions, characteristics, applications, and design implications. *IEEE Access*, **7**, pp. 167653–167671, 2019. DOI: 10.1109/ACCESS.2019.2953499.

[17] Zheng, Y., Yang, S. & Cheng, H., An application framework of digital twin and its case study. *Journal of Ambient Intelligence and Humanized Computing*, **10**, pp. 1141–1153, 2019. DOI: 10.1007/s12652-018-0911-3.

[18] Snijders, R., Pileggi, P., Broekhuijsen, J., Verriet, J., Wiering, M. & Kok, K., Machine learning for digital twins to predict responsiveness of cyber-physical energy systems. *Paper Presented at the 8th Workshop on Modeling and Simulation of Cyber-Physical Energy Systems, MSCPES 2020 – Proceedings*, 2020. DOI: 10.1109/MSCPES49613.2020.9133695.

[19] Geyer, P. & Singaravel, S., Component-based machine learning for performance prediction in building design. *Applied Energy*, **228**, pp. 1439–1453, 2018. DOI: 10.1016/j.apenergy.2018.07.011.

[20] Aivaliotis, P., Georgoulias, K. & Alexopoulos, K., Using digital twin for maintenance applications in manufacturing: State of the art and gap analysis. *Proceedings of the IEEE International Conference on Engineering, Technology and Innovation (ICE/ITMC)*, IEEE: Piscataway, USA, pp. 1–5, 2019.

[21] Alam, K.M. & El Saddik, A. C2PS: A digital twin architecture reference model for the cloud-based cyber-physical systems. *IEEE Access*, **5**, pp. 2050–2062, 2017. DOI: 10.1109/ACCESS.2017.2657006.

[22] Chhetri, S.R., Faezi, S., Canedo, A. & Faruque, M.A.A., QUILT. *Proceedings of the International Conference on Internet of Things Design and Implementation – IoTDI '19*, eds. O. Landsiedel & K. Nahrstedt, Montreal, Canada, ACM Press: New York, USA, pp. 237–248, 2019.

[23] Enders, M.R. & Hoßbach, N., Dimensions of digital twin applications – A literature review. *Proceedings of the 25th Americas Conference on Information Systems*, Cancun: Mexico, pp. 1–10, 2019.

[24] Otto, B., Hompel, M. ten & Wrobel, S. Reference architecture for the digitization of industries. *Digital Transformation*, ed. R. Neugebauer, Springer Vieweg: Berlin, Germany, pp. 109–128, 2019.

[25] Tao, F. et al., Digital twin-driven product design framework. *International Journal of Production Research*, **57**(12), pp. 3935–3953.

[26] Simplifying 5G with the Network Digital Twin, Spirent. https://www.spirent.com/assets/wp/wp simplifying-5 g-with-thenetwork-digital-twin. Accessed on: 16 Mar. 2020.

[27] Aivaliotis, P., Georgoulias, K. & Alexopoulos, K., Using digital twin for maintenance applications in manufacturing: State of the art and gap analysis. *IEEE International Conference on Engineering, Technology and Innovation (ICE/ITMC)*, pp. 1–5, 2019.

[28] Raza, M., Kumar, P.M., Hung, D.V., Davis, W., Nguyen, H. & Trestian, R., A digital twin framework for industry 4.0 enabling next-Gen manufacturing. *9th International Conference on Industrial Technology and Management (ICITM)*, Oxford, pp. 73–77, 2020.

[29] Barthelmey, A., Lee, E., Hana, R. & Deuse, J., Dynamic digital twin for predictive maintenance in flexible production systems. *IECON 2019 – 45th Annual Conference of the IEEE Industrial Electronics Society*, Lisbon, Portugal, pp. 4209–4214, 2019.

[30] Liang, Y.C., Li, W.D., Lu, X., & Wang, S., Fog computing and convolutional neural network enabled prognosis for machining process optimization. *Journal of Manufacturing Systems*, **52**, pp. 32–42, 2019.

[31] Xu, L.D., He, W. & Li, S., Internet of Things in industries: A survey. *IEEE Transactions on Industrial Informatics*, **10**(4), pp. 2233–2243, 2014. DOI: 10.1109/TII.2014.2300753.

[32] Telikani, A., Gandomi, A.H. & Shahbahrami, A., A survey of evolutionary computation for association rule mining. *Information Sciences*, **524**, pp. 318–352, 2020. DOI: 10.1016/j.ins.2020.02.073.

[33] Bock, H.H., Clustering methods: A history of k-means algorithms. Selected contributions in data analysis and classification. *Studies in Classification, Data Analysis, and Knowledge Organization*, eds. P. Brito, G. Cucumel, P. Bertrand & F. de Carvalho, Springer: Berlin/Heidelberg, Germany, pp. 161–172, 2007. DOI: 10.1007/978-3-540-73560-1\15.

[34] Van de Moortel, K., *Multidirectional Regression Analysis*, 2021. DOI: 10.13140/RG.2.2.16703.64162.

[35] Wang, Y. & Wang, C., Computer vision-based color image segmentation with improved kernel clustering. *International Journal on Smart Sensing and Intelligent Systems*, **8**, pp. 1706–1729, 2015. DOI: 10.21307/ijssis-2017-826.

[36] Negri, E., Fumagalli, L. & Macchi, M., A review of the roles of digital twin in CPS-based production systems. *Procedia Manufacturing*, **11**, pp. 939–948, 2017. DOI: 10.1016/j.promfg.2017.07.198.

[37] Napoleone, A., Macchi, M. & Pozzetti, A., A review on the characteristics of cyber-physical systems for the future smart factories. *Journal of Manufacturing Systems*, **54**, pp. 305–335, 2020. DOI: 10.1016/j.jmsy.2020.01.007.

[38] Marmolejo-Saucedo, J.A., Hurtado-Hernandez, M. & Suarez-Valdes, R., Digital twins in supply chain management: A brief literature review. *International Conference on Intelligent Computing and Optimization*, pp. 653–661, 2019.

COMPLEX WORKS PROJECT MANAGEMENT ENHANCED BY DIGITAL TECHNOLOGIES

FRANCESCO RUPERTO[1] & STEFANIA STRAPPINI[2]
[1]Department of Planning, Design and Technology of Architecture (PDTA), University of Rome La Sapienza, Italy
[2]ENEL Task Force, Italy

ABSTRACT

The digital transformation of the construction sector is aimed at increasing the levels of effectiveness and efficiency of construction and operating project management, improving, the expected levels of sustainability and favoring risk management with a data-based knowledge. The paper describes, in methodological and operational terms, the main research and development activities of the ENEL HQVRM – Digital Twin that is aimed at project management objectives of ENEL HQVRM Task Force for the new ENEL Head Quarters in Rome Viale Regina Margherita. The paper mainly illustrates the experimentation, conducted by a joint ENEL–CITERA working group, of modalities for platforming the project's organizational model, which allows for the collection, sharing and measurement in streaming of the main parameters and the simultaneous verification of compliance with contractual provisions. The HQVRM (Head Quarters Viale Regina Margherita) – Digital Twin is based on a common data environment (CDE) of the organizational model and from the point of view of technologies is enabled by cloud computing, artificial intelligence, machine learning, big data and building information modeling (BIM). HQVRM–CDE represents the connective of the digital ecosystem in which information models are organized depending on the need and the objectives to be pursued in particular for the temporal, economic, safety and sustainability aspects. From the client's perspective, the research investigates and experiments by implementing a CDE, replicable and scalable Digital Twin, in order to make the potential of ICT technologies available to project stakeholders in managing the complexities of a building for office use high architectural and representative value, with significant structural interventions and high plant integration for a total amount of €150 million in more than 3 years long works. An intervention that summarizes in itself the meanings of complexity as referable to those works characterized by particular complexity in relation to the type of works, the use of innovative materials and components, the execution in places that present logistical difficulties or particular environmental issues.
Keywords. project management, digital twin, BIM, artificial intelligence, cognitive services.

1 INTRODUCTION

The digital transformation involves companies in every sector and it defines a perimeter that is still being defined and that comes to involve the actions of people in the work, personal and social sphere. With digital technologies, organizations change, jobs change, individuals change. Innovative technologies disrupt established practices and disciplines, reshaping new ways of operating by forcing a project that also involves organizational models, work and skills integrated with each other. In this context, the management of the transition makes many managers of organizations oscillate between relying completely on technology suppliers or choosing the path of resistance to change in the name of a comfort zone that is reassuring but whose inefficiency is equally well known. This unstable equilibrium is also experienced in the AECO sector and involves large client companies, designers and companies each called their own way of interpreting BIM, an acronym that is too synthetic to represent a multiplicity of meanings.

This paper describes the effort put in place by ENEL S.p.A., Sapienza University of Rome in digitally transforming the operating methods of the construction site of the new Rome headquarters in Viale Regina Margherita. The project called HQVRM CDE is still in progress

WIT Transactions on The Built Environment, Vol 205, © 2021 WIT Press
www.witpress.com, ISSN 1743-3509 (on-line)
doi:10.2495/BIM210201

and is aimed at creating complex systems for risk management and for a better quality of the work. A work in progress that involves a complex organizational model that also includes the companies responsible for both the PMO and the construction of the work including the entire supply chain. This research which constitutes a useful opportunity to experiment a methodological approach of a construction site digital twin [1]–[3] implementation, potentially able at the same time to evaluate changes in project efficiency, to measure the attitude of the operators and to foresee further developments also in the transposition of the project information model (PIM) into an asset information model (AIM).

2 NEW ENEL HEADQUARTER IN ROME PROJECT

The project concerns the renovation of the ENEL headquarters located in Rome in Viale Regina Margherita. The building consists of seven buildings connected to each other and spread over a total area of about 70,000 square meters, of which 55,000 square meters above ground. Following the renovation works, in addition to the traditional workstations, it will also have focus stations (for informal meetings, sharing, concentration, etc.) and various spaces and ancillary services (auditorium, meeting rooms, areas for informal meetings, company bar, kindergarten, gym, etc.). The redevelopment of the property will take place in autonomous functional lots; this aspect will require particular attention to the management of the workings and interferences, in order to guarantee the operational continuity of the portions of the complex from time to time not affected by the interventions carried out on the single functional lot.

Figure 1: New ENEL Headquarters: Perspective view.

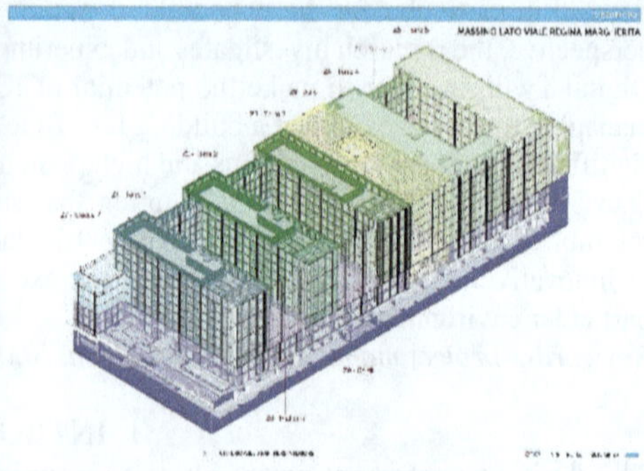

Figure 2: New ENEL Headquarters: Axonometric view.

2.1 Organisational model

ENEL has chosen to manage the construction of the new headquarters by entrusting the design and verification services to the architecture company Citterio Viel and Partners in 2018. Subsequently, in December 2019, the PMO and works management services were entrusted to an RTI formed by the companies Jacobs Italia S.p.A. and Artelia Italia S.p.A. In October 2020, the construction of the building to Colombo Costruzioni S.p.A. as main contractor.

2.2 ENEL Task Force and CITERA Group

Enel (Italy's national entity for electricity) was founded in 1962 with the fusion of energy producers. With a solid bedrock of hydroelectric power as our starting point, the task was to power the country's rapid growth. The energy crisis of the following decade spurred the Company to explore alternative sources of power how renewable energy and ENEL pioneered a number of renewable energy plants around the world: hydroelectric plants, photovoltaic power station, and wind farm. Innovation and sustainability were become for ENEL two biggest drivers. Now that climate change has become increasingly evident, ENEL are fully aware of the impact the new energy era can have, with investments in digitalization, e-mobility, circular economy, and decarbonization.

Today ENEL sustainable approach is to create long-term shared value is guiding the energy transition, and ENEL aim to carry on finding new, innovative tools to contribute to positive growth, without ever losing sight of the future of Company. The digital revolution is one of the cornerstones of Enel's industrial strategy.

In May 2019, ENEL set up the Task Force Headquarters, in order to manage in a coordinated and integrated manner the execution of the renovation works of the Viale Regina Margherita complex, with the aim of:

- implement the tender project by directly managing the activities related to the execution of the works (site management, safety management, accounting, payments and testing);
- ensure compliance with the specifications, deadlines and costs budgeted, by monitoring the progress and managing all those involved, both internal and external to the company.

The structure was divided into four functional areas as listed below:

- Project health, safety and environment to ensure the application of policies, procedures and operating instructions on safety and the environment, supervise the management of safety and the environment on site, supervise, construction site with those inside and outside the building.
- Project design for the management of the integrated executive design and monitoring of the development of building information modeling, assessment of any non-conformities. In particular for the collection of data relating to the certification / homologation of the works, the use and maintenance manuals, all related to the as-built information model of the building, to be made available to the Services Italy unit for the management of the facilities management.
- Project execution for the execution of constructability analyzes of construction activities, verifying their feasibility with reference to the project requirements, coordination of construction site activities ensuring the achievement of objectives in terms of safety, environment, quality, time and costs.
- Project control for the definition, planning and monitoring of the activities necessary for the development of the project, implementation of the ordinary and extraordinary management of contracts and carrying out the associated risk analysis, ensuring the definition and execution of intercompany contracts, providing adequate transfer pricing criteria in collaboration with the other functions involved.

ENEL Task Force is supported for digital governance themes and by a Sapienza University of Rome CITERA Group.

2.3 Project's information requirements

For this specific project, general strategic needs have been established, aimed at defining the operational specifications for the information management of the building process.
In particular, the aim was to obtain:

Table 1: Main project information requirements.

Improvement of the level of knowledge of the properties	Lower construction costs
Optimization of the design and subsequent execution phases in compliance with contractual deadlines	Availability of reliable and useful information for the management of the works in the subsequent operational phase
Optimization of the bioclimatic balance of the choices referred to the materials and systems of materials present in it environment	Decision-making processes more supported by timely, updated and reliable information throughout the life cycle of the building
Improvement of the health and safety of workers employed on the construction site;	Optimization of work management with BIM methodology
Mitigation of the risks of variations in progress	Highlighting the breakdown of costs into their primary components
Greater control of the execution times of the works	Collect and manage some data through the model that can contribute to the economic/ financial evaluation of the general sustainability of the work
Reduction of legal claims	Implement a supervision of the sustainability and the bioclimatic system of the entire design and construction process of the building

3 CDE AS A PROJECT DIGITAL TWIN

Collaboration between the participants involved in construction projects and in the management of real estate assets has a fundamental importance to ensure their efficient provision and management. Organizations are increasingly working in the context of new collaborative environments in order to achieve higher levels of quality and greater reuse of existing knowledge and experience. A core component of these collaborative environments is the ability to communicate, reuse and share data efficiently without data loss, contradiction or misrepresentation. This approach does not require any more work, as it has always been necessary to produce this information. However, true collaborative work requires mutual understanding and trust within the team and a deeper level of standardized process than previously experienced if information is to be produced and made available in a consistent and timely manner.

Information requirements must reach even the low levels of supply chains till the point where information can be produced more efficiently; furthermore, information must be collected as it moves up the supply chains. Currently, significant resources are devoted to correcting non-standard data, training new staff on approved data creation techniques, coordinating the efforts of subcontractor groups, and solving problems related to data reproduction. All of this is considered wasteful and can be reduced if the concepts and principles of the CDE are adopted jointly. A CDE is a system for managing data and information. The CDE is not just a web-based or cloud-based "data room". It includes the processes and rules necessary to ensure that people are working on or using the current version of a file or template, as well as to let them know what purposes they can use them for. These processes have been well defined and managed in the context of a paper filing system, however, with the adoption of new electronic technologies and the huge increase in data produced regarding a typical construction project, the need for good management it has been neglected and the old systems have not been replaced.

According to ISO 19650-1:2018, a CDE solution and workflow should be used for managing information during asset management and project delivery. During delivery phase, the CDE solution and workflow support the information management processes. At the end of the project, information containers required for asset management should be moved from PIM to AIM.

3.1 Methodological approach to HQVRM CDE

HQVRM CDE is a project aimed at experimenting, researching and developing the implementation methods of a crucial part of the BIM methodology: the structuring of the CDE for the renovation of the ENEL Headquarters. of Viale Regina Margherita (HQVRM). In particular, the research activities are aimed at defining a digital management framework for the organization of the construction site, and is aimed at monitoring the construction processes, enabling a correct flow of information between the various subjects participating in the construction of the work. The research, currently underway and which is developed over a minimum period of 2 years, is therefore aimed at structuring a cloud computing solution, developed on ENEL corporate technologies capable of having the Digital Twin. Within the solution all accredited subjects can share the information produced, according to pre-established rules. HQVRM CDE project is planned as a work in progress in which, according to intermediate milestones, it is possible to monitor the progress of subsequent developments.

HQVRM CDE features is going to implement main digital twin features: the set of data and information in any way referable to the entities represented by the digital twin; the connection between the elements of the physical component with the corresponding virtual (cybernetic) part; the possibility of ubiquitous access to data and IT resources through the web, with the possibility of searching and analyzing information (big data, machine learning, artificial intelligence).; the exchange of data and information between the virtual component (cybernetics) and the physical one, with the use of sensors and actuators.

All areas that have been designed within the "Digital Twin" have been structured through the use of standards, policies and company procedures.

Macro areas have been identified, within which the respective data have been collected. For example safety area is the area within which all the management of the data of companies, workers, vehicles, equipment, scaffolding and vehicles is carried out, through the collection of data aimed at authorizing entry to the construction site and the relative monitoring of their validity. The connection between the physical and digital elements are entrusted to actuators that allow the high level of interaction between the construction site and HQVRM CDE that allows a structured knowledge of the information relevant to the pursuit of the objectives of the ENEL Task Force. These include the actuators that allow controlled access to the site only by personnel actually authorized by the system on the basis of personal and documentary checks.

The main development activities of HQVRM CDE are summarized below:

1. Analysis and evaluation of the client organization by verifying its strategic objectives, main functions, multi-year planning, regulatory and legislative constraints or related to particular technical/administrative acts. Structuring of the decision-making line, of roles and responsibilities and of the approval process and definition of the necessary needs and of the information flows in response to the achievement of the objectives.
2. Analysis and evaluation of the nature of the project and of the asset to be managed by defining the managerial, commercial and technical aspects of the production of information. The first two areas define the information standards and the production methods and procedures that must be implemented. The technical aspects define the pieces of information necessary to satisfy the information needs of the organization.
3. Analysis and evaluation of the methods of procurement of project and work services by identifying the information requirements useful for achieving specific strategic objectives in correspondence with each of the decision-making milestones in the various phases of the work.
4. Alignment of the management systems of the processes of the organization assessed in relation to the nature of the project and the asset and the procurement methods, through audit activities aimed at identifying information gaps and structuring information flows based on useful data and preferably structured in interoperable open formats necessary for the return of significant indicators.
5. Definition of contractual schemes that regulate the action of the parties in terms of methods and responsibilities, in the digital environment, in the supply chain, and depending on the phase, even structuring penalty metrics in the event of non-fulfillment.
6. Verification of enabling digital technologies, acquisition and implementation of the same, evaluating the nature of the work, the type of project, the competence of the staff and possibly structuring training plans consistent with the role held in the organization.
7. Surveillance, maintenance and continuous improvement of the data sharing platform and environment in the life cycle of the procedures and possible progressive development of artificial intelligence algorithms useful for improving the performance of the organization.

From what has been briefly explained, some first derivative considerations are therefore evident, which are summarized below:

A. HQVRM CDE is a cloud information resource available to the ENEL Task Force where the same manages the project in all phases, with rules previously shared and contracted through the information specifications/information management plan.
B. HQVRM CDE is the "place" for the execution of the contract, allowing the performance of the internal work group and/or external suppliers to be constantly monitored.
C. HQVRM CDE reflects the organizational structures of the ENEL Task Force and possibly enables better levels of it and on this it must be customized with scaled solutions and with costs previously estimated and calibrated on the objectives considered strategic.

3.1.1 Development principles and requirements

Collaborative work based on container objects, is midway along the maturity development path from documentation and paper drawings to server-based work, in which data is saved in centralized databases and multiple parties work simultaneously on a single model. The introduction of the concept of "work based on objects and their containers" or "work based on files", as a first step, introduces enough variation to create an impact; furthermore, this process is sufficiently close to current practice to be able to be implemented without requiring a substantial modification of the legal and contractual framework. It is also a process designed to make the adoption of this working methodology realistic also by SMEs.

The term “collaborative work based on objects and containers” was adopted from international standard ISO 19650-1:2018. An “object and its container” can be a three-dimensional model, a drawing, a document, a table or a calendar; it is also often referred to as a “file”. A database that contains multiple structured data tables is also a container object. Such objects can be categorized into document container objects, graphic information container objects, and even non-graphic information container objects. The collaborative working methodology based on container objects basically means two things:

1. The principle that the author or generator of a given information, for example a model or a design, has the competence and is responsible for the content and quality applied.
2. Certain rules have been defined regarding information management processes in order to allow a safe and efficient exchange of data and information.

3.1.2 Needs and goals

The activity is carried out in accordance with the principles that characterize the requirements for information management contained in the ISO 19650 series and in accordance with IT UNI 11337, based on the fact that the proposing subject (ENEL), the principal appointed subject (general contractor) and the other organizations in charge (PMO and GC) work together in a collaborative way and that all parties must participate in the implementation of the ISO 19650 series specific standards. The key requirements of HQVRM CDE are:

- Accessibility, according to pre-established rules, by all the actors involved in the process; Traceability and succession history of the revisions made to the data contained; Support for a wide range of data types and formats and their processing; High query flows and ease of access, recovery and extrapolation of data (open data exchange protocols); Conservation and updating over time; Confidentiality and security guarantee;

The objectives and advantages of HQVRM CDE are:

- Automation of information coordination between interested parties; Transparency of information also in terms of authorship and temporal availability of information; Automated management of revisions and data updates; Reduction of data redundancy; Reduction of risks associated with data duplication; Communication between interested parties through reference forms and interfaces (requests for information, requests, correspondence, etc.).

The solution was implemented using different technologies, in order to allow the development of an aggregate information model. The activity is monitored quarterly and, at the end of each quarter, an analysis phase is carried out by HQVRM CDE, aimed at assessing the results and results of the period relating to the quarter itself, validating the contents and possibly integrating the objectives.

3.1.3 Main activities

The activities are developed on the basis of ENEL’s needs and the related input data provided and are carried out by the HQVRM CDE team, through the following main tasks:

1. Analysis of the ENEL information needs with reference to the objectives to be pursued, the contractual constraints and other corporate policy tools (organization, project, property);
   a. Identification of the organization’s information needs

    i. Data driven decision support for the ENEL Task Force;
    ii. Structured knowledge of the progress of the work;
    iii. Streaming monitoring;
    iv. KPI performance measurement;
    v. Reliable corporate and institutional disclosure and communication.

2. Identification of the information needs of the project
    a. Supervise the execution of the project through checks (design reviews) carried out directly in the data sharing environment, or through reports extracted directly from the aforementioned environment or specific software that will be indicated by the Service Contractors according to the following paragraphs;
    b. Simplify the assessments and estimates that ENEL will have to make, giving maximum transparency and control over the works to be carried out and the final results to be obtained;
    c. Carry out simulations of the construction sequences on the model, aimed at validating the program, according to the availability of materials and the interference of the individual assembly activities;
    d. Favor the economic control of the investment during construction and subsequent management, with traceability of materials coordinated with the execution phases of the works and subsequent phases;
    e. Allow the analysis of the feasibility of the construction of the work, crossing the dates of availability of the materials on site, with the expected dates of assembly of the systems, collecting, in an organized form, all the relative certification documentation to be produced in accordance with the law and certifications subject to intervention (LEED, WELL, etc.);
    f. Track the progress of progress during the construction phase, in digital and visual form;
    g. Provide a basis that includes the minimum data and information necessary to ensure the correct setting and management of the subsequent management phases of the work;
    h. Prepare, at the same time as the work is completed, an "IT" tool, which can also be used remotely, which allows the work to be managed during its useful life (such as, for example, operation, monitoring, maintenance, maintenance of necessary certifications).
3. Analysis of the tender documentation elaborated by ENEL and of the offers and information management proposals, elaborated by the RTI and GC;
4. Support for the definition of the initial contractual information management plan (ENEL/RTI/GC) and subsequent editions;
5. Structuring of HQVRM CDE according to ENEL proprietary rules and customized according to the company's own needs and procedures, developed on Microsoft technologies (mainly Sharepoint®, Teams®, PowerBI®) and corresponding to the requirements of current main standards with particular reference to ISO 19650:2018 and UNI 11337:2017 implementation of HQVRM CDE, on the HQVRM redevelopment pilot project, by structuring sites, information sheets, metadata, taxonomies, authorization levels, etc.;
6. Continuous measurement of the efficiency and effectiveness of the rules established in order to gradually improve the process;
7. Progressive alignment of the contract documentation (information management plan) and simultaneous processing of what has been validated at the level of corporate strategic documentation (OIB-OIR-AIR-PIR).

HQVRM CDE Team conducted sessions dedicated to the digital management of the project mainly attributable to the definition of the methods of management and coordination of the project processes and information flows, as well as the identification of the main information modeling rules. Over time, the aforementioned activities have also involved PMO and GC (with its subcontractors and third party contractors) and have been carried out by individual professionals or specific groups of professionals, The information processes of digitization have been managed, defining and managing general rules of the organization, research and development, training, resource management (human and instrumental), contracts, relations with mandatory and voluntary standardization systems, relations with higher bodies.

CDE management main functions needs:

1. Management and maintenance of the CDE and its use, the data contained therein, the databases connected to it and information applications in general, relations with higher bodies and information management;
2. Care and management within the individual orders, in the specifically digital information field of: declination of the general rules; information coordination internal and external to the organization; optimization of the use of resources (human instruments, etc.); progress of activities; relations with higher bodies.

A consistent information process must be guaranteed during all stages and phases of the process, taking care of the integrity of the data.

A solution and workflow of the CDE environment will be structured for the management of information during the management of the design, construction, preparation for management and delivery of the property itself to management.

At the end of the management of the building renovation project, the CDE will be prepared for the transfer from the job order information model (PIM) to the property management information model (AIM). The remaining job order information containers that will not be used for the management of the building will be stored in the CDE set up for the job itself. For the management of the CDE, metadata will be structured for the various encodings.

3.1.4 Reference standards and regulations

Information management framework, including BIM, compliant with the following main regulations:

- UNI EN ISO 19650:2018 (Organization and digitization of information about buildings and civil engineering works, including building information modelling (BIM) – Information management using building information modelling);
- UNI 11337:2017 (Building and civil engineering works – Digital management of the informative processes);
- Ministerial Decree MIT 560/2017;
- ISO 27001(Information technology – Security techniques – Information security management systems – Requirements);
- ENEL standards and regulations.

The goals of the CITERA Team concern the analysis of the needs framework of the proposing part of the project, the definition of KPIs, the modeling, implementation and management of digital information processes, defining and managing general rules of the organization, research and development, support training, management of resources (human and instrumental), support for contracts, relations with mandatory and voluntary standardization systems, relations with higher bodies.

The management function of the ACDat includes:

- management and maintenance of the ACDat and its use, the data contained therein, the databases connected to it and information applications in general, relations with higher bodies and information management;
- care and management within the individual project, in the specifically digital information field, of: declination of the general rules;
- information coordination internal and external to the organization; optimization of the use of resources (human instrumental, etc.);
- progress of activities; relations with higher bodies.

Team CITERA also guarantees an information process during all stages and phases of the process, taking care of the integrity of the data. At the end of the management of the order, the ACDat is also prepared, if necessary, for the transfer from the information model of the order (PIM) to the information model of the management of the property itself (AIM) according to methods that will be communicated by ENEL. The remaining job order information containers that will not be used for the management of the building will be stored in the ACDat prepared for the job itself.

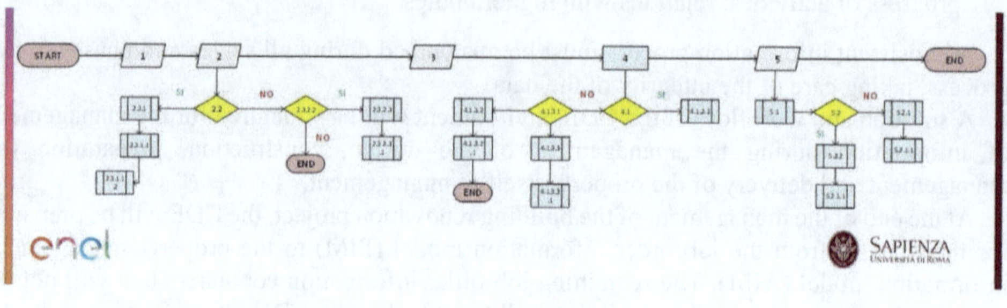

Figure 3: Workflow security.

3.1.5 HQVRM CDE uses definition

HQVRM CDE is intended as the set of tools and workflows used for the organization, management and sharing of data relating to the work. This set may vary over time in line with the evolution of the implementation process first and then management and the configuration of the subjects involved considering specific objectives and uses identified for HQVRM CDE.

HQVRM CDE is therefore structured according to the aggregation of multiple technological solutions each governed by internal workflows and related to each other through external workflows and is configured as P-CDE project CDE in the management phase of the redevelopment project able to pass through in an A-CDE asset CDE under asset management. The analysis of the information needs of the ENEL Task Force made it possible to identify the main uses of HQVRM CDE:

1. Field and management tracking. Intended as the process in which a digital resource is used during the construction, and commissioning of a building to manage and monitor activities, report on quality, safety, construction documents etc. The goal is to ensure compliance with contract documents, compliance with safety regulations, and with the customer's project requirements;
2. Design repository and versioning;

3. Design review, with issue resolution and request for information management;
4. 4D management;
5. 5D Management.

3.2 Digital ecosystem

The HQVRM – Digital Twin is based on a CDE of the organizational model and from the point of view of technologies is enabled by cloud computing, artificial intelligence, machine learning, big data and BIM. HQVRM–CDE represents the connective of the digital ecosystem in which information models are organized depending on the need and the objectives to be pursued in particular for the temporal, economic, safety and sustainability aspects.

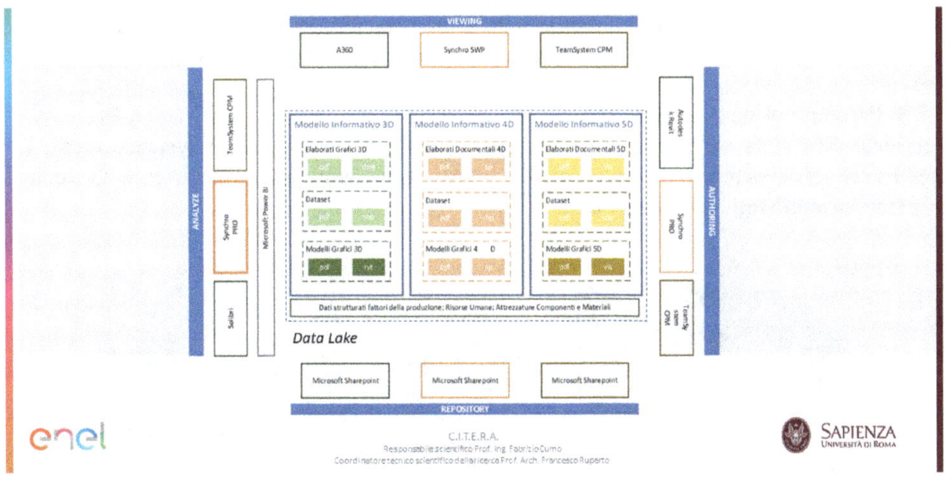

Figure 4: Environment structure.

3.2.1 Content management system (CMS)

Microsoft SharePoint® is a content management system (CMS) software platform developed by Microsoft, that is a program that, running on the server side, allows the creation and distribution of particular websites mainly for corporate use (intranet), but which can also be distributed on the net and therefore also be used as normal websites. The purpose of the software is to share information and/or documents in different ways. You can create lists, document repositories, calendars synchronized with Outlook® and more. SharePoint® is fully integrated with the Microsoft 365® package and the Power Platform and offers solutions such as document "versioning". Authentication is done by entering a username and password at login. This procedure is facilitated by single sign on, which in the context of Microsoft technologies is often inserted when turning on your PC. The system will therefore take care of automatically authenticating the user in the sites where he has access credentials. SharePoint® manages a system of permissions for its elements (site collection, site, list/ repository, page or folder/item). It is possible to associate a permission to each element, discriminating against any object mentioned above. This allows you to create particular websites in which, depending on the level of authorization a user has, it is allowed to see or not see certain elements by specifying the type of access (whether read-only or contributing).

3.2.2 App development and workflow automation

Power Apps is a suite of apps, services, connectors and data platform that offers a rapid development environment for building custom apps for business needs. With Power Apps, you can quickly build custom business apps that connect to business data stored either in the underlying data platform (Microsoft Dataverse) or across multiple online and on-premises data sources (SharePoint®, Microsoft 365®, Dynamics 365®, SQL Server®, etc.). Microsoft Power Automate is the tool used for the automation of information flows.

3.2.3 Business Intelligence

The Business Intelligence tool adopted is Microsoft Power BI®. PowerBI® is a collection of software services, apps, and connectors that work together to transform unrelated data sources into a coherent, visually engaging, and interactive set of information. The data can be structured in an Excel® spreadsheet or a collection of on-premises cloud-based or hybrid data warehouses. Power BI® allows to easily connect to data sources, view and discover important information, and share it with as many or as many users as you need.

3.2.4 BIM authoring platform

Autodesk Revit® is a CAD and BIM program for Windows operating systems, created by Revit Technologies Inc. and bought in 2002 by AutodesK, which allows design with elements of parametric modeling and drawing.

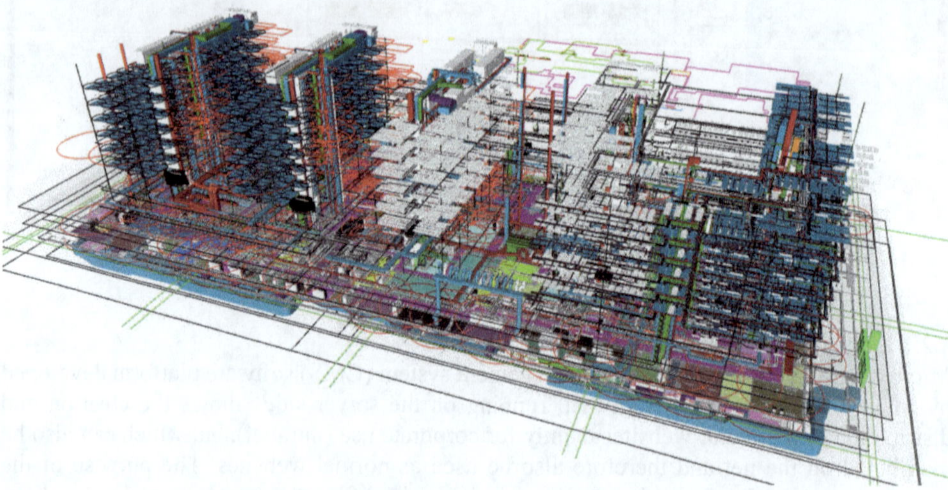

Figure 5: Axonometric view.

3.2.5 4D environment

SYNCHRO is a complete portfolio of integrated software and services for digital construction management that allows project management based on 4D simulations, improving the use of data to optimize decision-making, resource monitoring and critical processes.

3.2.6 5D environment

In order to ensure correct accounting of the works, the following requirements have been defined to be guaranteed:

1. Connect the cost coding system, and the WBS by means of shared parameters relating to the WBS COMPUTO items, associated with each element/category within the model;
2. Specify the rules that generate the surveys, especially for those that are not a direct reading of the parameters of the elements of the model but a re-elaboration carried out on the basis of further calculation conventions;
3. Identify the nature and type of price lists/items of reference;
4. Method of extracting and linking data between price lists and model objects.

To pursue the aforementioned objectives, a software dedicated to the management of construction site costs (field BIM) has been identified in order to ensure the verification of the times and costs of execution analysis in progress relating to the progress and progress of the construction site, traceability of materials, etc.

The software identified for the management of the above is team system construction project management, which allows you to record the progress of the work. The correspondence of these advances with the virtual BIM model allows us to have even the visual rendering of the actual work progress of the work which can also be associated with any detailed analysis of the data allowed by the business intelligence functions made available by CPM®.

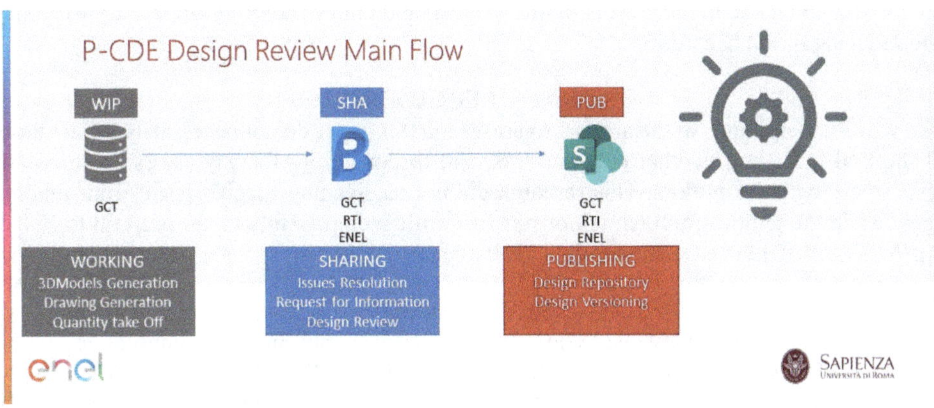

Figure 6: Project management workflow.

4 CONCLUSIONS

The information needs of the ENEL Task Force are aimed at monitoring the construction processes of the new headquarters, providing for the active collaboration of all the parties involved. To this end, the information flows concerning the project take place within a data sharing environment, in which the digital management of information processes takes place, made explicit through a process of correlation and optimization of digitized information flows and the decision-making processes that they concern every single activity. The objective of increasing the efficiency and effectiveness of the ENEL Task Force's monitoring action is pursued by establishing a framework of digital processes and technologies capable of enabling better levels of knowledge and control of each individual work phase in the main areas of safety and environment. A first evidence measured in carrying out the project, however, concerns the inevitable paradigm change required of all stakeholders; a transition that must gradually recognize its action which is less and less founded on documentary obligations and increasingly oriented towards managing and monitoring the construction site on data articulated in information containers. This digital transformation, on the one hand, therefore, allows an optimization of the Client's resources, on the other, imposes a general strategy aimed at increasing the reliability and availability of data on the basis of rules previously shared within the organization and contracted, also with collaborative approaches. in relations with suppliers.

In conclusion, whereas the main development activities of HQVRM CDE: (analysis and evaluation of the client organization, Analysis and evaluation of the nature of the project and of the asset to be managed, Analysis and evaluation of the methods of procurement of project and work services, Alignment of the management systems of the processes of the organization, definition of contractual schemes, verification, acquisition and implementation of enabling digital technologies and the surveillance, maintenance and continuous improvement of the data sharing) and his derivative considerations that we have previously discussed we can say HQVRM CDE constitutes a strategic asset of the organization capable of returning value in terms of efficiency, effectiveness and cost-effectiveness, to a greater extent depending on the specificity of the solution adopted and the medium / long-term vision capable of putting the governance apparatus of the project, recognizing its value also in terms of environmental sustainability (supporting the optimization of resources and reducing waste), economic (mitigating the risk of lengthening times and budget overruns), social (being able to enable higher levels of site safety), and ethics (favoring greater transparency and reducing moral hazard risk).

ACKNOWLEDGEMENTS

The views and opinions of the authors expressed in this paper do not necessarily reflect those of the ENEL S.p.A., or other collaborators, and responsibility for any errors or omissions rests solely with the authors. This research did not receive any specific grant from funding agencies in the public, commercial, or not-for-profit sectors. Authors are grateful to ENEL S.p.A. for making the case studies data available.

REFERENCES

[1] El Saddik, Digital twins: The convergence of multimedia technologies. *IEEE MultiMedia*, **25**(2), pp. 87–92, 2018. DOI: 10.1109/MMUL.2018.023121167.

[2] Chhetri, M.B., Krishnaswamy, S. & Loke, S.W., smart virtual counterparts for learning communities. *Web Information Systems – WISE 2004 Workshops, Lecture Notes in Computer Science*, **3307**, pp. 125–134, 2004. DOI: 10.1007/978-3-540-30481-4_12.

[3] Bacchiega, G., Creating an embedded digital twin: Monitor, understand and predict device health failure (PDF). *Inn4mech – Mechatronics and Industry 4.0 Conference Presentation*, 2018.

Author index

www.ingramcontent.com/pod-product-compliance
Lightning Source LLC
LaVergne TN
LVHW062228200926
848484LV00006B/1500